# STITCHING THE
# WEST BACK TOGETHER

# SUMMITS

**ENVIRONMENTAL SCIENCE, LAW, AND POLICY**

A series edited by Marc Miller,
Jonathan Overpeck,
and Barbara Morehouse

# STITCHING THE WEST BACK TOGETHER

*Conservation of Working Landscapes*

EDITED BY SUSAN CHARNLEY,
THOMAS E. SHERIDAN, AND
GARY P. NABHAN

The University of Chicago Press
Chicago and London

**Susan Charnley** is a research social scientist at the USDA Forest Service's Pacific Northwest Research Station. **Thomas E. Sheridan** is professor of anthropology at the University of Arizona and a research anthropologist at the university's Southwest Center. **Gary P. Nabhan** is a research scientist at the University of Arizona's Southwest Center.

The University of Chicago Press, Chicago 60637
The University of Chicago Press, Ltd., London
© 2014 by The University of Chicago
All rights reserved. Published 2014.
Printed in the United States of America

23  22  21  20  19  18  17  16  15  14      1  2  3  4  5

ISBN-13: 978-0-226-16568-4 (cloth)
ISBN-13: 978-0-226-16571-4 (paper)
ISBN-13: 978-0-226-16585-1 (e-book)
DOI: 10.7208/chicago/9780226165851.001.0001

SUMMITS PARTNERS AT THE UNIVERSITY OF ARIZONA

Institute of the Environment (www.environment.arizona.edu)
James E. Rogers College of Law (www.law.arizona.edu) and its Environmental Law,
Science and Policy Program (www.law.arizona.edu/environment)
Biosphere 2 and the Biosphere 2 Institute (www.b2science.org, www.b2science.org/institute)
Udall Center for Studies in Public Policy (www.udallcenter.arizona.edu)

SUPPORT FOR THE SUMMITS BOOK SERIES

The Law Firm of Girardi & Keese (Los Angeles; www.girardikeese.com)

SUPPORT FOR *STITCHING THE WEST BACK TOGETHER*

USDA Forest Service, Pacific Northwest Research Station (www.fs.fed.us/pnw/) and Cooperative
Forestry (www.fs.fed.us/spf/coop/)
Stanford University, Department of Anthropology (www.stanford.edu/dept/anthropology/cgi-bin/
web/) and The Bill Lane Center for the American West (http://west.stanford.edu/)
The Southwest Center at the University of Arizona (http://swc.arizona.edu/)

LIBRARY OF CONGRESS CATALOGING-IN-PUBLICATION DATA

Stitching the West back together : conservation of working landscapes / edited by Susan Charnley,
Thomas E. Sheridan, and Gary P. Nabhan.
pages cm — (Summits: environmental science, law, and policy)
ISBN 978-0-226-16568-4 (cloth : alk. paper) — ISBN 978-0-226-16571-4 (pbk. : alk. paper) —
ISBN 978-0-226-16585-1 (e-book) 1. Landscape ecology—West (U.S.) I. Charnley, Susan II. Sheridan,
Thomas E. III. Nabhan, Gary Paul IV. Series: Summits: environmental science, law, and policy.
QH541.15.L35S858 2014
577.0978—dc23
2014001560

# CONTENTS

Foreword       *Charles F. Wilkinson*  ix
Introduction   *Susan Charnley, Thomas E. Sheridan, and Gary P. Nabhan*  xiii

**PART 1**      THE IMPORTANCE OF CONSERVING
               WESTERN WORKING LANDSCAPES  1

Chapter 1      A Brief History of People and Policy in the West
               *Thomas E. Sheridan and Nathan F. Sayre*  3
Chapter 2      Status and Trends of Western Working Landscapes
               *Susan Charnley, Thomas E. Sheridan, and Nathan F. Sayre*  13
Chapter 3      The Biodiversity That Protected Areas Can't Capture: How
               Private Ranch, Forest, and Tribal Lands Sustain Biodiversity
               *Gary P. Nabhan, Richard L. Knight, and Susan Charnley*  33

**PART 2**      COLLABORATIVE CONSERVATION  49

Chapter 4      Beyond "Stakeholders" and the Zero-Sum Game: Toward
               Community-Based Collaborative Conservation in the
               American West
               *Thomas E. Sheridan, Nathan F. Sayre, and David Seibert*  53
Spotlight 4.1  Historic Precedents to Collaborative Conservation in
               Working Landscapes: The Coon Valley "Cooperative
               Conservation" Initiative, 1934
               *Curt Meine and Gary P. Nabhan*  77

v

Chapter 5     The Quivira Experience: Reflections from a "Do" Tank
*Courtney White* 81

Spotlight 5.1     Grass-Fed and Grass-Finished Livestock Production: Helping to Keep Working Landscapes Intact
*Gary P. Nabhan, Carrie Balkcom, and Amanda D. Webb* 95

Chapter 6     Place-Based Conservation Finds Its Voice: A Case Study of the Rural Voices for Conservation Coalition
*Maia Enzer and Martin Goebel* 101

**PART 3**     CASE STUDIES OF WORKING FORESTS 119

Chapter 7     Swan Story
*Melanie Parker* 123

Spotlight 7.1     Arcata Community Forest
*Mark Andre* 137

Chapter 8     Taking a Different Approach: Forestland Management in the Redwood Region
*Mike Jani* 141

Spotlight 8.1     The Conservation Fund's Garcia River Forest, California
*Chris Kelly* 155

Chapter 9     Stewardship Contracting in the Siuslaw National Forest
*Shiloh Sundstrom and Johnny Sundstrom* 159

Spotlight 9.1     Stewardship Agreements: The Weaverville Community Forest, California
*Pat Frost* 177

**PART 4**     CASE STUDIES OF WORKING RANCHES 181

Chapter 10     Lava Lake Land & Livestock: The Role of Private Landowners in Landscape-Scale Conservation
*Michael S. Stevens* 185

Spotlight 10.1     Country Natural Beef
*Susan Charnley and Sophia Polasky* 203

Chapter 11     Conservation and Development at Sun Ranch: The Search for Balance in the U.S. West
*Roger Lang, William H. Durham, and Josh Spitzer* 209

Spotlight 11.1     The Madison Valley Ranchlands Group
*Thomas E. Sheridan* 223

Chapter 12    Integrating Diversified Strategies on a Single Ranch: From
              Renewable Energy and Multiple Breeds to Conservation
              Easements
              *Dennis Moroney* 227
Spotlight 12.1 Private Land Conservation Trends in the Western United
              States
              *Jon Christensen, Jenny Rempel, and Judee Burr* 241


PART 5      EMERGING APPROACHES TO CONSERVING
            WORKING LANDSCAPES 247


Chapter 13    The Sonoran Desert Conservation Plan and Ranch
              Conservation in Pima County, Arizona
              *Thomas E. Sheridan* 251
Spotlight 13.1 Ranching and the "Death Tax": A Matter of Conservation as
              Well as Equity
              *Thomas E. Sheridan, Andrew Reeves, and Susan Charnley* 267
Chapter 14    Payments for Ecosystem Services: Keeping Working
              Landscapes Productive and Functioning
              *Gary P. Nabhan, Laura López-Hoffman, Hannah Gosnell, Josh
              Goldstein, Richard Knight, Carrie Presnall, Lauren Gwin, Dawn
              Thilmany, and Susan Charnley* 275
Spotlight 14.1 The Conservation Reserve Program
              *Steven E. Kraft* 295


Conclusions and Policy Implications
*Thomas E. Sheridan, Gary P. Nabhan, and Susan Charnley* 301


Acknowledgments 311
Contributor Biographies 315
Summits Board of Advisers 325
Index 327

# FOREWORD

The pages that follow argue for applying the theory and practice of the "radical center" to the longstanding, engrained divisions over land and water in the American West in order to conserve working landscapes and the biodiversity they contain. Such a project is edgy. There are many camps and circumstances, and all too often the positions of the opposing sides are starkly different and come equipped with white-hot, disparaging views about each other. Searching for a center can generate loud cries of sell-out.

The heart of this fresh and ambitious book is that when you talk long enough and listen enough, and dare to open up your mind enough, you realize that most people and sides are making ethical and moral arguments that deserve high respect. Yes: achieving a vibrant community economic footing is a worthy—and moral—objective. Likewise, embedding communities in healthy, lasting landscapes is a moral imperative.

These are large and multifaceted propositions and *Stitching the West Back Together* takes them on in the right way. The diverse and impressive contributors recount efforts, already in place on the ground, to conserve the West's working forests and ranchlands using a consensus-based radical-center philosophy, revealing the content of this brand of collaboration. Most of all, they show that these ideas need to be lived, not just asserted. Employing them must be practical and workable. These accounts and case studies present practices that are transferrable in that they, or aspects of them, can be put in place elsewhere when people are willing to break from their past intransigence to achieve a broader community good.

One of the cases highlights the Weaverville Community Forest in the mountainous Klamath Knot country of northern California. In the late 1990s, a co-alition of local citizens and businesses in Weaverville and Trinity County rose up in opposition when the Bureau of Land Management (BLM) was consider-ing a land swap that would have given a prime block of forest land, a favorite for its recreational and scenic values, to a large, northern California timber company that would probably log the land intensively. To establish a form of comanagement, a citizen-supported organization negotiated a steward-ship agreement with the BLM instead that established the community for-est and then expanded the land area by entering into a second stewardship agreement with the Forest Service to add 12,000 acres to the community for-est. Today, the area is managed conservatively for ecological and recreational objectives. The logs from harvesting small-diameter trees for fuel reduction are used for firewood collection and milling two-by-fours at the local mill in Weaverville. Stewardship contracts and agreements can be clunky and overly bureaucratic, but collaborative stewardship, which is being improved and put to increasing use, holds great potential for infusing public-land forest management with informed community values.

Another example, one of several on western ranching, originates with Doc and Connie Hatfield, two western originals. In the 1980s, facing relentless financial hardship, they made a last-ditch effort to save the ranch that they loved. Determined to ranch in environmentally sensitive ways, they reached out to other ranchers and environmentalists. What emerged was Country Natural Beef (CNB), a cooperative that reflected their values.

Ever since, CNB has epitomized the "green ranching" movement. Today it includes more than 100 ranches across the West that graze cattle on millions of acres, including public lands. All the members, who are certified by the independent Food Alliance, must also meet CNB's rigorous environmental standards. The cooperative successfully markets its members' beef to grocery and restaurant chains, and their high-end product returns reasonable profits.

Needless to say, none of this comes easy. The people of Weaverville and Trinity County had to attend grueling meetings to come up with an approach that represented the whole community. Then came tedious negotiations with the two land management agencies. That got them better management of a cherished landscape. It also got them more work, the job of managing the community forest. In the case of the Hatfields, the meetings of ranchers, environmentalists, and agency officials, all sitting in a circle at their ranch in Sisters, Oregon, is now the stuff of legend, and the participants wrought significant progress. Yet many meetings of many groups still lay ahead, and many were contentious. But the ranches and the public lands are better for it.

Can we continue to do the hard work of coming together in the name of both community and the land? Can we do even more of it? My guess is that we can because, if we pause to listen and honestly reflect, most of us believe that community and the land are both sacred, that both must be honored. And I will further guess that, in the years to come, people will look at this volume as a valuable contribution, a place where stories were gathered together to mark down what we have done to conserve the landscapes that we care about and to inspire us to do more by showing us that the center isn't so radical after all.

*Charles F. Wilkinson*

# INTRODUCTION

*Susan Charnley, Thomas E. Sheridan, and Gary P. Nabhan*

In January 2003, Courtney White of the Quivira Coalition brought together twenty ranchers, environmental activists, and conservation biologists at the Hotel Albuquerque to forge what Courtney called a Declaration of Interdependency. For forty-eight hours, these veterans of the so-called range wars of the late twentieth century put aside their differences and crafted a new vision of the West for the next century—"to find a way to take the West back from decades of divisiveness and acrimony that jeopardizes what we love," in Courtney's words. The group focused on their shared aspirations for the public lands grazed by livestock, lands that some environmental groups wanted to make cow-free.

After exhaustive debate, men and women like Bill McDonald, Bill deBuys, Teresa Jordan, Ed Marston, Paul Johnson, Gary Nabhan, and Heather Knight emerged from their hotel cocoon with a consensus statement as clear and bracing as the cold New Mexican winter air. That statement—the "Invitation to Join the Radical Center"—highlighted the importance of cooperation to promote the ecologically sound use of working landscapes, to preserve their biodiversity, and to halt their conversion to ecologically, aesthetically, and culturally destructive land uses. Since then, the document has been signed by thousands of individuals, including a who's who of the land stewardship movement: Stewart Udall, Theodore Roosevelt IV, Nina Leopold Bradley, Wendell Berry, Alvin Josephy, Curt Meine, Bill McKibben, Charles Wilkinson, Linda Hasselstrom, Peter Forbes, Patricia Limerick, and Dan Daggett, among many others. The initiative not only helped cool the western range wars; its principles also spread eastward to stimulate discussions in Wisconsin, Kentucky, Vermont, and Iowa. The manifesto remains the most elegant and visionary statement of why Americans need to find common ground over working landscapes and their conservation, and what is at stake if we fail to do so.

This book—written by ranchers and foresters as well as academics—is about why the struggle to find common ground in the rural West is so important, and about how more and more people are trying to do so. It makes two major interrelated arguments. One is that sustainable working landscapes are critical to the conservation of biodiversity in the American West and the cultures of rural ranching and forestry that depend on them. The West is a patchwork of private, tribal, and public lands. The only way to achieve conservation at a large landscape scale is to maintain or restore ecosystem health across the jurisdictional boundaries—private, tribal, and public; municipal, county, state, and federal—that divide us. To do so, ranchers and foresters and the rural communities they help sustain have to be able to make a living from the lands on which they have depended for so long.

The second major argument is that collaborative conservation is the most effective way to conserve biodiversity and ecosystem function on a large landscape scale. This requires the long, slow, patient process of talking to one another and searching for common ground that builds the trust necessary to collaborate across fence lines and make good things happen on the ground. And on the ground is, after all, where conservation truly matters.

In using the term "working landscapes," we are referring to landscapes where people make their living by extracting renewable natural resources such as grass and trees and turning them, through ranching and forestry, into wool, meat, and wood products. By sustainable—a notoriously difficult term to define—we mean both that (1) levels of extraction allow natural resources to recover and reach similar or higher levels of productivity for future generations and (2) the methods of extraction do not diminish the biodiversity and ecological integrity of terrestrial and aquatic ecosystems. But just why is it important to conserve these lands as working lands, in a manner that does not entail the retirement of all economic uses?

The reasons are both ecological and social and are described in many variations in the chapters that follow. They include the role of working landscapes in conserving biodiversity and providing ecosystem services such as clean water, wildlife habitat, and carbon sequestration outside of officially designated protected areas; in yielding products for everyday uses; in fostering relations between people and nature; in generating diverse revenue streams for residents of rural communities; in providing natural amenities such as open space, scenic values, and recreation opportunities; and in supporting social and cultural diversity. When rural producers can no longer survive financially, their private lands often end up in the hands of real estate speculators who carve those lands into subdivisions that destroy wildlife habitat, sever wildlife corridors, and fragment the wide open spaces. And because watercourses often run through or along the borders of private lands, both terrestrial and aquatic ecosystems are threatened (see chap. 3).

Our focus is on landscape-scale conservation, rather than on the conservation of small, individual properties per se. We are therefore talking about conservation that transcends jurisdictions: conservation across ownership and management types, and often across public and private boundaries to overcome historic processes of administrative and ecological fragmentation (see chap. 1). Although conservation frequently starts on individual parcels, it will ultimately fail to sustain pollinators or predators, livelihoods, or production infrastructures if it remains at that scale. For this reason, we are interested in the question of how to scale up within the broader geographic regions of which individual ownerships are a part so that landscapes can be stitched back together into functional, ecologically and economically healthy wholes. Doing so typically means that rural producers, government land managers, and environmentalists must act together to achieve common goals.

The chapters in this book highlight themes and raise questions that are worth drawing attention to at the outset. One theme is that conservation often happens in fits and starts and through processes of trial and error. The conservation entrepreneurs whose stories appear here have taken risks to innovate and experiment, sometimes succeeding, sometimes failing, yet always persisting and showing resilience in the face of challenges and change. They have tested different ideas and adapted their strategies as they went along, seeing what worked and what did not. The role of key individuals such as the case-study authors in this book cannot be underestimated; many conservation innovations come about because of an individual or a small number of people who have a vision and act on it, pushing the boundaries of regulatory and policy frameworks and social norms that often operate as constraints to innovation. Many times such individuals and groups are rural producers whose livelihoods depend, at least in part, on access to federal and state trust lands.[1] For their experiments to be successful, they need the assistance of federal, state, county, and municipal land managers who are willing to push such boundaries within their own agencies. Agency personnel can be bureaucratic gatekeepers who resist innovation, or they can be creative problem solvers who encourage flexibility and experimentation and who are sources of innovation themselves. Problem solvers—private, public, and tribal—are key to working landscape conservation.

Another theme is that the willingness to collaborate is critical when conservation initiatives involve different livelihoods, cultures, ownerships, and stakeholders. Social relationships between diverse interest groups who care about the rural West have, in the past, broken down over contentious forest and range management issues, in many cases creating perceived contradictions between what is good for the environment and what is good for the economy (see chap. 4). Much of the rural West consists of state and federal lands interspersed with tribal and private lands, meaning that working landscape conservation calls for understanding the ecological and socioeconomic interdependencies that operate among public, tribal, and private lands and landowners in specific places. It also means devising strategies that acknowledge and support these interdependencies. For example, many ranchers have depended on access to federal and/or state lands as well as private lands to run their operations, and still do today. Public, private, and tribal forestlands are important sources of wood and biomass for the wood products and bioenergy industries. The ability to maintain local processing infrastructure to make timber and biomass production economical requires sustaining a supply of forest products from these lands. Effective landscape-scale conservation thus calls for stitching the management of public, tribal, and private lands together using collaborative processes to achieve mutual social and ecological objectives.

The chapters that follow demonstrate that the land stewards who are striving to restore and sustainably manage working forests and rangelands today include a broad range of actors: private forest and ranch owners, public lands grazing permittees, tribal members, private contractors, natural resource crews affiliated with nonprofit organizations, civic groups and community members who work as volunteers, and local, state, and federal land managers. Thus, who plays what role in land management and stewardship has changed in the past few decades as a result of collaborative conservation and increasing partnerships, the outsourcing of public land management due to government downsizing and budget cuts, the acquisition of large landscapes by private investors, and other trends. It is worth considering who the managers now are of western working landscapes and the conservation implications of this shift. And so we ask readers to think about the following questions as they move through these case studies: (1) Who should be doing the work of maintaining ecosystem health? (2) How do landscape-scale conservation efforts that try to stitch the West back together challenge conventional notions of property rights and administrative boundaries and of the appropriate uses of public lands? (3) How can policy makers institutionalize flexibility and innovation in the management of working lands while still protecting those lands from exploitation and abuse?

We argue that partnerships among agency land managers, private land owners, rural producers, and informed citizens can stitch the West back together. This means breaking down the rural-urban divide and reengaging western urban dwellers in the conservation of the wide open spaces that surround them. It is a daunting endeavor because that divide has deepened in recent decades as the relationship between many Americans and the natural world has frayed. The connection between landscapes and the people who work them, in contrast, has not been broken. Multigenerational ranchers and foresters know their natural environments with a depth and intimacy few others can match. When they have the incentives to do so, they make long-term investments on the landscapes from which they make their livelihoods. This fosters a stewardship ethic and generates local ecological knowledge that complements and supplements the knowledge of experts in agencies and universities. Both types of knowledge, as well as practice, are needed to conserve biodiversity and ecosystem integrity.

As the following chapters attest, rural producers can also serve as the nuclei around which urban scientists, environmentalists, sportsmen, "foodies," and agency personnel can coalesce. Whether they are consumers of locally produced grass-fed beef, hunters who develop bonds of trust with rural landowners, volunteers who work on forest or watershed restoration projects, or environmentalists who join ranchers and foresters to advocate for policies that promote conservation, the collaborative conservation movement works to weaken the rural-urban divide through hundreds of local projects. It helps both urban and rural people realize that nature and biodiversity are not "out there" somewhere or limited to protected areas. On the contrary, a holistic view of conservation encompasses both the country and the city, the national park and the farm or ranch. Conserving working landscapes will produce benefits not only for the people who live and work there but for the urban dwellers who share their watersheds, buy their products, and engage in recreational activities on them as well.

A critical question for working landscape conservation is how to make innovations work economically. The cases in this book describe a range of creative financing strategies and business models that stitch together different mixes of private capital, government support, market economics, and funding from nongovernmental organizations. The complicated economics of conservation can sometimes create unexpected and seemingly contradictory alliances involving private investors, donors, public agencies, long-term, multigenerational landowners, and well-to-do newcomers, among others. Identifying the appropriate role of government in the mix is an important question for policy. For example, what kinds of tax policies, grants, landowner assistance programs, and other incentives make a difference? Figuring out how to make working landscapes financially viable may be the biggest challenge of all in this new conservation paradigm. As a case in point, the jury is still out on whether there will be buyers of the ecosystem services provided by working landscapes. We do not take it as a given that all of the strategies described in this book will succeed in improving the bottom lines of ranchers or foresters; the economic landscapes around them are forever changing, and the growing impact of climate change is already affecting them.

A final question we hope readers will consider is this: What are the alternatives to working landscape conservation, and what are the social, economic, and ecological consequences of these alternatives? One alternative is the explosive suburbanization and exurbanization of the rural West, fueled by the conversion of private forest and rangelands, as well as state trust lands, into commercial and residential developments.[2] The extraction of forest and rangeland products can be, and often are, sustainable economic activities. Real estate development, in contrast, frequently introduces a cascading chain of impacts that reduce biological diversity and undermine ecosystem function and ecological integrity.

Another alternative is to expand existing protected areas and create new ones. But that would require enormous amounts of money from either taxpayers or private donors. When Teddy Roosevelt and Gifford Pinchot created the national forest system in the early twentieth century, they carved national forests out of the federal public domain. U.S. taxpayers did not have to buy millions of acres of land, as they would have to do today. Does the political will exist to incur such expenses at the national level? Fights over federal budgets and growing concern about the national debt suggest not. And most environmental organizations would agree that there are not enough private funds to purchase all of the areas that need to be conserved.

A third alternative is to reduce or eliminate grazing and timber cutting on federal and state trust lands. Commercial logging on national forests has already dwindled, with severe economic impacts on many rural communities across the West (see chap. 2). A similar reduction of grazing on national forests or Bureau of Land Management districts could result in a severe contraction of the western livestock industry. We have found no compelling evidence that tourism or telecommuting will rejuvenate rural economies. Nor will tourism and telecommuting provide food, fiber, wood products, or bio-energy. The ecological, economic, and social health of the rural West will suffer as a consequence. This means that hard work must be done to sustain and enhance the social and ecological values embodied in working landscapes, values that we believe are in everybody's interest to conserve.

The book begins with a history and overview of working landscapes in the West, and why their conservation is important for protecting biodiversity (pt. 1). Part 2 goes on to explore collaboration as an approach to working landscape conservation. Parts 3 and 4 contain a series of case studies and spotlights of working landscape conservation in the West—one section on forestry and one on ranching—that have been written by key players in these efforts (see fig. I.1). Tribal lands are an important component of the West's working landscapes, making a significant contribution to the protection of biodiversity and ecosystem services (see chaps. 3 and 14). None of the cases or spotlights focus on tribal lands, however, because the status of tribal lands as working lands is not threatened the way that the status of private and public lands is. Private lands are subject to markets and the finances of private landowners; public lands are subject to political pressure from multiple interest groups and government budgets. Part 5 focuses on emerging approaches that attempt to bridge the rural-urban divide and make working landscape conservation a partnership between rural and urban Westerners. Closing comments in the book's conclusions identify opportunities for policy making to promote working landscape conservation across the West.

This volume is not an exhaustive review of all the tools being used to conserve western working landscapes today. We discuss many of those tools—conservation easements, sale of ecosystem services, stewardship contracts, and the like—but we do not provide a how-to manual or a cookbook of conservation recipes. The West is simply too dynamic a region for an encyclopedic approach, and the encyclopedia would be obsolete before it ever got published. Instead, the topical chapters and case studies illustrate the dynamic processes at play in working landscape conservation and highlight critical issues that must be addressed if such conservation is to be successful. We also hope the case studies provide inspiration and ideas to others working to conserve their own corners of the West. Increasing numbers of western watersheds are incubators of innovation, despite institutional structures and jurisdictional divisions that stifle creativity. We hope to encourage more innovation, more flexibility, and more experimentation. Above all, we hope to transform interest groups into partners by inviting them to

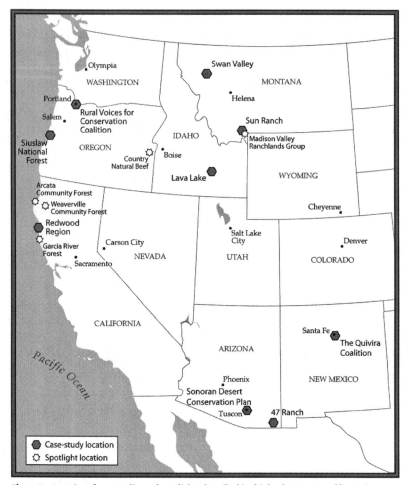

**Figure I.1.** Location of case studies and spotlights described in this book. Map created by Darin Jensen and Syd Wayman, 2013.

sit down at the same table together to solve problems rather than stake out positions.

The tone reflected in most of the chapters in this book is hopeful but cautious. Their authors celebrate important accomplishments and point out challenges and detours along the way. They show us that there is no template, no one model, no cookie-cutter approach for success. Instead, there are multiple pathways and mechanisms for stitching the West back together, each unique to the particular places and people involved. Moreover, as these cases portray, conservation is an adaptive process that is continually evolving, not a static, end state of affairs. This is to be expected, given the fact that western landscapes are dynamic, both socially and ecologically. What is needed, then, are enabling institutions and policies that support innovation, adaptive management, flexibility, and resilience. We hope this book will point toward such institutions and policies, helping to create a West that works.

## NOTES

1. State trust lands are lands held in trust by individual states and managed for specified beneficiaries, especially the K–12 public school systems in those states. The federal government granted each state portions of the federal public domain when it became a state to support public schools. Twenty-two states currently manage state trust lands, which total 135 million acres. Most of those acres are in the American West. See Jon A. Souder and Sally K. Fairfax, *State Trust Lands: History, Management, and Sustainable Use* (Lawrence: University Press of Kansas, 1996), 48, fig. 2-3, which provides the best overview of the subject.
2. State trust lands can and are being auctioned off to private real estate speculators and developers.

# PART ONE

## THE IMPORTANCE OF CONSERVING

## WESTERN WORKING LANDSCAPES

This book contends that sustainable working landscapes are critical to the conservation of biodiversity in the American West and the cultures of rural ranching and forestry that depend on them. Part 1 lays the groundwork for this argument, beginning in chapter 1, which provides historical context for how the West became the patchwork of private, tribal, and public lands that exist today. The authors of this chapter argue that two centuries of American land-use policy have left a legacy of contradictions and dichotomies (e.g., economy vs. environment, urban vs. rural, corporations vs. rural producers) that divide people having an interest in the West's working forests and rangelands. Today the collaborative conservation movement is trying to overcome these divides.

Chapter 2 highlights more recent trends in western working forests and rangelands and their current status. It shows that although public forests and rangelands are protected from land-use change, timber production and grazing on public lands have been declining for a number of reasons. This trend affects foresters and ranchers because nearly half of all western land is in federal ownership, because many of them depend on federal lands for their livelihoods, and because federal and private lands are economically and ecologically intertwined. At the same time, private forests and rangelands are being converted to agricultural and developed land uses and to investment properties, and are threatened by parcelization and fragmentation. The risk of this happening increases when their value for forestry and ranching cannot keep pace with their value for alternate land uses, especially real estate development.

Chapter 3 focuses on the question of why private forests and rangelands and tribal lands are so important for biodiversity conservation in the West. Protected areas are simply too small and poorly sited to support viable populations of area-sensitive and human-sensitive species, and a significant component of the biodiversity in different ecoregions occurs on private and tribal lands, which lie outside of protected areas. Furthermore, mixed-ownership landscapes generally contain more diverse habitat conditions than single-ownership landscapes. These findings suggest that environmental organizations and government agencies would do well to focus more attention on and investment in private ranch, private forest, and tribal lands and to work to promote biodiversity conservation at the landscape scale across land ownership boundaries.

Together the chapters in part 1 point to the need for stitching the West back together using the kinds of conservation strategies and tools illustrated in subsequent chapters. These strategies and tools are a response to the negative trends described here and offer an optimistic counterpoint that we hope will inspire more such efforts.

# 1

# A BRIEF HISTORY OF PEOPLE AND POLICY IN THE WEST

*Thomas E. Sheridan and Nathan F. Sayre*

IN BRIEF

- Today's checkerboard of public, tribal, and private land in the American West arose from contradictory land policies in the nineteenth and early twentieth centuries and now is at the heart of conflict among competing interests.
- The mid-nineteenth century brought western migration, clear-cut forests, and overstocked ranges and pushed Native Americans to ever-shrinking reservations, but in subsequent decades the federal government began establishing national forest preserves that regulated ranching and logging.
- The early twentieth century was characterized by government as both the source of conservation knowledge and the enforcer of management prescriptions, but soon tensions arose over managing for the "public" interest versus the interests of local residents dependent on public lands for livelihoods.
- Today finds ranchers and loggers pitted against environmentalists and urban recreationists, while federal agencies attempt to manage their lands for multiple uses; however, some collaborative conservation efforts have begun to bridge these divides.

Resource management in the West today reflects the contradictions in U.S. land policy during the nineteenth and early twentieth centuries. Beginning with the 1803 Louisiana Purchase of 828,000 square miles of land west of the Mississippi River, the aggressive young nation swallowed up huge chunks of

North America through a mixture of diplomacy and military conquest. Native American claims to the continent were largely ignored, justified in part by the belief that the land belonged to those who "improved" it under the civil right of private ownership, rather than to those who had only natural rights, "when men held the earth in common every man sowing and feeding where he pleased."[1] The Treaty of Guadalupe-Hidalgo, which ended the war between Mexico and the United States in 1848, acknowledged U.S. claims to Texas, drawing its boundary at the Rio Grande, and ceded California and most of the American Southwest between California and Texas to the United States. The treaty was supposed to safeguard the property rights of Mexicans, but of the 38 million acres bestowed under Spanish and Mexican land grants, only two million were confirmed by the U.S. Court of Private Land Claims. And of the confirmed grants, more than 80 percent of the land fell into Anglo-American hands.[2]

Historian Frederick Merk described the process as "the greatest real estate transaction in modern history" as the United States more than tripled in size in its march from the Atlantic to the Pacific.[3] During most of the 1800s, the primary goal of the federal government was to encourage the settlement and development of these vast acquisitions by granting portions of the public domain to corporate interests, particularly canal companies and railroads, and to individuals. But visions of a Jeffersonian democracy of small farmers civilizing the frontier were often at odds with the demands of an emerging industrial capitalism that hungered for the natural resources of the trans-Mississippi West.[4]

In the Midwest, land policies like the General Homestead Act of 1862 did encourage the proliferation of small farms rather than plantations, even if speculators often made huge profits in the transaction.[5] In the more arid regions west of the 100th meridian, however, farming homesteads only made sense along rivers and streams, which were few and far between. After the Civil War, the "free" land of the western public domain attracted capital from the British Isles and the eastern United States. Forests were clear-cut, prairies plowed, minerals plundered, and ranges overstocked.[6] It was a tragedy, not of the commons as Garrett Hardin and others argued, but of an open-access regime where no one, not even the federal government, regulated resource use.[7]

Another enlargement of the public domain came at the expense of American Indians. During the early years of the republic, the United States pushed Native Americans ever westward beyond the advancing frontier. It was a brutal form of primitive accumulation despite the numerous treaties signed between tribes and the federal government along the way. And even though

Chief Justice John Marshall declared the tribes "dependent domestic nations," Andrew Jackson and his supporters evicted the Five Civilized Tribes (Cherokee, Chickasaw, Choctaw, Creek, and Seminole) from their treaty lands in the Southeast and force-marched them on the so-called Trail of Tears to Indian Territory west of the Mississippi. But as U.S. pioneers crossed the Mississippi and settled the Great Plains, Indian Territory shrank. In 1849, Indian affairs were transferred from the War Department to the Department of the Interior, and the government began establishing reservations for Plains tribes on the portions of their homelands deemed least desirable to settlers. Under the Treaty of Guadalupe-Hidalgo, Spanish and Mexican land grants to Pueblo peoples in New Mexico had to be recognized. The first stages of a reservation "system" were beginning to take shape.[8]

Then, in 1872, Congress prohibited further treaties with Indians as conflicts with Plains tribes and the Apaches intensified. Military conquest, not negotiation with "domestic nations," became national policy. Reservations continued to be created, but their management was turned over to Indian agents appointed by different Protestant denominations. Native Americans often were not allowed to practice their own religions or follow their own cultural patterns. Instead, they were to be "civilized" and assimilated into capitalist, Christian society. This involved turning them into yeoman farmers as well. The Dawes Act of 1887 divided reservations into individual allotments that became the private property of Indian households. "Surplus" lands were sold to non-Indians. Tribes lost an estimated 91 million acres of reservation lands between 1887 and 1934, when the Indian Reorganization Act terminated the allotment program and gave tribes a little more control over their natural resources. They've been struggling to assert their sovereignty ever since.[9]

Beginning in the 1890s, the federal government began to reconsider its laissez-faire attitudes toward the public domain.[10] It set aside forest reserves to protect watersheds and timber resources. Under President Theodore Roosevelt, conservationist Gifford Pinchot transformed these scattered reserves into the national forest system, permanently removing millions of acres from the public domain, largely by executive order rather than Congressional legislation.[11] Pinchot won his battle with preservationists like John Muir, who wanted to eliminate grazing and logging on the national forests. Instead, ranchers and loggers could continue to harvest grass and timber but under federal regulation. When the Taylor Grazing Act of 1934 extended such regulation to the remaining public domain, the rural West of the twentieth century was born.[12]

Pinchot, Roosevelt, and other leaders of the so-called Progressive Era also institutionalized a particular vision of the relationship between science, politics, conservation, and management.[13] Renewable natural resources such as grass and timber were understood mechanistically, and the task of science was to reveal the "laws" that governed these mechanisms. The resulting knowledge was construed as universal and "objective"—and therefore apolitical—even when the science itself was in its infancy.[14] The resulting policies were implemented by bureaucratic national agencies, acting as both sources of information and enforcers of management prescriptions. The agencies viewed local people as lacking the knowledge needed for conservation—after all, most locals had only arrived very recently, often with little or no familiarity with their new surroundings. This, of course, was not true of Native Americans, and is no longer true for long-time ranchers and foresters, but "science" continues to be politically powerful precisely because it appears to operate above or outside of politics. These technocratic conceits are among the central legacies to which the current collaborative conservation movement responds.

Apart from the scientific shortcomings of the resulting policies, the progressive era model did not resolve the underlying tension that it was intended to address: how to balance the "public" interest, defined at the national or state scale, with the interests of local residents who depended on public lands for their livelihoods. Pinchot's Forest Service and the Taylor Grazing Act recognized local interests in certain ways (through rules about who could hold grazing permits, e.g., or institutions such as grazing districts), often to the detriment of marginalized groups, including Hispanics, Native Americans, African Americans, and subsistence producers.[15] In so doing, however, federal agencies opened themselves up to charges of co-optation, such as those leveled by Bernard DeVoto, who launched the half-century long "rangeland conflict" in the pages of *Harper's* beginning in the 1930s.[16] Whether faulty science or "agency capture" was more to blame in any particular instance, one clear result was to cast resource-dependent local communities as inherently opposed to environmentalism as it developed in the second half of the twentieth century.[17]

Today, both the U.S. Forest Service and the Bureau of Land Management manage their federal lands for multiple uses, including resource extraction, recreation, and the conservation of biodiversity. The priority given to these uses varies according to presidential administration and political pressure. As historian Paul Hirt has shown, the Forest Service privileged resource extraction, especially timber cutting, from World War II through the 1980s.[18] But as the West became an overwhelmingly urban region during the postwar period, more and more city dwellers turned to federal public lands for recreational opportunities. That profound demographic shift, along with the growth of the modern environmental movement beginning in the 1960s, challenged the political power of loggers and ranchers—the "lords of yesterday" in Charles Wilkinson's memorable phrase.[19] The passage of strong federal environmental statutes, especially the National Environmental Policy Act in 1969 and the Endangered Species Act in 1973, gave environmentalists powerful legal tools to force agencies to change entrenched management practices.

As C. Klyza and Sousa point out, however, this so-called golden era of bipartisan environmental legislation did not eliminate the lords of yesterday: "Instead, each new movement layered new institutions and agendas atop the old and empowered new interests even while preserving many of the legal and institutional bases of the claims made by old interests."[20] In their words, the lords of yesterday are now battling the "'lords of a little while ago'"—groups like the Center for Biological Diversity that employ environmental legislation adopted in the 1960s and 1970s to advance their interests and values. "This reality energizes all politics, certainly modern environmental politics, and there is no escape from the crashing and grinding of multiple orders in this field."[21]

Many of today's loggers and ranchers have experienced this legal, political, economic, and cultural sea change in their own lifetimes. It is easy to understand why many of them feel that their livelihoods and ways of life are under assault, especially when some environmental groups advocate the end of grazing and logging on public lands. The grassroots collaborative conservation movement sprang to life across the American West in the 1990s in part as a defensive response to these pressures. Ranchers and foresters saw their access to grass and timber on federal lands threatened by environmentalists who portrayed them as ecological scourges.[22] In the Pacific Northwest, logging operations and sawmills shut down because of lawsuits to preserve the habitat of the Northern spotted owl (*Strix occidentalis caurina*). Environmentalists did not succeed in eliminating grazing on federal lands, but ranchers watched environmental groups sue the Forest Service and the Bureau of Land Management for noncompliance with the Endangered Species Act or National Environmental Policy Act on grazing allotments on which their livelihoods depended. Unless they countersued, those ranchers were not even at the legal table. Most ranchers didn't have the deep pockets to hire lawyers and fight such battles on their own.[23] But they could reach out to potential allies among scientists, more moderate environmental groups, and local people to build alliances around specific landscapes.

The other and much deeper reason for the emergence of the collaborative conservation movement was the conservationist ethic of ranchers and foresters themselves.[24] Many ranchers wanted to heal watersheds and restore fire as a natural disturbance in order to halt the encroachment of woody plants and to grow more grass. When they tried to do so, however, they were frustrated by the crazy quilt of land tenure and jurisdictions—federal, state, county, private—that had carved up the West over the past century. Water and fire rarely confine themselves to land ownership boundaries, but bureaucracies have only recently begun to "think" like mountains or watersheds. The result was gridlock instead of collaboration, confrontation instead of cooperation, polemics instead of a search for common ground.[25] If anything, the battle between ranchers and environmentalists played into the hands of developers.[26]

At the beginning of the twentieth century, when preservationists like John Muir and conservationists like Gifford Pinchot were fighting for the support of Theodore Roosevelt, the West was a sparsely populated region dominated by the extractive industries of mining, ranching, forestry, and agriculture. Today it contains large urban areas, with most people concentrated in megalopolises like Seattle, Denver, Los Angeles, and Phoenix.[27] Those cities spin off suburbs and exurbs in every direction developers can find private land. Meanwhile, population is growing in rural western counties as a result of amenity migration by people from urban areas who seek open space, beautiful scenery, clean air and water, and easy access to outdoor recreation.[28] As a result, the value of private lands in the West has skyrocketed at a time when making a living as a rancher or forester on intermixed public and private lands is becoming increasingly difficult. Lands that once produced beef, mutton, wool, timber, and wood products now sprout second homes. Subdivisions degrade wildlife habitat, disrupt wildlife migrations, and accelerate the spread of exotic species.[29] Wildcat (or illegal) dumping and off-highway vehicle trails proliferate. The restoration of fire's necessary role as a natural disturbance in fire-adapted ecosystems, desperately needed because of the encroachment of woody species on grasslands and the buildup of fuel loads in forests, becomes politically challenging because the protection of private property usually trumps ecosystem health, and the prevailing policy of fire suppression is difficult to change.

One outcome of this history is that land tenure in the rural West involves a bewildering array of jurisdictions, especially across elevational gradients between the mountains (which are often administered by the Forest Service) and the lower, more fertile valley bottoms.[30] Boundaries between federal, state, and private lands tend to carve up ecosystems with little regard for the flow of water or other natural processes. One great challenge in the twenty-first century, then, is to stitch the West back together again through the creation of enduring partnerships that crosscut these boundaries. As this volume demonstrates, today's collaborative conservation movement is attempting to do just that.

## NOTES

1. John Winthrop, quoted in W. Cronon, *Changes in the Land: Indians, Colonists, and the Ecology of New England* (New York: Hill and Wang, 1983), 56.

2. R. Bradfute, *The Court of Private Land Claims: The Adjudication of Spanish and Mexican Land Grant Titles, 1891–1904* (Albuquerque: University of New Mexico Press, 1975); R. G. del Castillo, *The Treaty of Guadalupe-Hidalgo: A Legacy of Conflict* (Norman: University of Oklahoma Press, 1990); M. Ebright, *Land Grants and Lawsuits in Northern New Mexico* (Albuquerque: University of New Mexico Press, 1994); H. Lamar, *The Far Southwest, 1846–1912: A Territorial History* (Albuquerque: University of New Mexico Press, 2000); T. E. Sheridan, *Landscapes of Fraud: Mission Tumacácori, the Baca Float, and the Betrayal of the O'odham* (Tucson: University of New Mexico Press, 2006).

3. F. Merk, "Foreword," in *The Frontier in American Development: Essays in Honor of Paul Wallace Gates*, ed. David M. Ellis (Ithaca, NY: Cornell University Press, 1969), ix.

4. D. Worster, *Rivers of Empire: Water, Aridity, and the Growth of the American West* (New York: Pantheon Books, 1985); R. White, *"It's Your Misfortune and None of My Own": A History of the American West* (Norman: University of Oklahoma Press, 1991); W. Cronon, *Nature's Metropolis: Chicago and the Great West* (New York: W. W. Norton & Company, 1991).

5. White, *It's Your Misfortune*, 137–54; P. Gates, *The Jeffersonian Dream: Studies in the History of American Land Policy and Development*, ed. A. G. Bogue and M. B. Bogue (Albuquerque: University of New Mexico Press, 1996), esp. his seminal essay, "The Role of the Land Speculator in Western Development," 6–22; Sheridan, *Landscapes of Fraud*, 139–43.

6. N. F. Sayre, *Ranching, Endangered Species, and Urbanization in the Southwest: Species of Capital* (Tucson: University of Arizona Press, 2002).

7. G. Hardin, "The Tragedy of the Commons," *Science* 162, no. 3859 (1968): 1243–48; White, *It's Your Misfortune*, 222–27; D. Worster, "Cowboy Ecology," in *Under Western Skies: Nature and History in the American West* (New York: Oxford University Press, 1992), 34–52; T. E. Sheridan, *Arizona: A History*, rev. ed. (Tucson: University of Arizona Press, 2012), 131–65.

8. C. F. Wilkinson, *American Indians, Time, and the Law* (New Haven, CT: Yale University Press, 1985), 7–31; White, *It's Your Misfortune*, 85–118.

9. Wilkinson, *American Indians and Law*, 7–31; White, *It's Your Misfortune*, 85–118; D. E. Wilkins and K. T. Lomawaima, *Uneven Ground: American Indian Sovereignty and Federal Law* (Norman: University of Oklahoma Press, 2001).

10. For general overviews of federal land management up to the present, see T. Koontz, T. Steelman, J. Carmin, K. S. Korfmacher, C. Moseley, and C. Thomas, *Collaborative Environmental Management: What Roles for Government?* (Washington, DC: Resources for the Future, 2004); S. Mullner, W. Hubert, and T. Wesche, "Evolving Paradigms

for Landscape-Scale Renewable Resource Management in the United States," *Environmental Science and Policy* 4 (2001): 39–49.

11. D. Brinkley, *The Wilderness Warrior: Theodore Roosevelt and the Crusade for America* (New York: Harper Perennial, 2009), 751–91; T. Egan, *The Big Burn: Teddy Roosevelt and the Fire That Saved America* (Boston: Houghton Mifflin Harcourt, 2009), 53–72.

12. White, *It's Your Misfortune*, 477–81; Worster, "Cowboy Ecology."

13. S. P. Hays, *Conservation and the Gospel of Efficiency* (Cambridge, MA: Harvard University Press, 1959); F. Cubbage, J. O'Laughlin, and C. Bullock III, *Forest Resource Policy* (New York: John Wiley and Sons, 1993); P. W. Hirt, *A Conspiracy of Optimism: Management of the National Forests since World War II* (Lincoln: University of Nebraska Press, 1994); Mullner et al., "Evolving Paradigms."

14. N. F. Sayre, "Climax and 'Original Capacity': The Science and Aesthetics of Ecological Restoration in the Southwestern USA," *Ecological Restoration* 28 (2010): 23–31.

15. White, *It's Your Misfortune*, 431–57; J. Kosek, *Understories: The Political Life of Forests in Northern New Mexico* (Durham, NC: Duke University Press, 2006).

16. B. DeVoto, "The West: A Plundered Province," *Harper's Magazine* 179 (1934): 355–64.

17. R. White, "'Are You an Environmentalist or Do You Work for a Living?' Work and Nature," in *Uncommon Ground: Rethinking the Human Place in Nature*, ed. William Cronon (New York: Norton, 1996), 171–85.

18. Hirt, *Conspiracy of Optimism*.

19. C. F. Wilkinson, *Crossing the Next Meridian: Land, Water, and the Future of the West* (Washington, DC: Island Press, 1992), 3–27.

20. C. Klyza and D. Christopher, *American Environmental Policy, 1990–2006* (Cambridge, MA: MIT Press, 2008), 9.

21. Ibid., 10

22. Wilkinson, *Crossing the Next Meridian*, 3–27; T. Power, *Lost Landscapes and Failed Economies: The Search for a Value of Place* (Washington, DC: Island Press, 1996), 131–90; D. Donahue, *The Western Range Revisited: Removing Livestock from Public Lands to Conserve Native Biodiversity* (Norman: University of Oklahoma Press, 1999); G. Wuerthner and M. Matteson, eds., *Welfare Ranching: The Subsidized Destruction of the American West* (Washington, DC: Island Press, 2002); S. P. Hays, *Wars in the Woods: The Rise of Ecological Forestry in America* (Pittsburgh: University of Pittsburgh Press, 2007).

23. T. E. Sheridan, "Cows, Condos, and the Contested Commons: The Political Ecology of Ranching on the Arizona-Sonora Borderlands," *Human Organization* 60, no. 2 (2001): 141–52, and "Embattled Ranchers, Endangered Species and Urban Sprawl: The Political Ecology of the New American West," *Annual Review of Anthropology* 36 (2007): 121–38.

24. R. Knight, W. Gilgert, and E. Marston, *Ranching West of the 100th Meridian: Culture, Ecology, and Economics* (Washington, DC: Island Press, 2002); C. White, *Revolution on the Range: The Rise of the New Ranch in the American West* (Washington, DC: Island Press, 2008).

25. J. M. Wondolleck and S. L. Yaffee, *Making Collaboration Work: Lessons from Innovation in Natural Resources Management* (Washington, DC: Island Press, 2000); P. Brick, D. Snow, and S. Van de Wetering, eds., *Across the Great Divide: Explorations in Collaborative Conservation and the American West* (Washington, DC: Island Press, 2001); Koontz et al., *Collaborative Environmental Management*; Klyza and Sousa, *American Environmental Policy*; C. Wilmsen, W. Elmendorf, L. Fisher, J. Ross, B. Sarathy, and G. Wells, *Part-*

*nerships for Empowerment: Participatory Research for Community-Based Natural Resource Management* (London: Earthscan, 2008); R. D. Margerum, *Beyond Consensus: Improving Collaboration to Solve Complex Public Problems* (Cambridge, MA: MIT Press, 2011).

26. Sayre, *Ranching, Endangered Species, Urbanization* 105–25 (n. 6 above, this chapter), and "Climax and 'Original Capacity'" (n. 14 above, this chapter); see chap. 13 for how Pima County is conserving working ranches to preserve open space and biodiversity.

27. Robert E. Lang, Andrea Sarzynski, and Mark Muro, *Mountain Megas: America's Newest Metropolitan Places and a Federal Partnership to Help Them Prosper* (Washington, DC: Brookings Institution, 2008).

28. D. G. Brown, K. M. Johnson, T. R. Loveland, and D. M. Theobald, "Rural Land-Use Trends in the Conterminous United States, 1950–2000," *Ecological Applications* 15 (2005):1851–63.

29. A. J. Hansen, R. L. Knight, J. M. Marzluff, S. Powell, K. Brown, P. H. Gude, and K. Jones, "Effects of Exurban Development on Biodiversity: Patterns, Mechanisms, and Research Needs," *Ecological Applications* 15 (2005): 1893–1905.

30. J. M. Scott, F. W. Davis, R. G. McGhie, R. G. Wright, C. Groves, and J. Estes, "Nature Reserves: Do They Capture the Full Range of America's Biological Diversity?" *Ecological Applications* 11 (2001): 999–1007.

# 2

# STATUS AND TRENDS OF WESTERN WORKING LANDSCAPES

*Susan Charnley, Thomas E. Sheridan, and Nathan F. Sayre*

### IN BRIEF

- Although western forests supply almost half the nation's softwood lumber in addition to other wood products, timber harvest has dropped by roughly one-half since 1990, with most of the decrease occurring on federal lands.
- Even though permitted livestock use on federal lands has decreased over the past two decades, western rangelands still support about one-fifth of the cattle and half of the sheep in the United States.
- Rural land in the West is being converted to development at an average rate of 2.32 percent annually as human population growth there far exceeds the national average.
- Private nonindustrial forestlands are shrinking as the number of owners is increasing, leading to parcelization (fragmented ownerships of small parcels) and forest fragmentation; meanwhile private industrial forestlands are being sold and managed as investment properties, making their future conservation and land-use status uncertain.

This book is about the conservation of working landscapes in the West today. In order to provide context for saving these wide open spaces, it is helpful to look at where they occur, their current status, what has been happening to them over the past few decades, and how this has affected the people who depend on them for their livelihoods. Toward that end, we provide, in this chapter, an overview of the status and trends in the West's working forests and rangelands on both public and private lands.

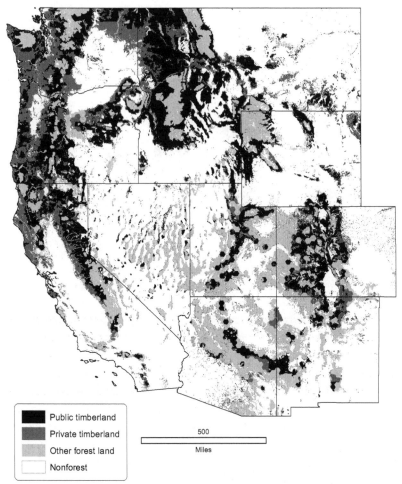

**Figure 2.1.** Extent and ownership of western U.S. timberlands. Map authored by John M. Chase, 2013 (U.S. Department of Agriculture, Forest Service, Pacific Northwest Research Station, Portland, OR). Data sources analyzed in the creation of the map: "Protected Areas Database of the US, PAD-US (CBI Edition)," version 2, Conservation Biology Institute, June 2009–October 2012, http://consbio.org/products/projects/pad-us-cbi -edition; Mark D. Nelson, Greg C. Liknes, and Brett J. Butler, 2010. "Forest Ownership in the Conterminous United States: ForestOwn_v1 Geospatial Dataset" (U.S. Department of Agriculture, Forest Service, Northern Research Station, Newtown Square, PA); USDA Forest Service 2007, "Unproductive Timberland of the Conterminous United States Summarized across the US Environmental Protection Agency Environmental Monitoring and Assessment Program Hexagon Sampling Framework," RPA2007_UnproductiveForest geospatial dataset, part of 2007 updates to Resource Planning Act Assessment of 2000 (U.S. Department of Agriculture, Forest Service, Northern Research Station, Newtown Square, PA). Methodological references: two nondataset references speak to the methodology behind the Forest Service's Forest Inventory and Analysis ownership dataset and its unproductive timberland dataset—W. Brad Smith, Patrick D. Miles, Charles H. Perry, and Scott A. Pugh, *Forest Resources of the United States, 2007*, General Technical Report WO-78 (Washington, D.C.: Washington Office, Forest Service, U.S. Dept. of Agriculture, 2009); and D. White, A. J. Kimerling, and W. S. Overton, "Cartographic and Geometric Components of a Global Sampling Design for Environmental Monitoring," *Cartography and Geographic Information Systems.* 19 (1992): 5–22.

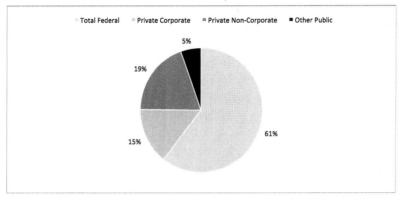

**Figure 2.2.** Division of ownership of western timberlands in 2007. Source: W. B. Smith, P. D. Miles, C. H. Perry, and S. A. Pugh, *Forest Resources of the United States, 2007: A Technical Document Supporting the Forest Service 2010 RPA Assessment*, General Technical Report WO-78 (Washington, DC: U.S. Department of Agriculture, Forest Service, Washington Office, 2009). Figure created in 2013.

## FORESTLANDS: CURRENT STATUS

The West's working forests from which wood products are produced lie primarily in the timberlands located in the Northwest, California, and the Rocky Mountains (fig. 2.1).[1] Nearly two-thirds of these timberlands are in public ownership (with 57 percent of the total on national forestlands); the rest are privately held by corporate owners (forest industry, forest management companies, timber investment management organizations, and other companies) or by noncorporate owners such as individuals, families, trusts, nongovernmental organizations, and other unincorporated groups (fig. 2.2).[2] Western timberlands comprise just over half of the region's forestlands (56 percent) and cover about 128.3 million acres.[3]

From an economic standpoint, the working forests of the West supplied roughly 45 percent of the nation's softwood lumber in 2011 (11.9 billion board feet).[4] The majority of this production came from Oregon and Washington. In addition, Northwestern states supplied about 16 percent of the nation's plywood and other structural panel board in 2011 (3,031 million sq. ft., based on 3/8-inch thickness).[5] In 2011, employment in the forest products industries was estimated at 137,200 people in Oregon, Washington, California, Montana, and Idaho, where these industries are concentrated.[6] Although these jobs comprise a small part of total employment in these states, they are nonetheless important. Unemployment in rural western counties is typically higher than in urban areas, and jobs in the wood products industry tend to be higher paying than jobs in the services sector, the major employment sector in many rural western counties.[7]

Working forests are also a source of nontimber forest products such as mushrooms, berries, and floral greens that support a multimillion dollar industry in the Pacific Northwest and provide a host of additional products that are valued for food, medicine, basketry, decoration, and other cultural uses.[8] Harvesting of commercial nontimber forest products can be an important source of supplemental income for rural community residents, as well as recent immigrants, and offers economic diversification opportunities in rural communities.[9] In addition, these forests hold a large supply of biomass that is increasingly being utilized as a domestic source of renewable energy and for manufacturing wood products for niche markets, creating economic development opportunities in forest communities.

## RANGELANDS: CURRENT STATUS

There are an estimated 426.7 million acres classified as rangelands in the U.S. West (fig. 2.3).[10] Of these rangelands, about 205 million acres (48 percent) occur on federal lands (primarily Forest Service and Bureau of Land Management), and the remainder are found on nonfederal lands (mostly private). Arizona, New Mexico, Montana, Nevada, Wyoming, and California are the western states with the largest acreages of rangelands (fig. 2.4). Grazing, of course, does not necessarily occur across all of these rangelands, but no one has quantified the extent of grazing lands in the West in a reliable way.[11] Grazing can also occur on pasturelands (land used primarily for the purpose of producing introduced forage grasses for livestock) and forestlands (in the understory).[12] Because grazing takes place on deserts, grasslands, and forestlands, parts of the West support both ranching and the timber industry.

Despite their aridity and rugged terrain, the rangelands of the eleven western states support about one-fifth of the cattle and half of the sheep in the United States, with California having the greatest numbers of both (table 2.1). In 2012, according to the USDA National Agricultural Statistics Service, the West supported 20.2 million cattle including calves, 21.8 percent of the national total (92.7 million; table 2.1). The West's proportion of sheep including lambs is more than twice as high: 2.7 million, or 51.5 percent of the total (5.3 million) in 2012 (table 2.1). The number of cattle in the West has remained fairly stable over the past two decades, while the number of sheep has dropped by almost half. Below, we look more closely at trends in forestry and ranching on public and private working lands in the West.

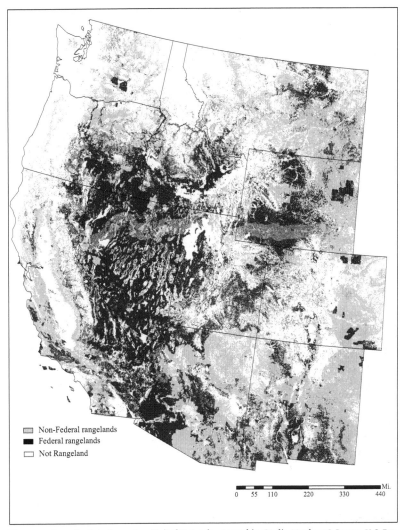

Non-Federal rangelands
Federal rangelands
Not Rangeland

Mi.
0   55   110        220        330        440

**Figure 2.3.** Rangelands in the western United States, by ownership. Credit: Matthew C. Reeves, U.S. Forest Service, Rocky Mountain Research Station. Data source: M. C. Reeves and J. E. Mitchell, "Extent of Conterminous US Rangelands: Quantifying Implications of Differing Agency Perspectives," *Rangeland Ecology and Management* 64, no. 6 (2011): 585–97. Figure created in 2013.

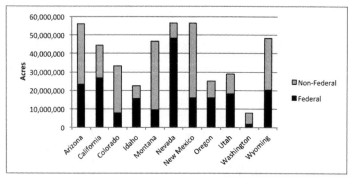

**Figure 2.4.** Acres of federal and nonfederal rangelands in the western states. Source: Nonfederal—National Resources Inventory Data, 2007; Federal—2001 data derived from M. C. Reeves and J. E. Mitchell, *A Synoptic Review of U.S. Rangelands: A Technical Document Supporting the Forest Service 2010 RPA Assessment*, General Technical Report RMRS-GTR-288 (Fort Collins, CO: U.S. Department of Agriculture, Forest Service, Rocky Mountain Research Station, 2012). Figure created in 2013.

TABLE 2.1. Cattle and Sheep in the 11 Western States, 2012

| State | Cattle including Calves (Per 1,000 head) | Sheep including Lambs (Per 1,000 head) |
|---|---|---|
| Arizona | 920 | 140 |
| California | 5,350 | 570 |
| Colorado | 2,750 | 460 |
| Idaho | 2,220 | 240 |
| Montana | 2,500 | 225 |
| Nevada | 470 | 70 |
| New Mexico | 1,390 | 100 |
| Oregon | 1,300 | 200 |
| Utah | 800 | 305 |
| Washington | 1,110 | 52 |
| Wyoming | 1,360 | 370 |
| Total West | 20,170 | 2,732 |
| West as percent of U.S. total | 21.8 | 51.5 |
| U.S. Total | 92,700 | 5,300 |

Source: USDA National Agricultural Statistics Service (for cattle: http://usda01.library.cornell.edu/usda/current/Catt/Catt-02-01-2013.pdf; for sheep: http://usda01.library.cornell.edu/usda/current/SheeGoat/SheeGoat-02-01-2013.pdf). 2013.

## FEDERAL LANDS AS WORKING LANDS

About 48 percent of all western lands are in federal ownership, ranging from nearly 85 percent in Nevada, to about 27 percent in Washington:

| | |
|---|---|
| Nevada | 84.6 |
| Utah | 63.1 |
| Idaho | 62.7 |
| Oregon | 50.3 |
| California | 45.9 |
| Wyoming | 45.9 |
| Arizona | 41.7 |
| Colorado | 35.7 |
| New Mexico | 33.9 |
| Montana | 28.8 |
| Washington | 27.1[13] |

These figures do not include Native American reservations, where tribes and the Department of the Interior's Bureau of Indian Affairs are engaged in complex, ongoing negotiations to determine where federal oversight ends and tribal sovereignty begins. Historically, most federal lands, including Indian reservations, were managed as working forests and rangelands, as described in chapter 1. They have been, in fact, critical to the ranching and timber industries of the West, and continue to be so today, in part because they are protected from residential development.

Multiple uses, such as timber harvesting, grazing, mining, or off-highway vehicle use, are allowed on roughly 69 percent of federal lands in the western states.[14] However, since the 1990s, the focus of federal land management has shifted to recreation, biodiversity conservation, and ecological restoration in response to changing public values and to comply with the National Environmental Policy Act, Endangered Species Act, Healthy Forests Restoration Act, and other federal environmental laws, particularly where lawsuits brought by environmentalists have forced the issue.[15]

As a result, the pendulum has swung in the other direction. Timber harvest from western timberlands (fig. 2.5) and employment in the forest products industries (fig. 2.6) have dropped by 50 percent and 48 percent, respectively, since 1990. Most of this harvest decrease occurred on federal lands. Permitted livestock use on federal lands has also decreased, from about 10 to 7.9 million animal unit months over the past two decades on Bureau of

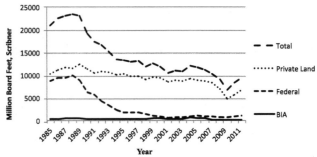

**Figure 2.5.** Total timber harvested in Washington, Oregon, California, Montana, and Idaho, by ownership, 1985–2011. Timber harvests from state and other public lands are not broken out from the total. Annual harvests under one million board feet from an individual ownership are not included in the totals.Sources: D. D. Warren, *Production, Prices, Employment, and Trade in Northwest Forest Industries, All Quarters 2000*, Research bulletin PNW-RB-236 (Portland, OR: U.S. Department of Agriculture, Forest Service, Pacific Northwest Research Station, 2002); X. Zhou and D. D. Warren, *Production, Prices, Employment, and Trade in Northwest Forest Industries, All Quarters 2011*, Research Bulletin PNW-RB-264 (Portland, OR: U.S. Department of Agriculture, Forest Service, Pacific Northwest Research Station, 2012). Figure created in 2013.

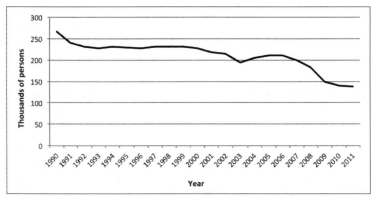

**Figure 2.6.** Total employment in the forest products industries in Washington, Oregon, California, Montana, and Idaho, 1990–2011. Sources: D. D. Warren, *Production, Prices, Employment, and Trade in Northwest Forest Industries, All Quarters 2000*, Research bulletin PNW-RB-236 (Portland, OR: U.S. Department of Agriculture, Forest Service, Pacific Northwest Research Station, 2002); X. Zhou and D. D. Warren, *Production, Prices, Employment, and Trade in Northwest Forest Industries, All Quarters 2011*, Research Bulletin PNW-RB-264 (Portland, OR: U.S. Department of Agriculture, Forest Service, Pacific Northwest Research Station, 2012). Figure created in 2013.

Land Management (BLM) lands, and from about 9.5 to 8.3 million animal unit months on Forest Service lands in the last decade.[16] At the same time, the majority of outdoor recreation activities in the West take place on public lands. Recreational use of federal lands nationwide has been stable or increasing (with the exception of Forest Service lands), even as traditional hunting

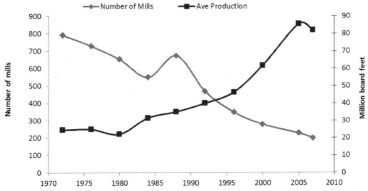

**Figure 2.7.** Change in number of mills and their average production, 1971–2009. Source: R. W. Haynes and T. M. Quigley, "The Western Forest Products Industry: Status, Issues, and Implications for the Future" (report prepared by Management and Engineering Technologies International, El Paso, TX, for the Western Forestry Leadership Coalition, 2009). Figure created in 2013.

and fishing uses are flat or in decline.[17] Mechanized recreation in particular has increased nationwide: between 1995 and 2006, retail sales of off-highway vehicles in the United States nearly tripled, from 368,600 to more than one million. The number of off-highway vehicle operators climbed from about 27.3 million in 1994–95 to more than 40 million in 2007.[18] Such recreation can potentially degrade federal watersheds and wildlife habitat.

Reductions in timber production and grazing on federal lands have been detrimental for many residents of nearby communities who have historically made a living from ranching and forestry on federal and state lands. The dramatic drop in timber production on federal lands that has occurred since the late 1980s (fig. 2.5) has been a major contributor to the drop in timber-based employment, though reductions in federal timber harvests are only one of several factors contributing to these employment trends. Other factors include technological changes in the wood products manufacturing industry that have increased efficiency (see fig. 2.7) and changes in market demand for wood products.[19] Reduced federal timber harvesting affected forest communities differently, depending on local characteristics and relations to national forests. Communities located close to federal forestlands, and whose economies were highly dependent on those lands, experienced greater employment declines and exhibited lower socioeconomic well-being rankings following the declines (in 2000) than communities farther away.[20]

The loss of mill infrastructure throughout the West to support forest products manufacturing (fig. 2.7) has important implications for the ability of federal agencies to produce forest products in a cost-effective manner because mills constitute markets for those products. Fewer mills mean less competition and lower stumpage prices for logs. And, the farther the haul distance from the harvest site to the processing facility, the higher the transport costs and the less economical the timber sale. In the absence of nearby infrastructure, sawlogs, small-diameter trees, and biomass lose commercial value and become expensive to remove, making it harder for agencies to produce timber, accomplish forest restoration goals such as fire risk reduction, and support jobs in the wood products industries.[21] Mill infrastructure is not only important for federal land management; it also provides a market for wood products harvested on private and tribal lands, which is also less economical in the absence of local processing infrastructure. If the economic value of private lands for timber production drops relative to their value for other land uses, they are more likely to be sold and converted to agricultural uses or to development.[22] The remaining mill infrastructure in the West is concentrated along major interstate highways in western Washington, western Oregon, California, western Montana, and northern Idaho (fig. 2.8). Because the West's working forests are highly dependent on this remaining infrastructure, it is important to maintain what is left of it.

Ranchers are also highly dependent on public lands.[23] The way in which ranching developed in the West means that today, many ranches consist of a core of private land, and grazing allotment(s) or leases on federal and/or state lands.[24] Such ranches are mosaics of private and public land tenure, but they are managed together as a ranch unit.[25] National forest grazing allotments provide important summer pasture, while BLM and state trust lands, generally found at lower elevations, are often grazed year-round. Rangeland health on BLM lands is good and for the most part stable; the Forest Service does not monitor rangeland health consistently.[26]

The Forest Service and BLM are mandated to manage public lands for multiple uses. The power of other stakeholders like environmental groups and recreation interests has grown in recent decades, often at the expense of ranchers and foresters. As mentioned in chapter 1, grazing on public lands is controversial, and a host of environmental regulations constrain grazing activities and demand more attention and expense from ranchers. These factors together create a state of uncertainty around public lands grazing and threaten ranching in the West because ranching on private lands is often highly dependent on the public lands component: without secure access to grazing leases or allotments, many ranchers would not be able to maintain viable operations.

One alternative to public lands grazing for these ranchers is leasing private range. However, the cost of leasing private rangelands to supplement grazing on their deeded lands can be prohibitive, especially where compet-

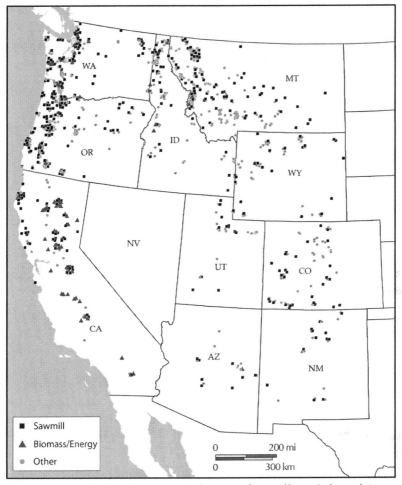

**Figure 2.8.** Location of mills in the western United States, as of 2006–8 (date varies by state). Source: Data come from the University of Montana, Bureau of Business and Economic Research. Credit: GIS map produced by Jean Daniels, U.S. Forest Service, Pacific Northwest Research Station; presentation modified by Darin Jensen and Syd Wayman in 2013.

ing land uses, such as farming or real-estate development, create a shortage of range.[27] The predominance of public rangelands in many rural western counties, and in states like Nevada and California, means leasing private range may not be an option. Ranching is a difficult way to make a living under the best of circumstances. Frequent and prolonged droughts, and meat prices that rise more slowly than the price of fuel, feed, and other inputs, shave profit margins down to the bone. The loss or reduction of animal unit months on federal grazing allotments could drive many ranchers out of busi-

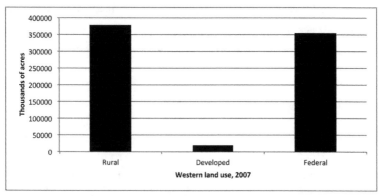

**Figure 2.9.** Use of western land in 2007. Source: 2007 National Resources Inventory data (U.S. Department of Agriculture, *Summary Report: 2007 National Resources Inventory* (Washington, DC: Natural Resources Conservation Service; Ames, IA: Center for Survey Statistics and Methodology, Iowa State University, 2009). Figure created in 2013.

ness and force them to sell their private lands to developers or to develop it themselves.

## PRIVATE LANDS AS WORKING LANDS

The majority of nonfederal land in the West remains rural and undeveloped (fig. 2.9).[28] Nevertheless, between 1982 and 2007, the amount of land converted to development in the western states increased at an average rate of 2.32 percent annually, effectively removing it from natural resource–related uses.[29] Population growth is one driver of development: between 1990 and 2010, the population of the western states increased 37 percent, from 51.1 million to 69.9 million people (the national average was 24 percent).[30] Nationwide, land conversion to development occurs primarily as a result of the conversion of cropland and forestland, with forests being the largest source of land conversion to developed uses.[31]

Private forestlands can be divided into two categories: corporate and noncorporate. Family forestlands are a subset of noncorporate private forestlands and are owned by families, individuals, trusts, estates, family partnerships, and other unincorporated groups of individuals.[32] The conversion of forestland to development occurs primarily on family forest (as opposed to timber industry) land.[33] However, rapidly changing industrial timberland ownership trends have meant increasing sales of these lands for development in some places where their value as real estate exceeds their value for timber production.[34] Timber production from private forestlands in the West has been more stable than from federal lands, though harvest levels have declined in recent years in response to the economic downturn (fig. 2.5).

As shown in figure 2.2, 20 percent of the timberlands in the West are owned by private noncorporate owners (most of whom are family forest owners), who own forestland for a variety of reasons. Not all family forest owners in the West are economically dependent on commercial timber production, though it does provide a supplementary source of income for many. Nearly half of the family forest owners in the West have harvested timber from their lands, and about 18 percent have engaged in commercial timber harvests from their lands (of sawlogs, veneer logs, or pulpwood).[35] There is a strong relationship between ownership size and management for timber: the larger the ownership, the more likely a family forest owner is to manage for timber production.[36]

The private noncorporate ownership category has been declining. For example, the amount of timberland owned by private noncorporate owners in the Pacific Northwest dropped by 4.4 million acres (34 percent) between 1953 and 2002, and by 1.5 million acres in California during this same period (25 percent).[37] At the same time, the number of private noncorporate owners has been increasing, implying increasing fragmentation and parcelization of the landscape.[38] Parcelization results from a reduction in average forest parcel size and an increase in the number of forest landowners.[39] In the eleven western states, 65 percent of the family forest owners owned parcels between one and nine acres in size in 2006 (fig. 2.10), though this accounted for only 6.6 percent of family forestland in the region. Most family forestland consists of tracts over 200 acres in size, owned by a minority of the owners (fig. 2.10). The number of owners of several size classes of forestland under 500 acres in size has been increasing, however—an indicator of parcelization.[40]

Parcelization and low-density rural home development have a number of negative environmental impacts. These include forest fragmentation and associated habitat loss or alteration, changes to ecosystem processes and biotic interactions, higher likelihood of invasion by exotic species, and increased human disturbance.[41] Parcelization does not necessarily lead to forest fragmentation, depending on how individual forest owners manage their lands.[42] But it becomes increasingly difficult to maintain continuous habitat when the number of owners proliferate because each additional owner brings her or his own needs, management preferences, and levels of knowledge to the landscape. Robles et al. found that several coastal watersheds in the West have high numbers and densities of at-risk species associated with private forestlands, as well as private forestlands that are predicted to experience increasing housing densities in the coming decades.[43] These trends suggest a need

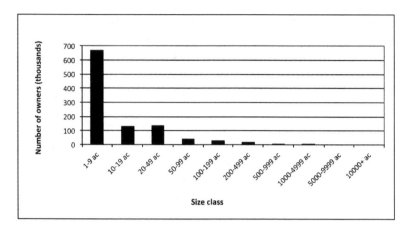

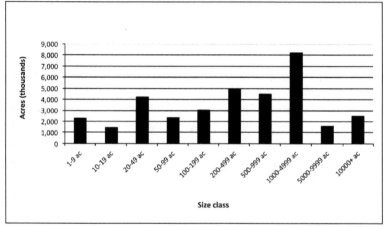

**Figure 2.10.** Number of owners (*top*) and acres owned (*bottom*) by family forest owners according to size class in the 11 western states, 2006. Source: 2006 data from "National Woodland Owner Survey," Forest Inventory and Analysis National Program, U.S. Department of Agriculture Forest Service, http://www .fia.fs.fed.us/nwos/. Figure created in 2013.

for targeting private forestlands and timberlands with conservation efforts using the kinds of strategies discussed in this book.

Private corporate timberlands are also an important component of the West's working forests. Historically, forest products companies were vertically integrated, owning timberlands to have control over the supply of timber to their manufacturing facilities. In the mid-1990s, however, many forest products companies began selling off their timberlands for economic reasons to timber investment management organizations (TIMOs) and real-estate investment trusts (REITs), as well as to individual private owners.[44] Timber investment management organizations buy, manage, and sell forestland and timber on behalf of institutional investors; they do not typically own forestland. In contrast, REITs do own land; many are former forest products companies that have restructured for tax purposes to separate legal ownership and control of their forestland and timber from that of their manufacturing facilities, with the land and timber being acquired by the REIT. Such trusts buy, manage, and sell real estate and related assets on behalf of private investors.

In 1992 there were approximately 28.5 million ha of industrial timberland in the United States.[45] Between 1996 and 2007, timberland sale transactions for sales over 10,000 acres in size (4,047 ha) totaled nearly 24 million ha (some parcels were sold more than once). Forest industry offered nearly 76 percent of the total sales, while TIMOS, REITs, and private buyers purchased 77 percent of these sales. Though much of this shift in timberland ownership from forest industry to TIMOs and REITs is taking place in the southern and northeastern United States, it is also occurring in the Northwest.[46]

This trend is of concern for working landscape conservation because it creates uncertainty around the future ownership and management of large blocks of timberland. Organizations like TIMOs and REITs manage their lands as investments. Limited evidence suggests that TIMOs and REITs manage these lands more intensively than private industry, are less concerned with long-term management horizons, sell land in parcels that are smaller than what they acquired, and convert timberlands to developed uses if they deem this to be the most profitable use of the land.[47] Nevertheless, as the authors of chapter 7 of this volume demonstrate, there can also be positive conservation outcomes from forest industry divestitures of working forestlands.

*Rangelands*

Although the extent of federal rangeland nationwide has remained fairly constant over time and is not expected to change much in the future, the extent of nonfederal rangeland declined at an average rate of 350,000 acres annually between 1982 and 2007 as a result of conversion to cropland and developed land uses and is projected to continue to decline slowly in the future.[48] In the West, the greatest decreases occurred in New Mexico, Montana, and California.[49] Rangeland fragmentation is also occurring as a result of agricultural uses, subdivision, and urbanization.

The loss of rangeland has occurred primarily as a result of conversion to crop production rather than to residential development, though the dynamics of land-use changes vary by place.[50] Residential development is projected to contribute to future rangeland conversion, particularly in California, followed by Arizona and Colorado. Ranchlands are also undergoing a process of parcelization and fragmentation owing to development and, in some cases, because of estate tax demands (see spotlight 13.1). Much of this parcelization occurs near major urban centers such as Denver, Boulder, and Fort Collins, in Colorado, Salt Lake City and the Wasatch Front in Utah, and the greater Phoenix and Tucson areas of Arizona (see chap. 13).[51] The conversion of ranchlands to real-estate development also occurs in areas of the West having high amenity values. For example, in the Madison Valley of Montana, part of the greater Yellowstone ecosystem, the average price per acre rose from $563 to $1,251 during the 1990s, triple the agricultural value of the land (see spotlight 11.1).[52] Much of this land was subdivided into "ranchettes." Some large ranches, like large tracts of private forest near more developed resort destinations, have not been carved up. Nonetheless, the new owners may be less interested in livestock production and more drawn by the amenity and investment values of the ranch, taking it out of production.[53]

A rangeland health inventory system developed in the early 2000s for nonfederal rangelands examined 17 different rangeland health indicators pertaining to soil and site stability, hydrologic function, and biotic integrity to assess their degree of departure from reference conditions representing healthy rangelands. Arizona and New Mexico contained the largest percentage of nonfederal rangeland acres that had departed from reference conditions for all three attributes, indicating the most severe degradation.[54] Nonnative species and invasive native species (such as juniper and mesquite) are also widespread on western rangelands, though the degree to which they pose a problem for grazing is variable.[55] These findings underscore the need for collaborative conservation efforts to restore rangeland health.

## SUMMARY

In sum, western working forests and rangelands are threatened on both public and private lands, for different reasons but with the same outcome: it has become increasingly difficult for rural ranchers and foresters to maintain natural resource-based livelihoods. And, although biodiversity conservation and ecological restoration have become a central focus of federal land management over the past two decades, trends on private lands have often worked against these goals. These trends make it challenging to stitch the West back together, especially in places having checkerboard ownership patterns or in watersheds having mixed ownerships. Nevertheless, doing so is important for maintaining and restoring the integrity of working forests and rangelands, as well as the biodiversity and rural communities they support.

## NOTES

1. For purposes of the statistics reported here, "timberland" is defined as forestland that is capable of producing at least 20 cubic feet per acre per year of industrial wood from natural stands and that is not withdrawn from timber utilization by statute or administrative regulation. "Forestland" is defined as land that is at least 120 feet wide and one acre in size and that has at least 10 percent live tree cover, including land that had such tree cover in the past that will be regenerated. W. B. Smith, P. D. Miles, C. H. Perry, and S. A. Pugh, *Forest Resources of the United States, 2007: A Technical Document Supporting the Forest Service 2010 RPA Assessment*, General Technical Report WO-78 (Washington, DC: U.S. Department of Agriculture, Forest Service, Washington Office, 2009).

2. Ibid.

3. Ibid.

4. X. Zhou and D. D. Warren, *Production, Prices, Employment, and Trade in Northwest Forest Industries, All Quarters 2011*, Resource Bulletin PNW-RB-264 (Portland, OR: U.S. Department of Agriculture, Forest Service, Pacific Northwest Research Station, 2012), 1, 6.

5. Zhou and Warren, *Production, Prices, Employment*, 10.

6. Ibid., 24, 27, 28.

7. Susan Charnley, Rebecca J. McLain, and Ellen M. Donoghue, "Forest Management Policy, Amenity Migration, and Community Well-Being in the American West: Reflections from the Northwest Forest Plan," *Human Ecology* 36, no.5 (2008): 743–61.

8. E. T. Jones, R. J. McLain, and J. Weigand, eds., *Nontimber Forest Products in the United States* (Lawrence: University of Kansas Press, 2002), xvii–xviii.

9. R. Hansis, "A Political Ecology of Picking: Non-timber Forest Products in the Pacific Northwest," *Human Ecology* 26, no. 1 (1998): 67–86; E. T. Jones and K. A. Lynch, "Integrating Commercial Nontimber Forest Product Harvesters into Forest Management," in *Forest Community Connections: Implications for Research, Management, and Governance*, ed. E. M. Donoghue and V. E. Sturtevant (Washington, DC: Resources for the Future Press, 2008), 143–61.

10. The Natural Resources Conservation Service's *National Resource Inventory* defines "rangeland" as land on which native grasses, grasslike plants, and forbs or shrubs suitable for grazing and browsing, and introduced forage species that are managed like rangeland, make up the climax or potential plant cover (U.S. Department of Agriculture [USDA], *Summary Report: 2007 National Resources Inventory* [Washington, DC: Natural Resources Conservation Service and Center for Survey Statistics and Methodology, Iowa State University, 2009], http://www.nrcs.usda.gov/Internet/FSE_DOCUMENTS/stelprdb1041379.pdf). This is the definition adopted for purposes of the statistics reported in this chapter; M. C. Reeves and J. E. Mitchell, *A Synoptic Review of U.S. Rangelands: A Technical Document Supporting the Forest Service 2010 RPA Assessment*, General Technical Report RMRS-GTR-288 (Fort Collins, CO: U.S. Department of Agriculture, Forest Service, Rocky Mountain Research Station, 2012).

11. M. C. Reeves and J. E. Mitchell, "Extent of Conterminous US Rangelands: Quantifying Implications of Differing Agency Perspectives," *Rangeland Ecology and Management* 64, no. 6 (2011):585–97.

12. U.S. Department of Agriculture, *RCA Appraisal: Soil and Water Resources Conservation Act* (Washington, DC: U.S. Department of Agriculture, 2011).

13. The source for these figures is USDA, *Summary Report: 2007 National Resources Inventory.*

14. Multiple-use percentages were calculated from raw data found at National Gap Analysis (GAP), U.S. Geological Survey, http://gapanalysis.usgs.gov/data/.

15. Regarding the federal land management's shift in focus in response to changing public values, see B. A. Shindler, T. M. Beckley, and M. C. Finley, *Two Paths toward Sustainable Forests: Public Values in Canada and the United States* (Corvallis: Oregon State University Press, 2003).

16. An animal unit month is the amount of forage it takes to sustain one cow and calf, one horse, or five sheep or goats for one month. Permitted use is the maximum number of animal unit months that can be used on U.S. Forest Service or Bureau of Land Management lands in a given year under a permit document. This number is generally higher than the number of animal unit months that are actually authorized for grazing use in a given year; Reeves and Mitchell, *A Synoptic Review of U.S. Rangelands.*

17. H. K. Cordell, *Outdoor Recreation Trends and Futures: A Technical Document Supporting the Forest Service 2010 RPA Assessment,* General Technical Report SRS-150 (Asheville, NC: U.S. Department of Agriculture Forest Service, Southern Research Station, 2012); K. Calvert, C. H. Vincent, and R. W. Gorte, *Recreation on Federal Lands,* CRS Report for Congress RL33525 (Washington, DC: Congressional Research Service, 2010).

18. H. K. Cordell, C. J. Betz, G. T. Green, and B. Stephens, *Off-Highway Vehicle Recreation in the United States and Its Regions and States: A National Report from the National Survey on Recreation and the Environment (NSRE),* USDA Forest Service RECSTATS Research Report in the Internet Research Information Series, ([Athens, GA: Southern Research Station?], 2008), http://www.fs.fed.us/recreation/programs/ohv/IrisRec1rpt.pdf.

19. T. M. Power, "Public Timber Supply, Market Adjustments, and Local Economies: Economic Assumptions of the Northwest Forest Plan," *Conservation Biology* 20, no. 2 (2006): 341–50.

20. S. Charnley, E. M. Donoghue, and C. Moseley, "Forest Management Policy and Community Well-Being in the Pacific Northwest," *Journal of Forestry* 106, no. 8 (2008): 440–47.

21. M. Nielsen-Pincus, S. Charnley, and C. Moseley, "The Influence of Proximity to Sawmills and Biomass Facilities on National Forest Hazardous Fuels Treatments," *Forest Science* 59, no. 5 (2013): 566–77.

22. R. J. Alig, "A United States View on Changes in Land Use and Land Values Affecting Sustainable Forest Management," *Journal of Sustainable Forestry* 24, nos. 2–3 (2007): 209–27.

23. J. M. Fowler, "Historic Range Livestock Industry in New Mexico," in *Livestock Management in the American Southwest,* ed. R. Jemison and C. Raish (Amsterdam: Elsevier, 2000); G. B. Ruyle, R. Tronstad, D. W. Hadley, P. Heilman, and D. A. King, "Commercial Livestock Operations in Arizona," in *Livestock Management,* ed. Jemison and Raish; A. Sulak and L. Huntsinger, "Public Land Grazing in California: Untapped Potential for Private Lands?" *Rangelands* 29, no. 3 (2007): 9–12.

24. For an overview of how ranching developed in the West, see P. F. Starrs, *Let the Cowboy Ride: Cattle Ranching in the American West* (Baltimore, MD: Johns Hopkins University Press, 1998).

25. L. Huntsinger, L. C. Forero, and A. Sulak, "Transhumance and Pastoralist Resilience in the Western United States," *Pastoralism* 1, no.1 (2010): 9–36.

26. U.S. Department of Agriculture, Forest Service, *Future of America's Forests and Rangelands: 2010 Resources Planning Act Assessment* (Washington, DC: USDA Forest Service, 2012).

27. B. J. Gentner and J. A. Tanaka, "Classifying Federal Public Land Grazing Permittees," *Journal of Range Management*, 55, no.1 (2002): 2–11; Sulak and Huntsinger, "Public Land Grazing."

28. For purposes of the data reported here, developed land consists of a combination of large urban and built-up areas, small built-up areas, and rural transportation land. Rural land is land that does not fall into one of the three preceding categories. Definitions from USDA, *Summary Report: 2007 National Resources Inventory* (n. 10 above, this chap.).

29. Data from ibid.

30. Population figures based on data found at U.S. Census Bureau, U.S. Department of Commerce, http://www.census.gov/prod/2001pubs/c2kbr01-2.pdf, and http://www.census.gov/prod/cen2010/briefs/c2010br-01.pdf

31. R. J. Alig, A. J. Plantinga, S. Ahn, and J. D. Kline, *Land Use Changes Involving Forestry in the United States: 1952 to 1997, with Projections to 2050—a Technical Document Supporting the 2000 USDA Forest Service RPA Assessment*, General Technical Report PNW-GTR-587 (Portland, OR: U.S. Department of Agriculture, Forest Service, Pacific Northwest Research Station, 2003).

32. B. J. Butler, *Forest Ownership.* In W. B. Smith, P. D. Miles, C. H. Perry, and S. A. Pugh, *Forest Resources of the United States, 2007: A Technical Document Supporting the Forest Service 2010 RPA Assessment*, General Technical Report WO-78 (Washington, DC: U.S. Department of Agriculture, Forest Service, Washington Office, 2009), 19–21.

33. J. D. Kline and R. J. Alig, "Forestland Development and Private Forestry with Examples from Oregon (USA)," *Forest Policy and Economics* 7 (2005): 709–20.

34. J. C. Bliss, E. C. Kelly, J. Abrams, C. Bailey, and J. Dyer, "Disintegration of the U.S. Industrial Forest Estate: Dynamics, Trajectories, and Questions," *Small-Scale Forestry* 9 (2010): 53–66.

35. The data, which are from 2006, are from the National Woodland Owner Survey: http://www.fia.fs.fed.us/nwos/.

36. Butler, *Family Forest Owners*, 23.

37. R. W. Haynes, D. M. Adams, R. J. Alig, P. J. Ince, J. R. Mills, and X. Zhou, *The 2005 RPA Timber Assessment Update: A Technical Document Supporting the USDA Forest Service Interim Update of the 2000 RPA Assessment*, General Technical Report PNW-GTR-699 (Portland, OR: U.S. Department of Agriculture, Forest Service, Pacific Northwest Research Station, 2007), 22. Note that this publication uses the term "private nonindustrial" instead of "noncorporate"; we use the latter for consistency of terminology in the chapter. Since the time of this publication, the categorization of private forest lands has changed (see Butler, *Forest Ownership*). However, the categories "private nonindustrial" and "private noncorporate" are analogous.

38. Ibid.; B. J. Butler, J. J. Swenson, and R. J. Alig, "Forest Fragmentation in the Pacific Northwest: Quantification and Correlations," *Forest Ecology and Management* 189, nos. 1–3 (2004): 363–73; B. J. Butler and E. C. Leatherberry, "America's Family Forest Owners," *Journal of Forestry* 102, no. 7 (October–November 2004): 4–9.

39. S. R. Mehmood, and D. Zhang, "Forest Parcelization in the United States: A Study of Contributing Factors," *Journal of Forestry* 99, no. 4 (2001): 30–34.

40. Butler and Leatherberry, "America's Family Forest Owners," 6.

41. A. J. Hansen, R. L. Knight, J. M. Marzluff, S. Powell, K. Brown, P. H. Gude, and K. Jones, "Effects of Exurban Development on Biodiversity: Patterns, Mechanisms, and Research Needs," *Ecological Applications* 15, no. 6 (2005): 1893–1905; G. E. Heilman Jr., J. R. Strittholt, N. C. Slosser, and D. A. Dellasala, "Forest Fragmentation of the Conterminous United States: Assessing Forest Intactness through Road Density and Spatial Characteristics," *BioScience* 52, no. 5 (2002): 411–22.

42. Y. Zhang, D. Zhang, and J. Schelhas, "Small-Scale Non-industrial Private Forest Ownership in the United States: Rationale and Implications for Forest Management," *Silva Fennica* 39, no. 3 (2005): 443–54.

43. M. D. Robles, C. H. Flather, S. M. Stein, M. D. Nelson, and A. Cutko, "The Geography of Private Forests That Support At-Risk Species in the Conterminous United States," *Frontiers in Ecology and the Environment* 6, no. 6 (2008): 301–7.

44. Bliss et al., "Disintegration of the U.S. Industrial Forest Estate."

45. Ibid.

46. N. E. Block and V. A. Sample, *Industrial Timberland Divestitures and Investments: Opportunities and Challenges in Forestland Conservation* (Washington, DC: Pinchot Institute for Conservation, 2001).

47. Bliss et al., "Disintegration of the U.S. Industrial Forest Estate."

48. Reeves and Mitchell, *A Synoptic Review of U.S. Rangelands* (n. 10 above, this chapter), 14–16, 33–35, http://www.fs.fed.us/rm/pubs/rmrs_gtr288.html

49. Data from USDA, *Summary Report: 2007 National Resources Inventory* (n. 10 above, this chap.), 32–41, table 2: "Land Cover/Use of Non-Federal Rural Land, by State and Year."

50. On the sources of rangeland loss, see Reeves and Mitchell, *A Synoptic Review of U.S. Rangelands*, 24–27.

51. Robert E. Lang, Andrea Sarzynski, and Mark Muro, *Mountain Megas: America's Newest Metropolitan Places and a Federal Partnership to Help Them Prosper* (Washington, DC: Brookings Institution, 2008).

52. A. B. Pearce, "Uncommon Properties: Ranching, Recreation, and Cooperation in a Mountain Valley" (PhD thesis, Stanford University, 2004).

53. H. Gosnell and W. R. Travis, "Ranchland Ownership Dynamics in the Rocky Mountain West," *Rangeland Ecology and Management* 55, no. 1 (2005): 2–11.

54. J. E. Herrick, V. C. Lessard, K. E. Spaeth, P. L. Shaver, R. S. Dayton, D. A. Pyke, L. Jolley, and J. J. Goebel, "National Ecosystem Assessments Supported by Scientific and Local Knowledge," *Frontiers in Ecology and the Environment* 8, no. 8 (2010): 403–8; USDA, *RCA Appraisal* (n. 12 above, this chapter).

55. Regarding nonnative and invasive species on rangeland, see Herrick et al., "National Ecosystem Assessments." Concerning the degree to which such species are problematic, see USDA, *RCA Appraisal*.

# 3

# THE BIODIVERSITY THAT PROTECTED AREAS CAN'T CAPTURE

How Private Ranch, Forest, and Tribal Lands Sustain Biodiversity

*Gary P. Nabhan, Richard L. Knight, and Susan Charnley*

## IN BRIEF

- Formally designated protected areas in the western United States are not large enough, diverse enough, or located strategically enough to support viable populations of many area-sensitive species, endemic species, or endangered species; a broader, landscape-scale focus is needed to incorporate conservation on adjacent private and tribal lands as well.
- Many working tribal and private lands provide an ecological buffer zone between adjacent public lands and encroaching exurban development; their lower elevation relative to many protected public lands offers better soils, more water features, and often a significant component of the biodiversity in a region.
- The conversion of working lands to exurban development will cause a decline in human-sensitive species, in turn increasing the number of these species on the federal threatened and endangered species list and placing a greater burden on public-policy makers and land managers who must act to protect them.
- Working landscapes on private and tribal lands are core conservation areas in the West that deserve protection using tools such as zoning restrictions, tax incentives for establishing conservation easements, and market rewards for maintaining their ecological goods and services.

## INTRODUCTION

For decades, conservation biologists have conceded that the ecosystems protected within the largest national parks in the western United States are neither large nor diverse enough to support viable populations of area-sensitive species.[1] Less than 6 percent of the coterminous United States lies within formally designated protected areas such as national parks, the Nature Conservancy lands, or designated wilderness areas on federal lands.[2] In general, these protected areas are too small and not located where the highest biodiversity, most productive soils, or most accessible wetland areas for migratory wildlife occur.[3]

For terrestrial endangered species, in particular, the current location, size and management practices of national parks and other federally protected lands will not suffice to keep these species from extinction over the next century.[4] The outcome of numerous geographic analyses of biodiversity at risk in the United States has been summarized by Rodriguez et al.: "Recent government studies indicate that over half of the species on the federal endangered species list have more than 80 percent of their habitat on non-federal lands."[5] It is not merely the extent of those lands but also the diverse ways in which they are managed that make them into safe or unsafe sites for the conservation of at-risk biodiversity.

In response to the acknowledged limitations of the current U.S. system of national parks and other protected areas, Wilcove and May initially proposed three remedies: (1) make parks larger by adding selected ecosystems in order to make the populations of their most endangered species more viable; (2) buy up functional wildlife migration corridors between parks and other protected areas; and (3) constrain urban and suburban developments around parks in a manner that does not disrupt wildlife movements.[6] However, these conservation biologists neglected to consider another type of geography just as essential to continent-wide conservation strategies for biodiversity: the private and tribal lands in large working landscapes where forestry and ranching have long occurred.[7]

It is now clear that this "neglected geography" of private and tribal lands intermingled with public lands has always been critically important to sustaining, conserving, and restoring the wild biodiversity of western North America. Further, if Shafer is correct, the role of private lands in biodiversity conservation may become even more important, since "our opportunities for larger [reserve] sizes, or better designs, may have vanished in many cases and, in the conceivable future . . . regional cooperation [between public and

**Figure 3.1.** Location of ecoregions featured in this chapter. Map by Darin Jensen and Syd Wayman, 2013.

private sectors] offers [the best] chance to alleviate this problem."[8] Moreover, under global climate change, shifts in ecosystems could occur that alter the distribution of species across ownerships that protected areas were designed to preserve.

In this chapter we cover three regional case studies (see map, fig. 3.1) of how private and tribal lands support wild biodiversity, followed by lessons derived from each. These case studies exemplify how the hundreds of millions of acres of working landscapes in the West retain their productive capacity while also supporting biodiversity of conservation value.

## THE COLORADO PLATEAU

The Colorado Plateau is an ecoregion that spans the Four Corners region of the states of Colorado, New Mexico, Arizona, and Utah (fig. 3.1). Both the biological and the cultural diversity of the Colorado Plateau are so impressive that this ecoregion can be ranked, along with the Everglades and the Great Smoky Mountains, in the highest tier of the 110 North American ecoregions compared and analyzed by Ricketts et al.[9]

At last count, some 16,108 vascular plant species, 2,536 vertebrate species, and 1,163 invertebrate species have been documented on the Colorado Plateau, including 34 plant taxa and 19 vertebrate taxa that are federally listed as threatened or endangered.[10] A rather high proportion of this diversity is composed of species native to a very small area or soil type that can be found nowhere else in the world.[11] The Colorado Plateau ranks among the top three ecoregions in continental North America for total numbers of endemic and narrowly restricted species, particularly those of vascular plants, small mammals, fish, reptiles, tiger beetles, and leafhoppers.

Across the 130 million acres of the plateau, elevations range from 1,000 feet to well over 12,000 feet. This gradient produces, within ten- to sixty-mile distances, the entire range of ecological zones found between Mexico and Canada. When this range of terrestrial habitat types is classified using the classification scheme for biotic communities found in North America developed by Brown and Lowe, some 21 distinct biotic associations are harbored on the Colorado Plateau, compared to 15 in the adjacent Sonoran Desert ecoregion, 20 in the Rocky Mountain ecoregion, and six in the Great Plains grasslands.[12]

With nearly 2.3 million acres protected in the 11 largest national parks on the Colorado Plateau, and another 1.8 million acres protected through the U.S. National Wilderness Preservation System, many visitors to the Colorado Plateau perceive this ecoregion to be more fully protected than any other in the coterminous United States. In reality, however, less than 2 percent of the region occurs in National Park Service land, whereas 50 percent consists of Bureau of Land Management and U.S. Forest Service lands—most of which are open to extractive resource uses. Importantly, the 23 percent of the ecoregion that is in tribal lands and 26 percent in state and private ownership are also unprotected from resource extraction and from the conversion of plant communities.

So even when other categories of lands that are federally protected for their cultural, historical, recreational, or wildlife values are taken into account, only 11.2 million acres (8.6 percent) of the 130 million total acres on the Colorado Plateau have been set aside for protection of their natural or cultural resources by federal or state governments. (This figure does not include state and federal lands managed for resource extraction, or state trust lands that can be put up for public auction.)[13] In short, the acreage of wildlands under strict protection in this ecoregion is dwarfed by the acreage of wildlands in working landscapes on tribal lands and under private ownership and on federal lands managed for multiple uses, often through leases to families or small private corporations.

It is therefore not surprising that the majority of endemic plants and animals, as well as the majority of habitats critical for threatened and endangered species on the Colorado Plateau, occur on tribal and private lands in working landscapes—and not in national parks, wildlife refuges, or wilderness areas. For example, fewer taxa of mammals endemic to the Colorado Plateau have been found in the national parks and monuments of the region than on Indian reservations, even though the biota of tribal lands remains poorly surveyed by zoologists and dramatically underrepresented in regional biodiversity inventories.[14] The patterns that Scott et al. have noted for the coterminous United States in general appear to hold true for the Colorado Plateau in particular:

1. Formally designated protected areas are most frequently found at higher elevations and on less productive soils (or on barren slickrock).
2. The known distributions of native wild plants and animals suggest that the highest level of species richness is found at lower elevations, on floodplains or around springs and other water features.
3. A cursory assessment of the geographic occurrence of mapped habitat cover types suggests that 60 percent of the habitat types have less than 10 percent of their areas in formally designated protected areas.
4. Land ownership patterns show that areas at lower elevations, especially those on more productive and better watered soils, are most often on tribal or privately owned lands that qualify as "working landscapes."[15]

Given these prevalent patterns, it is problematic that most ecoregional assessments of the Colorado Plateau—formulated by the federal government, the Nature Conservancy, the Wildlands Project, and other entities—largely neglect the contributions of tribal and private lands to the maintenance of wild biodiversity. It seems that either pertinent data from these lands have not been factored into regional databases or the human dimensions required to deal with the complexities of working cross-culturally with tribal and other rural communities to achieve conservation goals is still lacking. This tendency to overlook tribal and private lands is institutionalized: when Sisk surveyed 80 environmental professionals from across the Colorado Plateau regarding which actions they considered to be the highest priorities for conservation in the ecoregion, they ranked supporting "private land stewardship" fourth, recognizing "tribal sovereignty" seventh, and building capacity "on and off reservations" ninth.[16]

In spite of this relative neglect of tribal and private lands by researchers and environmental professionals, the current scale of private-public collaboration to conserve and restore working landscapes on the Colorado Plateau is unprecedented. With Diablo Trust lands including more than 650,000 acres in northern Arizona, Babbitt Ranch lands covering another 700,000 acres on the South Rim of the Grand Canyon, and Grand Canyon Trust lands on the North Rim including another 850,000 acres of grazed private and federal lands, over two million acres of working landscapes are now being managed by ranchers involved in collaborative conservation experiments.

## THE SOUTHERN ROCKY MOUNTAINS ECOREGION

The ecoregion of the Southern Rocky Mountains includes some 48 counties in Colorado, Wyoming, and New Mexico (fig. 3.1).[17] Within this ecoregion, 4.69 million acres of private lands are owned by ranchers, who, altogether, also have grazing leases on 14.1 million acres of U.S. Forest Service and Bureau of Land Management lands. Some 1,456 ranches in the ecoregion have private land "bases" that are intermixed with some 2,217 allotments on federal lands. Despite this comingling of private and public lands in the Southern Rockies, what is revealing are the differences in landscape qualities found between the private and public components of these working landscapes:

1. The private lands of the base ranches are, on average, 600 feet lower in elevation than the federal grazing allotments, are more gently sloped, and are better watered, having twice the stream densities as that of the public lands.

2. Soil fertility and productivity are significantly higher on the private lands of the base ranches than on the U.S. Forest Service and Bureau of Land Management allotments where ranchers maintain their grazing leases.

3. In the portion of the Southern Rockies that falls within Colorado, all of the lands have been assessed by the state's National Heritage Program, located at Colorado State University, for the presence or absence of species ranked

for their significance to conservation efforts. Importantly, there were more occurrences of species of conservation value on the private ranchlands than on the public lands, even though the private land acreage in the total study area was only a third of the public land acreage.[18]

In short, Talbert et al. found that the geophysical and hydrological features that typically foster high levels of biodiversity in the Southern Rockies are more evident on private ranchlands than on public lands, which is consistent with the nationwide patterns reported by Scott et al.[19]

An important component of Talbert et al.'s study was the finding that private ranchlands with public-grazing leases provide a buffering effect to public lands.[20] Over 43 percent of the private lands abutting public lands in the Southern Rockies ecoregion are ranchlands with public-grazing leases. These private ranchlands provide critical ecological buffers for public lands, in essence shielding them from the harmful effects of the rural sprawl that increasingly encircles them.[21]

However, these results do not necessarily suggest that all of the 21 million acres of private lands in the Southern Rockies have the conservation value that the 4.69 million acres of private-grazing lands do. To the contrary, the working landscapes of these ranches retain certain elements of wild biodiversity that have been lost from those private lands converted to exurban residential and commercial development.[22] Over the last decade, an increasing number of scientists have reported on the dramatic differences in the landscape attributes and levels of biodiversity in exurban residential developments versus those in larger, unfragmented working landscapes where livestock grazing persists.[23]

Whereas ranching is synonymous with reduced human densities, roads, and houses, exurban development brings year-round human presence and elevated human activities that fragment ecosystems with roads, houses, vehicles, security lights, and pets. For example, when ranches in Larimer County, Colorado, were subdivided, there was an order-of-magnitude increase in roads and housing, which perforated and dissected the previously intact wildlands.[24]

The observation that exurban development alters the density and composition of biodiversity on private lands motivated studies examining biodiversity across different private land uses, from working ranches to exurban developments.[25] Studies specific to the Southern Rockies ecoregion found that ranchlands and protected areas supported birds and carnivores of conservation interest while land in exurban developments supported what were essentially the same songbird and carnivore communities found in urban suburbs (e.g., robins, cats, and dogs).[26]

Characteristics of plant communities differed on these three types of lands as well. Both the protected areas and the exurban developments were far weedier than the private ranchlands. These differences can be attributed to the ranchers' stewardship, judicious use of herbicides and livestock, and discernment. Ranchers apparently are doing what Aldo Leopold suggested when he wrote, "The central thesis . . . is this: game can be restored by the creative use of the same tools which have heretofore destroyed it—axe, plow, cow, fire, and gun."[27]

The biological changes associated with the conversion of ranchlands alter the natural heritage of an area. Exurban developments support biodiversity of mostly human-adapted species, whereas ranches and protected areas support biodiversity that is of conservation value.[28] In the years to come, as the Southern Rockies ecoregion is gradually transformed from rural ranches with low human densities to increasingly sprawl-riddled landscapes, the region will be altered forever, its once-rich natural diversity diminished. There will be more generalist species—species that thrive in association with humans—and fewer specialist species—those whose evolutionary histories failed to prepare them for elevated human densities and twenty-first-century technology. Rather than lark buntings (*Calamospiza melanocorys*) and bobcats (*Lynx rufus*), there will be starlings (*Sturnus vulgaris*) and striped skunks (*Mephitis mephitis*). Rather than rattlesnakes and warblers, garter snakes and robins (*Turdus migratorius*) will prevail. Conservationists devote their efforts to minimizing the harmful effects of generalist species on specialist species as these human-adapted species can be effective competitors, predators, and diseases and parasites on the human-sensitive species. And, from a public-policy perspective, with the continuing decline of human-sensitive species there will many more federally threatened and endangered species, an inheritance that brings evermore social and economic riddles to resolve.

### THE OREGON COAST RANGE

The Oregon Coast Range (fig. 3.1), part of the Pacific Northwest's temperate rainforest, contains some of the most productive forestlands in the West and is the source of roughly half of Oregon's annual timber harvest.[29] Spanning 5.75 million acres, the region is dominated by mixed conifer forests composed mainly of Douglas fir, western hemlock, Sitka spruce, and western red cedar, interspersed with hardwoods (bigleaf maple, red alder).[30] Land ownership in the Oregon Coast Range is a mix of federal (25 percent), state (12 percent), private industrial (41 percent), and nonindustrial private (NIPF; 22 percent) ownership.[31] Forest management policy is similarly diverse, ranging from wilderness protection on a portion of federal lands to intensive wood production on private industrial lands.[32]

Historically, timber production from both public and private forests in the region was high, with state and federal lands managed in a manner similar to that of private industrial lands. Since the early 1990s, however, timber production from federal lands has dwindled—from roughly half of the annual regional total to under 5 percent of that total—owing to concerns over the northern spotted owl (*Strix occidentalis caurina*) and marbled murrelet (*Brachyramphus marmoratus*) (both threatened bird species) and the environmental effects of clear-cutting and old-growth harvesting. Biodiversity protection and ecological restoration have since become the focus of forest management policy on federal lands in the Oregon Coast Range, while timber production remains the management emphasis on private lands. State lands are managed for multiple uses.[33]

The mix of land ownership and forest management approaches in the Oregon Coast Range offers an opportunity to assess how forest conditions and biodiversity vary by ownership. Scientists who analyzed the forest cover of 66 Oregon Coast Range watersheds found a significant correlation between land ownership and forest cover patterns: the more diverse the land ownership in a watershed, the more diverse the forest cover types there.[34] Federal lands had large patches of forest, high levels of forest connectivity, and low diversity of forest cover types (with mature forest being the dominant cover type). State lands had moderate levels of forest cover diversity and connectivity. Very large private industrial lands also had low forest cover diversity—dominated by young forest cover—but large patches of forest and high forest connectivity. Finally, NIPF lands had a wide diversity of forest cover types, though forest cover patterns were spatially fragmented. Only NIPF lands contained large amounts of nonforest and mixed woodland in addition to forest cover. The high degree of forest cover diversity and fragmentation among NIPF owners was due to relatively small ownership sizes and the diversity of management objectives among them. The researchers concluded that, in the Oregon Coast Range, the greater the mix of forest ownership in a watershed, the greater the likelihood that the watershed would contain a mix of forest cover classes. Federal, state, private industrial, and NIPF lands contribute to landscape structure in different ways because they contain different types of forest cover. It follows that, in western Oregon and possibly elsewhere, land-scapes with highly mixed ownership patterns may support a more diverse array of habitats and a broader array of ecological functions than single-owner landscapes and may more effectively support biodiversity at a watershed scale.[35] Of course, measures of biodiversity are highly variable, and no single measure is adequate; this assessment emphasized landscape structure and cover diversity.

Next, researchers used current forest conditions and management practices to build a spatial simulation model to predict change in biodiversity by ownership over a 100-year period in response to different forest management policy scenarios. Based on this model, they predict that under current forest management policy, federal and state forestlands will be dominated by stands of large and very large conifers; forest industry lands will be dominated by recently clear-cut open areas and small and medium-size conifers; and NIPF lands will have a mixture of forest size classes.[36] Most timber harvest will come from forest industry lands, with NIPF lands providing the rest; this represents an increase over harvest volume levels from NIPF lands since the mid-1990s, when federal harvests declined sharply. In the absence of natural disturbance (e.g., fire) and timber harvesting, the loss of young forest in the early stages of succession on public lands could reduce populations of big game species like elk and deer that favor this forest type. Hardwood stands are predicted to decrease across all ownerships—on public lands, because of the absence of disturbance due to fire suppression and reduced timber harvesting that promote old-growth conifer development, and on private lands, because management favors young conifer stands for commercial timber production.[37] Hardwood forests can support unique assemblages of species that do not occur in conifer forests.

Spies et al. used the simulation model to assess what these changes in forest conditions might mean for forest biodiversity in the Oregon Coast Range over the next 100 years.[38] They used several indicators of biodiversity in their assessment: habitat availability for a group of focal species, forest stand structure, and landscape structure and composition. They found that, in general, ownership explained little of the variation in habitat quality for most focal species. With respect to forest stand structure, the distribution of structurally diverse old-growth forest would be strongly concentrated on federal and state forestlands under current management policies. Regarding landscape structure, the remaining patches of hardwood forest would be found largely on private forestlands. State and private lands would provide the majority of young stands with remnant trees; large-diameter trees would occur mainly on federal and state lands; and private lands would have the majority of forests with small- and medium-diameter trees.

Landscape diversity on public lands is projected to decline strongly over the next 100 years, with federal lands exhibiting the lowest patch-type diversity. The greatest diversity is projected to occur on private nonindustrial and industrial forestlands. High diversity on private lands results from the mix of open (harvested) areas and forests in the earlier stages of succession, a mix that is uncommon on federal lands.[39] However, it is important to note that the presence or absence of habitat/landscape diversity alone is not sufficient to adequately characterize the degree of biodiversity. For example, some rare habitats may be missing from diverse landscapes; and large, relatively homogeneous blocks of similar habitat (e.g., old-growth forests) are often needed to support some species and ecological processes.

Biodiversity is a multifaceted concept—no single measure or small set of measures is adequate to capture it. These studies point out, however, that ownership diversity can contribute to landscape and habitat diversity—an important component of overall biodiversity. Forest management practices undertaken on different ownerships for different objectives can create a diversity of habitat conditions that support a variety of native species. The decline of landscape diversity on federal lands stems largely from the policy, articulated in the Northwest Forest Plan, of creating reserves to protect and increase an important component of biodiversity—old-growth habitat and its associated threatened species. Old-growth habitat was lost through decades of intensive timber harvesting on both public and private lands. However, old-growth reserves alone cannot provide or create the full range of biodiversity that is possible in the Oregon Coast Range. In the future, without "the biodiversity that protected areas can't capture" on private forestlands, there would be an overall loss of landscape diversity and associated biodiversity in the Oregon Coast Range. These findings illustrate the interdependence of public and private lands for biodiversity conservation at the landscape scale and the positive role that forest management can play in creating biodiversity at regional scales.

## CONCLUSIONS

The continent-wide analyses and regional case studies we have discussed suggest that (1) formally designated protected areas are neither sufficient in size and heterogeneity nor located ideally to capture the bulk of North America's wild biodiversity within their boundaries; (2) many elements of this bio-diversity are better represented on private and tribal lands than on the public lands managed by federal agencies, though it is unclear whether this is simply because the private lands often occur in richer landscapes or because tribal and private land and water management also favors habitat heterogeneity and species diversity; and (3) a mosaic of private and public forests and rangelands that includes protected areas, but is not limited to them, is more effective for maintaining biodiversity than are protected areas alone.

But it is not enough simply to repeat the clichés that effective conservation cannot occur without private-public partnerships or that wildlife populations, to remain viable, need corridors through the private lands that connect protected areas. Instead, it may be time to move beyond conventional wisdom, to consider whether working landscapes—rather than formally designated protected areas—are actually functioning as the "core" conservation areas of the West: those areas required for maintaining the continent's unique biodiversity.[40] If this is the case, we must ask why the public is reluctant to invest in the maintenance of our region's natural heritage through zoning restrictions, tax incentives for establishing conservation easements, and market-value enhancement for their ecological goods and services.[41]

Of course, many Americans have witnessed not only the conversion of ranches and private forests to development but also the degradation of these lands by poor management practices that have diminished their diversity. At the same time, as Imhoff and Dagget have documented, there is a vitality and variety to some private landowner's experiments with conserving, sustaining, and restoring biodiversity and productivity to private lands in the working landscapes of the West.[42] The very heterogeneity of these private land experiments may potentially offer a modicum of resilience in the face of uncertainty, but the data are not in and the jury is still out on whether such experiments collectively sustain landscape-level biodiversity over the long haul. Nevertheless, it could be argued that, over the last two decades, conservation biologists and land managers in parks and wildlife refuges have remained far more static in their approaches to habitat management than those in the private sector. This is in part because federal and state managers of public lands have explicit legal mandates to sustain or recover rare native species and face more regulatory, institutional, and social constraints than do private and tribal managers, giving them less leeway to experiment.

Much of the American public subscribes to the flawed notion that private landowners manipulate nature to their own (economic) ends, while parks, wildlife refuges, and other protected areas are not truly "managed" but are simply protected from people.[43] Perhaps acknowledging the importance of private and tribal lands to the maintenance of biodiversity in the American West will encourage people to think about why and how these lands have come to play this role. Is it possible that the century-plus of livestock grazing and forestry in large working landscapes hasn't necessarily been the categorical "subsidized destruction of the American West" that some activists have made it out to be?[44] The so-called subsidies that ranchers or foresters do gain for utilizing and co-managing federal lands are supported by the citizenry at large, just as the populace supports libraries, hospitals and walkways for the public good. The core issue may be how to harmonize the management of public, private, and tribal lands to maintain landscape heterogeneity through "additive effects," rather than seeing sharp discrepancies between adjacent land uses that effectively fragment habitats and deplete rare populations.

Clearly the solution is not found in them-versus-us approaches. Instead, we submit, it will be found in the "radical center," the place where people from different persuasions meet to work on things they all hold dear.[45] Rather than pitting energies and resources against one another in battles that always have a loser, more and more Americans are dropping their weapons, leaving their ideological trenches, and meeting on the common ground. The formation of the Rural Voices for Conservation Coalition and the Coalition for Conservation through Ranching—confederacies of dozens of community-based, grassroots collaborative conservation organizations—has opened up new opportunities to explore such issues on a larger scale. It is to be hoped that these opportunities will lead us into an era of both policy and practice that keeps elements of the West unique, dynamic, diverse, and resilient for many decades to come.

## NOTES

1. D. S. Wilcove and R. M. May, "National Park Boundaries and Ecological Realities," *Nature* 324 (1986): 206–7.
2. J. M. Scott, F. W. Davis, R. G. McGhie, R. G. Wright, C. Groves, and J. Estes, "Nature Reserves: Do They Capture the Full Range of America's Biological Diversity?" *Ecological Applications* 11 (2001): 999–1007.
3. B. A. Stein, L. S. Kutner, and J. S. Adams, eds., *Precious Heritage: The Status of Biodiversity in the United States* (New York: Oxford University Press, 2000); Scott et al., "Nature Reserves"; J. M. Scott, R. Abbit, and C. Groves, "What Are We Protecting? The U.S. Conservation Portfolio," *Conservation Biology in Practice* 2, no. 1 (2002): 18–19.
4. A. P. Dobson, J. P. Rodriguez, W. M. Roberts, and D. D. Wilcove, "Geographic Distribution of Endangered Species in the United States," *Science* 275 (1997): 550–53.

5. J. P. Rodriguez, W. M. Roberts, and A. P. Dobson, "Where Are Endangered Species Found in the United States?" *Endangered Species Update* 14, nos. 3–4 (1997), http://www.umich.edu/~esupdate/library/97.03-04/rodriguez.html.

6. Wilcove and May, "National Park Boundaries."

7. R. L. Knight, "Private Lands: The Neglected Geography, *Conservation Biology* 13 (1999): 223–24.

8. C. L. Shafer, *Nature Reserves: Island Theory and Conservation Practice* (Washington, DC: Smithsonian Institution Press, 1990), 99.

9. L. E. Stevens, and G. P. Nabhan, "Biodiversity: Plant and Animal Endemism, Biotic Associations and Unique Habitat Mosaics in Living Landscapes," in *Safeguarding the Uniqueness of the Colorado Plateau: An Ecoregional Assessment of Biocultural Diversity*, ed. G. P. Nabhan (Flagstaff: Center for Sustainable Environments, Northern Arizona University, 2002), 41–48; T. H. Ricketts, E. Dinerstein, D. Molson, and C. Loucks, "Who's Where in North America? Patterns of Species Richness and the Utility of Indicator Taxa for Conservation," *BioScience* 49 (1999): 369–81.

10. Stevens and Nabhan, "Biodiversity: Plant and Animal Endemism," 44.

11. Ricketts et al., "Who's Where in North America?"

12. D. E. Brown and C. H. Lowe, "Biotic Communities of the American Southwest—United States and Mexico," *Desert Plants* 4 (1982): 1–342; Stevens, and Nabhan, "Biodiversity: Plant and Animal Endemism."

13. Tony Joe, Gary Paul Nabhan, and Patrick Pynes, "Diversifying Conservation Strategies: Innovative Means of Protecting Traditional Cultural Places and Their Wildlands Settings," in *Safeguarding the Uniqueness of the Colorado Plateau*, ed. Nabhan, 66–68.

14. Wayne Armstrong, personal communication, cited in Stevens and Nabhan, "Biodiversity: Plant and Animal Endemism," 44.

15. Scott et al., "Nature Reserves."

16. T. D. Sisk, "Eliciting Perceptions of Biocultural Diversity on the Colorado Plateau: A Methodology for Defining Threats and Response Options," in *Safeguarding the Uniqueness of the Colorado Plateau*, ed. Nabhan, 13–26.

17. C. B. Talbert, R. L. Knight, and J. E. Mitchell, "Private Ranchlands and Public Land Grazing in the Southern Rocky Mountains," *Rangelands* 29 (2007): 5–8.

18. Ibid., 7.

19. Scott et al., "Nature Reserves"; Talbert et al., "Private Ranchlands and Public Land Grazing."

20. Talbert et al., "Private Ranchlands and Public Land Grazing."

21. P. B. Landres, R. L. Knight, S. T. A. Pickett, and M. L. Cadenasso, "Ecological Effects of Administrative Boundaries," in *Stewardship across Boundaries*, ed. R. L. Knight and P. B. Landres (Washington, DC: Island Press, 1998), 39–64.; A. A. Wade and D. M. Theobald, "Residential Development Encroachment on U.S. Protected Areas," *Conservation Biology* 24 (2010): 151–61.

22. A. J. Hansen, R. L. Knight, J. M. Marzluff, S. Powell, K. Brown, P. H. Gude, and K. Jones, "Effects of Exurban Development on Biodiversity Patterns, Mechanisms, and Research Needs," *Ecological Applications* 15 (2005): 1893–1905.

23. E. A. Odell and R. L. Knight, "Songbird and Medium-Sized Mammal Communities Associated with Exurban Development in Pitkin County, Colorado," *Conservation Biology* 15 (2001): 1143–50; J. D. Maestas, R. L. Knight, and W. C. Gilgert, "Biodiversity

across a Rural Land-Use Gradient," *Conservation Biology* 17 (2003): 1425–34; Hansen et al., "Effects of Exurban Development."

24. J. E. Mitchell, R. L. Knight, and R. J. Camp, "Landscape Attributes of Subdivided Ranches," *Rangelands* 24 (2002): 3–9.

25. J. D. Maestas, R. L. Knight, and W. C. Gilgert, "Biodiversity and Land-Use Change in the American Mountain West," *Geographical Review* 91, no. 3 (2001): 509–24.

26. Ibid.

27. A. Leopold, *Game Management* (New York: Charles Scribner's Sons, 1933), vii.

28. Hansen et al., "Effects of Exurban Development."

29. K. N. Johnson, P. Bettinger, J. D. Kline, T. A. Spies, M. Lennette, G. Lettman, B. Garber-Yonts, and T. Larsen, "Simulating Forest Structure, Timber Production, and Socioeconomic Effects in a Multi-Owner Province," *Ecological Applications* 17 (2007): 34–47.

30. T. A. Spies, B. C. McComb, R. S. H. Kennedy, M. T. McGrath, K. Olsen, and R. J. Pabst, "Potential Effects of Forest Policies on Terrestrial Biodiversity in a Multi-Ownership Province," *Ecological Applications* 17 (2007): 48–65.

31. Here, "private industrial lands" refers to forestlands owned by a corporate entity; "private nonindustrial lands" are nonpublic forestlands that are not owned by a corporate entity (ibid.).

32. T. A. Spies, K. N. Johnson, K. M. Burnett, J. L. Ohmann, B. C. McComb, G. H. Reeves, P. Bettinger, J. D. Kline, and B. Garber-Yonts, "Cumulative Ecological and Socioeconomic Effects of Forest Policies in Coastal Oregon," *Ecological Applications* 17 (2007): 5–17.

33. Ibid.

34. B. J. Stanfield, J. C. Bliss, and T. A. Spies, "Land Ownerships and Landscape Structure: A Spatial Analysis of Sixty-Six Oregon (USA) Coast Range Watersheds," *Landscape Ecology* 17 (2002): 685–97.

35. Ibid.

36. Johnson et al., "Simulating Forest Structure."

37. Ibid.

38. Spies et al., "Potential Effects of Forest Policies."

39. Ibid.

40. Knight and Landres, eds., *Stewardship across Boundaries* (n. 21 above, this chapter).

41. A. Anella and J. B. Wright, *Saving the Ranch: Conservation Easement Design in the American West* (Washington, DC: Island Press, 2005).

42. D. Imhoff, *Farming the Wild: Enhancing Biodiversity on Farms and Ranches* (San Francisco, CA: Watershed Media/Sierra Club Books, 2003); D. Dagget, *Gardeners of Eden: Discovering Our Importance to Nature* (Santa Barbara, CA: Thatcher Charitable Trust, 2005).

43. Dagget, *Gardeners of Eden.*

44. G. Wuerthner and M. Matteson, *Welfare Ranching: The Subsidized Destruction of the American West* (San Francisco: Foundation for Deep Ecology, 2002).

45. R. L. Knight, "Bridging the Great Divide: Reconnecting Rural and Urban Communities in the New West," in *Home Land: Ranching and a West That Works,* ed. L. Pritchett, R. L. Knight, and J. Lee (Boulder, CO: Johnson Books, 2007), 13–25.

# PART TWO

COLLABORATIVE CONSERVATION

As the authors of chapter 1 of this volume make clear, the United States launched a revolution in the American West a century ago. The result was a mosaic of private, tribal, state, and federal lands across the West—what historian Donald Worster called "a hybrid of capitalist and bureaucratic regimes, each assuming it knows what is best for the nation's pocketbook and for nature."[1]

Ranchers, loggers, and miners spent much of the twentieth century either fighting those "bureaucratic regimes" (i.e., government agencies) or attempting to co-opt them. But the contradictions and tensions within the West were exacerbated as more interest groups—environmentalists, sportsmen, recreationists—demanded to impose their visions on public lands as well. The proliferation of multiple uses on federal lands engendered a zero-sum game that often promoted gridlock rather than progress.[2] By the late twentieth century, conflict intensified as interest groups became ever more polarized. Federal land managers were caught in the middle, beset by lawsuits and debilitated by unfunded mandates and shrinking budgets.

To retain a voice in this increasingly challenging arena, rural producers began to form grassroots community-based collaborative conservation groups (CBCCs) throughout the West. These groups pursue a variety of political and ecological goals (see chap. 4). Many producers, especially ranchers and foresters, depend on federal or state trust lands for at least part of their livelihoods. There are now so many ranching and forestry-related CBCCs that a number of umbrella organizations, including the Quivira Coalition based in Santa Fe, New Mexico (see chap. 5), and the Rural Voices for Conservation Coalition based in Portland, Oregon (see chap. 6), have emerged to bring these groups together and provide regional forums for their concerns.

Ranching and forestry CBCCs are fundamentally different from the traditional organizations that represent commodity producers, like cattlegrowers' or forestry associations. Rather than acting like another interest group, CBCCs attempt to bring individuals from different interest groups together to promote the long-term ecological and economic health of particular landscapes. Neither are they temporary political coalitions. In many cases, CBCCs avoid taking stands on controversial issues. Instead, the successful ones serve as nodes that bring producers, environmentalists, scientists, recreationists, tribal members, and agency personnel together. Their goal is to search for the common ground that unites people rather than remain trapped within the issues that divide them. Rural CBCCs are trying to stitch the West back together in a fundamental way: by bringing people with different interests, ideologies, and backgrounds together to speak with, rather than at, each other.

The common ground they discover usually turns out to be literal as well as metaphorical: a watershed, valley, mountain range, or county. And common goals often revolve around ecological processes that cut across the jurisdictional boundaries that carve up the West: the movement of wildlife, the spread of invasive species, the flow of water, the management of fire as a natural disturbance in fire-adapted ecosystems. As such, CBCCs serve as vehicles to conserve those regions by organizing or participating in specific projects. They develop burn plans and help carry out prescribed fires. They restore eroded gullies. They improve wildlife habitat or build and maintain rural roads that capture runoff and reduce erosion.

Another shared goal is economic: a desire to keep ranch and forestlands from being converted into subdivisions and strip malls. "It's land fragmentation, stupid!" became the mantra of the Arizona Common Ground Roundtable, an early effort to mobilize ranchers, environmentalists, and sportsmen to find ways to slow the explosive urban, suburban, and exurban growth devouring Arizona's wide open spaces (chap. 4). To prevent that, however, rural producers have to be able to make a living off the land. It is not easy to be a rancher or a forester in the modern West, especially during an economic recession. If a second revolution sweeps across the West in which federal and state agencies develop more inclusive and collaborative decision-making processes to manage public lands, CBCCs may provide the critical institutional framework needed to partner with agencies in stitching the West back together, one watershed at a time.

## NOTES

1. D. Worster, *Under Western Skies: Nature and History in the American West* (New York: Oxford University Press, 1992), 44.
2. R. M. Cawley and J. Freemuth, "A Critique of the Multiple-Use Framework in Public Lands Decision Making," in *Western Public Lands and Environmental Politics*, ed. C. Davis (Boulder, CO: Westview Press, 1997), 32–44.

# 4

# BEYOND "STAKEHOLDERS" AND THE ZERO-SUM GAME

## Toward Community-Based Collaborative Conservation in the American West

*Thomas E. Sheridan, Nathan F. Sayre, and David Seibert*

### IN BRIEF

- A growing trend in attempts to manage land for multiple interests is the formation of community-based collaborative conservation groups (CBCCs); they are successfully facing the challenges posed by invasive species, watershed deterioration, wildfires, economic decline, and loss of open space.
- Often created and led by rural ranchers, forest owners, and other producers, CBCCs include alliances with scientists, conservationists, and agency resource managers all seeking common ground and pragmatic solutions to resource management issues.
- Examples of successful CBCCs include the Altar Valley Conservation Alliance in southern Arizona, the Diablo Trust in northern Arizona, the Malpai Borderlands Group in southern Arizona and New Mexico, and the Laramie Foothills Group in Colorado.
- Collectively, the greatest accomplishments of CBCCs may be the formation of partnerships based on trust in the face of legitimate differences and presenting a unified voice in natural resource management and decision making.

### INTRODUCTION

During the past two decades, scores of community-based collaborative conservation groups (CBCCs) have sprung up across the West. They are often created and led by rural producers—particularly ranchers and foresters—frustrated by bureaucratic regulations that restrict their ability to make a living from the land, especially if much of that land belongs to states or the federal government. Usually place-based and place-specific, these groups focus much of their attention on achieving tangible conservation goals like erosion control, removal of invasive species, or the restoration of fire to forest and grassland ecosystems.

Unlike traditional industry or advocacy organizations, however, most CBCCs have forged strong alliances with scientists, conservationists, agency resource managers, and other stakeholders. These allies are just as frustrated by the zero-sum game that often dominates resource politics on public lands in the modern West.[1] As legal scholar Cass Sunstein points out, positions grow more extreme and internal diversity diminishes when members of interest groups meet and talk among themselves because like-minded people reinforce rather than challenge one another's positions. In a society marked by the balkanization of "interest groups" who only speak to their own members, CBCCs and their partners represent a countertrend: the search for common ground. Together they form the so-called radical center, a place where people of diverse interests meet and work together to achieve common goals (see the introduction to this volume).[2]

The collaborative conservation movement is not limited to producer-led groups. There are government-led collaborative endeavors such as the Animas River Stakeholder Group in Colorado, and "hybrid" efforts like the Laramie Foothills Group in Colorado, initiated by the Nature Conservancy (see chap. 13 on Arizona's Sonoran Desert Conservation Plan).[3] There are also umbrella organizations like the Quivira Coalition (chap. 5) and the Rural Voices for Conservation Coalition (chap. 6) that bring these different groups together. Westerners are experimenting with an ever-increasing, ever-evolving number of political and economic strategies to stitch the West back together, as the case studies in this volume attest.

Social scientists have made several preliminary attempts to analyze this emerging phenomenon. R. D. Margerum surveys the typologies that researchers have developed to categorize these groups by different criteria, including the outcomes these groups seek to achieve; whether they are directed by governments, citizens, or NGOs; or the scale at which they work.[4] Margerum also presents a typology of his own, which focuses on institutional analysis and design.

In this volume, we are less interested in classifying collaborative conservation organizations than in exploring their successes and failures on the ground, especially for those led by rural producers. Our approach is more ethnographic than analytic. We want to capture the contingent, experimental vitality of these groups as they strive to create new ways of fostering conservation across the boundaries that divide the rural West. And even though all three authors of this particular chapter are academics, our insights are derived primarily from our participation in two rancher-led CBCCs, the Altar Valley Conservation Alliance and the Malpai Borderlands Group, both located in the Arizona-Sonora (Mexico) borderlands. We write not as detached, "objective" scholars but as practitioners learning as we go.

Most of the collaborative efforts discussed in the present volume have not been profiled in earlier publications on collaborative conservation.[5] No single volume, including this one, could ever encompass more than a small subset of collaborative groups in the West. What distinguishes this volume from others about collaborative conservation is that its purpose is not to describe and evaluate CBCCs as political institutions, how they work, and what makes them successful or not. Rather, it is to draw attention to the critical role they play in working landscape conservation by examining the innovative tools and strategies they have developed to keep forests and rangelands producing in an ecologically sound and economically sustainable manner. In this chapter we focus on CBCC in the context of rangeland conservation and ranching to illustrate the many reasons why CBCCs have formed in the West and some of the ways in which they are conserving working landscapes.

Collaborative conservation is a subset of a much larger effort to solve public-policy problems through processes that build consensus among stakeholders rather than resorting to litigation or lobbying.[6] As Donald Snow points out, "Efforts that evolved into collaborative conservation probably got their start in the arena of alternative dispute resolution, as it was applied to environmental issues beginning in the 1970s."[7] Environmental alternative dispute resolution often limited itself to the resolution of disputes between two parties through the forging of formal agreements. Collaborative conservation, in contrast, usually involves numerous stakeholders.[8] Moreover, many CBCCs, including those profiled in this volume, seek to create networks of partners that endure through time. Their goal is not just to resolve a single dispute but, rather, to establish a process based on trust that addresses continuing issues of resource management.

In that respect, community-based collaborative conservation offers a new model of resource management in the twenty-first-century West. Moving beyond a political process in which stakeholder interest groups engage in a zero-sum game to advance their own agendas by attacking the agendas of oth-

ers, the collaborative conservation movement strives to bring rural producers, environmentalists, government managers, and others together to find
common ground and common solutions to social, economic, and ecological
problems. No one knows how many of these CBCCs exist across the West. One
of the authors of this chapter (Sheridan) distributed an analytic tool called
the "Grassroots Collaborative Conservation Survey" at the annual meeting of
the Quivira Coalition in 2009. Eight organizations ranging from the Madison
Valley Ranchlands Group in Montana to the Malpai Borderlands Group of the
United States–Mexico border responded. Based on those surveys as well as
the published literature, we have identified four key themes that many grassroots CBCCs led by rural producers seek to address in the American West.
These themes are not meant to be exhaustive, but we feel they capture some
of the major political challenges to which these groups are responding.

Community-based collaborative conservation groups:

- react to, and try to overcome, the flaws and limitations of top-down "command and control" regulation and its associated confrontation and gridlock
  by searching for common ground and pragmatic solutions to problems,
  especially between environmentalists, rural producers, recreationists, and
  land management agency personnel;
- contribute to the formulation of site-specific goals and management
  practices for landscapes that cross jurisdictional and other boundaries that
  often hamper effective ecosystem management, with some also contributing to the formulation of policy that affects the management of these
  landscapes;
- involve local people in the definition, design, implementation, and evaluation of conservation efforts, both large and small scale, over time; and
- recognize and build on the long histories of local people in specific places,
  with an eye to extending those histories into the future—that is, maintaining long-term commitments to place by enhancing both economic and
  ecological conditions on intermixed private and public lands.

## STITCHING THE WEST BACK TOGETHER, ONE WATERSHED AT A TIME

More than a century ago, John Wesley Powell argued that the West should be
divided into "irrigation districts" where settlers would "establish local self-
government by hydrographic basins" and share water, forests, and rangelands.[9] Powell tried to convince Congress that instead of 160-acre homesteads,

it should award grants of 80 acres for irrigated farmland and at least 2,560 acres (four sections) of pasture. The rest of the watershed would remain public domain managed by the local irrigation districts themselves: "I say to the Government: Hands off! Furnish the people with institutions of justice, and let them do the work for themselves," Powell growled.[10]

Powell's vision of western society made immanent ecological sense. Communities of ranchers, farmers, and foresters would be organized according to the flow of water—the most critical resource in arid and semiarid lands. But one look at a land tenure map of almost any western watershed reveals that such logic has rarely, if ever, been followed. Historical and current patterns of ownership in western watersheds do not conform to topography or hydrology, with the partial exception of the higher mountain ranges, which the U.S. Forest Service often controls. The basins between mountains, in contrast, look like puzzles with different-colored pieces—rectilinear polygons in seemingly random patterns of private, state trust, or Bureau of Land Management possession. Rules and regulations governing one parcel may not apply to its neighbors. Decision making is fractured along lines that have little or nothing to do with the flow of water or the lay of the land.

Most of that decision making is vested in different levels of government, with the federal government dominating resource management because it controls so much of the West. Because of the National Environmental Policy Act and other laws, federal agencies are mandated to inform the public and solicit public opinion whenever they review or change the administration of public lands under their jurisdiction. But those solicitations were not designed to develop consensus or achieve compromise. Instead, they were informational rather than deliberative; their purpose was to convey and gather information, not to bring interest groups together to make decisions. As such, the process often encouraged posturing and adversarial behavior by pitting interest groups against one another as each struggled to advance its agenda, often at the expense of others (the zero-sum game). Interest groups may have formed temporary political coalitions with one another, but the process rarely forced them to sit down with one another and hash out compromises. On the contrary, decision making resides with the agencies, not the public.

The grassroots collaborative conservation movement has been trying to reverse this top-down, command-and-control flow of power and information in the rural West. Rather than being clients of government agencies, CBCCs often define their own agendas and then invite agencies and nongovernmental organizations to join them as partners. If successful, the result is an ongoing dialogue about resource issues and conservation goals that reduces the friction between regulators and producers and increases the communication between them. It also recasts their roles to a certain extent by creating a metaphorical roundtable around which producers, agencies, and nongovernmental organizations can gather as equals to discuss common problems and find solutions to them. Differences of mission—or of opinion—do not disappear, but the collaborative process concentrates the attention of all parties on the common ground that unites them rather than the lines of ideology or jurisdiction that divide them.

This is not an easy process, even within CBCCs themselves. As this chapter reveals, CBCCs sometimes take no position on controversial issues that divide their own members. Members of a group need to reach consensus among themselves before they try to reach consensus with their partners or other stakeholders. Not all issues can be addressed through collaboration. Litigation, lobbying, and regulation will probably never disappear in the rural West.

Nonetheless, CBCCS are trying to piece together a modern version of Powell's vision by providing an enduring framework for conservation that cuts across jurisdictions and land ownership boundaries. Echoing Powell, many CBCCs organize themselves by watershed and are often labeled "watershed groups."

One example of a watershed group is the Altar Valley Conservation Alliance (AVCA) in Arizona, which began in 1995 and became a 501(c)3 not-for-profit organization in 2000.[11] The Altar Valley consists of more than 600,000 acres of desert grassland drained by Altar Wash and its tributaries, including Arivaca Creek. In part because it is located southwest of Tucson, the birthplace of Earth First! and headquarters of the Center for Biological Diversity (CBD), the Altar Valley used to be ground zero in the western range wars. Jim Chilton, one of AVCA's founding members, was engaged in a bitter battle with the CBD and its allies in the Forest Service and U.S. Fish and Wildlife Service over grazing on the Montana Allotment of the Coronado National Forest. When the Forest Service renewed Chilton's grazing lease, the CBD appealed, contending that Chilton's cows endangered the threatened Sonora chub (Gila ditaenia). It also published 21 photos on its website that purported to show degradation of the allotment due to grazing. When Chilton found out about the photos, he sued the CBD for malicious libel in 2003. He and his lawyers also rephotographed all 21 locations. Four were not even on the allotment. The others were either hunters' campsites, old mining roads, or tight shots that, when shown in panorama, revealed healthy landscapes. The coup de grâce was a photo of a dry lake bed where several hundred people, including the CBD's photographer, had celebrated May Day a few days before the shot was taken. In 2005, a jury in Pima County Superior Court concluded that the CBD had made "false, unfair, libelous and defamatory statements" against Chilton on its website and in a press release. It awarded Chilton $100,000 in actual and $500,000 in punitive damages. The CBD appealed but both the Arizona Court of Appeals and the Arizona Supreme Court upheld the verdict.[12]

The long battle with the CBD understandably embittered Chilton and his family. Tensions also simmered between AVCA ranchers and the nearby Buenos Aires National Wildlife Refuge, which in 1985 had converted a huge ranch at the south end of the Altar Valley into a 118,000-acre preserve for the endangered masked bobwhite (*Colinus virginianus ridgwayi*).[13] During the same period, Pima County was launching its ambitious Sonoran Desert Conservation Plan (see chap. 13). Even though the county identified ranch conservation as one of its five major goals and pledged that it would only acquire land from willing sellers, many ranchers were deeply suspicious of the county's intentions. Polarization, paranoia, and interest-group politics made collaboration exceedingly difficult during the first decade of AVCA's existence. Angry confrontations occasionally erupted at AVCA meetings, creating an atmosphere of mistrust.

To reverse these negative dynamics, AVCA turned to Dr. Kirk Emerson, the first director of the U.S. Institute for Environmental Conflict Resolution of the Morris K. Udall Foundation in Tucson. A professional facilitator, Emerson had worked with several members of AVCA on the Arizona Common Ground Roundtable, a statewide forum that began in 1997 when the Arizona chapter of the Nature Conservancy asked the Udall Center for Studies in Public Policy at the University of Arizona to help it improve relations with Arizona ranchers. Emerson knew the ranchers and understood their issues. With her guidance, AVCA established ground rules and learned how to run meetings where agency officials and representatives from conservation nongovernment organizations felt welcome, not threatened. Meanwhile, U.S. Fish and Wildlife Service removed the manager of the Buenos Aires refuge and replaced him with leaders who reached out to AVCA. Finally, the county demonstrated its commitment to working landscapes by signing management agreements with ranchers who sold their spreads to it, including two members of AVCA (the Rowley family of Rancho Seco and the Chilton family of Diamond Bell Ranch). Meetings became civil and productive. Slowly but surely, bonds of trust began to develop among ranchers, scientists, conservationists, and agency personnel.

Today AVCA and Buenos Aires National Wildlife Refuge are partners, not antagonists, in the valley they share. The refuge routinely cooperates with AVCA on conservation projects on private and state trust lands. When Buenos Aires received federal economic stimulus funds in 2010, its new manager asked former AVCA restoration coordinator David Seibert, one of the authors of this article, to conduct a gully restoration project on refuge lands. Today AVCA and the refuge utilize the project's four coordinated restoration sites and others as demonstration areas for students, agency personnel, and private interest groups. Completed in 2011, perhaps no other project better symbolizes the spirit of collaboration that now characterizes conservation in the Altar Valley.

## NATURAL ELEMENTS: FIRE

Political concerns such as endangered species issues are not the only reasons ranchers come together to form CBCCs. For those in business to raise cattle or sheep, not speculate in real estate, the health of the land they work is of fundamental importance. Like ancient philosophers, they focus on the natural elements, especially earth (soil), water, and fire. Grass cannot grow unless soil is held in place and water captured. Preventing and reversing erosion therefore become major goals, especially in the arid and semiarid Southwest. Another problem is the invasion of woody shrubs because of fire suppression. Thus two major goals of many western CBCCs are the restoration of eroded watersheds and the reintroduction of fire into desert grasslands and forests.

A paradigm shift that began around 1990 has revolutionized fire management across the West. Once considered a monster that destroyed both property and nature, fire is now seen as a natural process in certain ecosystems, as necessary to their long-term health as rain or snow.[14] Properly used, fire is also a tool to restore and maintain healthy landscapes—one that indigenous peoples around the world have employed for thousands of years.[15] Primarily because of fire suppression, shrubs have encroached on more than 84 percent of the grasslands in the southwestern United States.[16] "Had fires continued to sweep the grasslands down through the years to the present with their original frequency," ecologist Robert Humphrey has observed, "the desert grassland would probably occupy about the same area today as it did prior to the white settlement of the Southwest."[17]

Fire led to the formation of the Malpai Borderlands Group, the granddaddy of ranching CBCC organizations.[18] The Malpai planning area encompasses a

1,250-square-mile triangle of ranchlands in southeastern Arizona and south-western New Mexico. On July 2, 1991, a blaze started next door to Warner and Wendy Glenn's Malpai Ranch near the United States–Mexico border. The fire was not threatening any structures, so the ranchers urged the Forest Service to let it burn. A Forest Service crew put it out anyway because that's what fire crews did in those days. In response, the Glenns and their neighbors be-gan meeting in September of that year. They invited others as well, including scientists like ecologist Ray Turner and political activists like Jim Corbett.[19] Corbett and a few like-minded landowners in the San Pedro River Valley had formed the Saguaro-Juniper Association in 1988, which envisioned using livestock to live in harmony with, rather than taming, wildlands. The Mal-pai's idea of "working wilderness" sprang in part from Corbett's covenant be-tween humans and "untamed communities of plants and animals."[20]

By July 1992, the group had written the "Malpai Agenda for Grazing in the Sonoran and Chihuahuan Bioregions." One of the goals of the Malpai agenda was to reverse the polarization between ranchers and environmentalists by identifying "the conservational common ground that unites all of us who love the land, then to create programs in which we can work together to im-plement the values we share." Another was to recognize the environmental contributions of ranchers. "In the Sonoran and Chihuahuan bioregions and most of the arid West, ranching is now the only livelihood that is based on human adaptation to wild biotic communities . . . conservationists who are ranchers are divided from many other conservationists by their belief that ranching can be stewardship that preserves the health and unreduced diver-sity of the native biotic community."[21]

The development of a fire management plan was one of the primary ways the Malpai ranchers put their principles into action. Gathering several times in 1993, they invited representatives from the state and federal agencies that managed public lands in the region. After a two-day meeting, the agencies agreed to let their fire control policies be "informed and guided by the man-agement goals of the ranchers." Everyone then pledged to engage in "a coor-dinated, comprehensive ecosystem management approach" that would "en-hance and restore the use of natural processes in these ecosystems, to improve their renewable resources, to provide for wildlife habitat and productivity of grasslands, and to sustain rural and grazing livelihoods." The Radical Center was being born.[22]

In 1994, the ranchers formalized their loose network as the Malpai Bor-derlands Group, a 501(c)3 not-for-profit corporation. A year later, the Malpai Group and the Forest Service conducted their first prescribed burn, Baker I, on 6,000 acres in the Peloncillo Mountains. The next was the Maverick Burn (9,014 acres), where concerns about the endangered lesser long-nosed bat (*Leptonycteris curasoae*) and New Mexico ridge-nosed rattlesnake (*Crotalus willardi obscurus*) triggered consultation with U.S. Fish and Wildlife. After a year of discussion, Fish and Wildlife authorized an incidental take permit that allowed the burn to take place in June 1997.[23] In return, the Forest Service agreed to monitor the effects of fire on the two species.

The study concluded that fire had little effect on nectar and pollen production of the agave (*Agave palmeri*) on which the bats fed.[24] But planning the third and largest prescribed burn, Baker II, was considerably more contentious, because concerns about the ridge-nosed rattlesnake persisted. Only one of nine rattlesnakes radio-tagged before the Maverick Fire had been killed by the burn (and it wasn't a New Mexico ridge-nosed), but snake biologists argued that fuel buildup due to fire suppression had created conditions for fires so hot they might damage the snakes' habitat. The biologists squared off against the Forest Service and the ranchers, and accusations of bad faith flew from both sides. It took more than five years of planning and debate before Baker II was set in June 2003. The controversy revealed how concern over a single species could hold ecosystem management hostage. "A fire that kills an individual of a listed species constitutes 'take' regardless of long-term benefits to the species as a whole," geographer Nathan Sayre observed, "whereas activities that have no direct effect on listed individuals (e.g., fire suppression) do not constitute take even if their indirect effects may be significantly detrimental. The Fish and Wildlife Service must somehow resolve these trade-offs and potential contradictions."[25]

When Baker II did finally burn, it spread across 46,000 acres in the Peloncillos, reputedly the largest successful prescribed fire ever in the western United States. Between 1989 and 2005, prescribed and naturally ignited fires covered more than 300,000 acres in the Malpai Borderlands, including 250,000 acres on the Gray Ranch, which the Animas Foundation had purchased from the Nature Conservancy using funds from the Hadley family. Patient collaboration has returned fire to the landscape, overcoming dissension and bureaucratic infighting along the way.

## NATURAL ELEMENTS: EARTH AND WATER

Both the Malpai and the Altar Valley Conservation Alliance have also conducted extensive erosion control projects in their watersheds. Most of the work has concentrated on the restoration of small upland gullies using rock check dams and other simple, inexpensive structures. One principle guiding these efforts is, "first, do no harm." Too often in the past, land managers have unintentionally triggered even greater erosion by installing large gabions or concrete dams that were subsequently overwhelmed or undercut by heavy floods. Instead, AVCA and Malpai have been guided by retired Forest Service biologist Bill Zeedyk and his colleagues, independent consultants who have developed low-tech, low-impact techniques to "let the water do the work" (see chap. 5). The principles that underlie these restoration techniques mirror those of successful CBCCs in the West—working with existing elements (soil, water, vegetation), rather than attempting to alter or remove them, and in coordination with existing ecosystem processes (e.g., sediment deposition and the movement of water in gullies) rather than against them.[26]

What CBCCs cannot accomplish, however, is the restoration of major arroyos that were carved through the alluvium of Southwestern basins in the late nineteenth and early twentieth centuries.[27] One such arroyo is Altar Wash. Before 1900, Manuel King, the founder of the Anvil Ranch, could ride his horse faster than any flood because stands of giant sacaton grasses (*Sporobolus wrightii*) slowed the waters down and spread them across the valley floor, where they recharged valley aquifers. The sacaton floodplain also sheltered masked bobwhite and other species of birds and animals. But overgrazing and fuel wood cutting in the late 1800s denuded the uplands while the construction of a wagon road and reservoirs like Aguirre Lake removed riparian vegetation. When heavy rains in 1904–5 poured into the drainage, Aguirre Lake burst its dam and sliced a channel down the wagon road.[28] Today that channel is twenty feet deep and more than 1,440-feet wide in places. Floodwaters now roar out of the valley, and the sacaton is long gone. Because the grade of the main trunk has dropped, Altar Wash causes head cutting up its tributaries as well, a dynamic that today affects the wildlife refuge and working ranches alike. The Natural Resource Conservation Service estimates that about seven acres of land and 100 acre-feet of sediment wash away each year.[29]

Overgrazing is largely a thing of the past in the Altar Valley. Stocking rates are 65–90 percent lower than they were when Altar Wash created. Most ranchers have implemented rotational grazing systems that allow pastures to recover. They also have installed wells and pipelines to provide additional sources of water for their cattle and spread them across the landscape. But Altar Wash remains an open and erosive wound. For more than 30 years, the Kings and other ranchers have advocated the construction of grade control/sediment retention structures along Altar Wash itself. They point to 19 such structures built by the Civilian Conservation Corps, Natural Resource Conservation Service, and Bureau of Land Management in the San Simon Valley of southeastern Arizona between the 1930s and the 1950s. According to a 1992 report by the Soil Conservation Service (predecessor of Natural Resource Conservation Service), those retention dams have aggraded 10 miles of the San Simon floodplain and captured about 19 million tons of sediment.[30]

But neither individual ranchers nor AVCA have the funds or expertise to construct such structures on their own. The Altar Valley Conservation Alliance therefore turned to Pima County's Natural Resources, Parks and Recreation Department for help. The department, which manages all the Pima County ranches purchased with open-space bond funds under the Sonoran Desert Conservation Plan (see chap. 13), wrote a $1.5 million proposal for the next open-space bond election, which was supposed to occur in 2008 but has been postponed because of the economic recession. Such partnerships are critical for CBCCs like AVCA to engage in landscape-level conservation endeavors.

Before any structures are built, however, a hydrological study of the Altar watershed will have to be carried out. Many hydrologists and geomorphologists have grown increasingly skeptical about manipulating drainage systems. Western landscapes are littered with failed experiments to modify the flow of water across them. In some cases, the experiments have done far more harm than good. That is why it is necessary for CBCCs to consult with appropriate scientists in their areas before they spend precious time, labor, money, and social capital on restoration projects. They also need to monitor the experiments they do carry out. Even one-rock check dams need to be evaluated over time. The Civilian Conservation Corps built thousands of similar structures during the 1930s, but the impacts of its work have never been systematically investigated. Anecdotes are poor substitutes for rigorous empirical research.

## ECONOMIC DIVERSIFICATION

Another goal of some CBCCs is economic development and diversification. Ranching is a notoriously difficult way to make a living.[31] Stock raisers from Montana to Arizona tell variations of the same joke: "How do you make a small fortune in ranching? Easy: you start with a large fortune."[32] The operating costs of beef production—fuel, feed, veterinary care, and so forth—have risen much faster than the price of beef. Small wonder that so many ranchers have sold their private lands to real-estate speculators and left the business.[33]

Economic diversification is not a new trend in the ranching industry.[34] Many "dude ranches" started out as working outfits that ran a few guests on the side. In states that allow it (Arizona does not), other ranchers make some of their income by selling deer and elk tags to hunters who want to hunt on their private lands. Others are exploring leases to energy companies to install wind- and solar-energy projects on their properties.

Perhaps the most widespread initiative, however, is to capture more value for their beef or lamb by selling it directly to consumers (see spotlight 5.1). Vertical integration is not as complete in the beef industry as it is among poultry and pork producers. Most cattlemen remain independent business-men in the sense that their cow-calf or steer operations are not owned by the big feedlots and packing companies that dominate beef production in the United States. But those conglomerates form an oligopsony—a small num-ber of middlemen who can dictate prices to large numbers of both producers and consumers.

Betty Fussell exposes the nature of that oligopsony in her *Raising Steaks: The Life and Times of American Beef* (2008). During the 1960s, there were 200,000 feedlots in the United States. By 2005, their number had shrunk to 800, and only 2 percent of them were processing 85 percent of the approximately 34 million cattle slaughtered each year.[35] The Big Four packing companies—Tyson Foods; Cargill Meat Solutions, a subsidiary of Cargill; JBS USA, a subsidiary of Brazilian-based JBS S.A., the largest beef packer in the world; and National Beef Packing Company—control more than 83 percent of U.S. slaughtering and processing.[36] Five retailers—Wal-Mart, Kroger, Safeway, Albertsons, and Ahold USA—control 46 percent of the marketing. "Despite their endless business talk and cowboy costumes, the majority of the 800,000 ranchers today who raise cattle cannot make a living out of cattle alone," Fus-sell points out. "Only big-time feedlots in conjunction with packers in con-junction with retailers—in other words, the industrial beef chain, supported by government agencies—make big-time money."[37]

It's not easy to evade the clutches of the conglomerates. In most areas of the West, the infrastructure to slaughter, process, and market one's own beef or lamb is not accessible or has very limited capacity to accommodate many ranchers. United States Department of Agriculture–certified slaugh-terhouses in particular are few and far between. Even when ranchers live near major metropolitan centers like Phoenix, they are able to slaughter only a fraction of their calves, steers, or lambs. Then, they have to rely on farmers' markets and online sales to reach consumers. The lack of infrastructure en-sures that locally produced grass-fed beef and lamb remains a niche market, not a mainstream food choice. There are a few exceptions, like Country Natural Beef (see spotlight 10.1) or the Diablo Burger restaurants in Flagstaff and Tucson, Arizona (discussed below). But they are the happy exceptions that prove the rule.

## SLOWING THE SUBURBS: THE FIGHT AGAINST SPRAWL

Diablo Burger is an independent offshoot of the Diablo Trust, a collaborative group created in 1993 by the Prosser family, who own the Bar T Bar Ranch, and the Metzger family, who own the Flying M Ranch, in northern Arizona.[38] The Diablo Trust provides most of the beef used by Diablo Burger, thereby pioneering retail and direct marketing of beef in Coconino County, where livestock constitutes 93 percent of the agricultural sector, 99.5 percent of which gets exported for processing and consumption elsewhere.[39] Located southeast of Flagstaff, Arizona's third largest city, the two ranches comprise 426,000 acres ranging from desert grasslands to ponderosa pine forests. About one-third is part of Coconino and Apache-Sitgreaves National Forests. The rest is private and state trust lands, much of it locked into a checkerboard of alternating sections that the U.S. government originally granted to the Santa Fe Railroad in the nineteenth century.

The Diablo Trust's greatest success has been to preserve the Bar T Bar and Flying M as working ranches. During the real-estate booms of the 1980s and early 2000s, the price of land in and around Flagstaff skyrocketed faster than in other Arizona markets. The private parcels of the Bar T Bar and Flying M were therefore hot commodities in the land-starved speculative frenzy. As the Prosser and Metzger families surveyed their domain in the early 1990s, selling those parcels seemed very attractive at times. Their cattle competed for forage with Arizona's exploding elk herds. Their management came under increasing scrutiny from environmentalists, especially on their Forest Service allotments. But instead of cashing in, the Prossers and Metzers formed the nonprofit Diablo Trust and held a public meeting to ask for help in 1993. The response was overwhelming. Scientists, environmentalists, and other citizens as well as state and federal agency personnel joined the two ranching families to create the Northern Arizona Collaborative Grassroots Management Team. Together they pursued the trust's mission: "To ensure the long-term economic, social, and ecological sustainability of the Diablo Trust land area by providing a forum for active community participation in a collaborative land stewardship process."

The trust met monthly and formed numerous working groups to achieve its key goals: "Sustaining open space (preventing land fragmentation); sustaining biological diversity; sustaining multiple-generation stewards working on the land; producing high-quality food; protecting watersheds with stable living soils; restoring historic grasslands; enhancing wildlife corridors; achieving community of place." Because of these efforts, the trust has won numerous state and national awards and was designated a National Partnership for Reinventing Government Laboratory in 1998. More important, Diablo Trust lands remain undeveloped despite real-estate prices that approach California levels. With its land and watershed improvement projects, its community outreach including annual Artists' Days on the Land, it is a living laboratory to promote its motto: "Learning from the land and sharing our knowledge, so there will always be a West."

Similar CBCCs have sprung up in other development hotspots throughout the West. It is difficult to select a textbook example of sprawl in the region because the textbook is being rewritten every day. *Mountain Megas*, a 2008 report from the Brookings Institution, calls the intermountain West the "New American Heartland." In that new heartland, the report identifies "five emerging 'megapolitan' areas—vast, newly recognized 'super regions' that often combine two or more metropolitan areas into a single economic, social, and urban system."[40] The two largest are Phoenix-Tucson and Denver, the core city of Colorado's Front Range. "At the peak of land conversion, Colorado was losing the size of Rocky Mountain National Park in farm and ranch land each year," conservation biologist Rick Knight points out. "Interstate 25 was called 'Main Street,' and towns from Colorado Springs to Fort Collins were referred to as neighborhoods. Colorado was well on its way to copying the trajectory of Atlanta, Georgia and Southern California."[41]

To prevent that megalopolis from swallowing northern Colorado, the Colorado chapter of the Nature Conservancy partnered with the "last ranching community located on Colorado's North Front Range" to create the Laramie Foothills Mountains to Plains Conservation Project in 1987. According to one of the founders, the project "lies across an ecotone, hence we work in rangelands, foothills shrublands, forests, wetlands, and riparian corridors."[42] Such ecotones are among the most threatened of the West's wide open spaces because they consist of mosaics of public and private lands. To stitch the Laramie Foothills back together, the Colorado chapter of the Nature Conservancy and its partners forged a "remarkable consortium of rural and urban constituencies" to "ensure that land beyond city limits stayed open and productive, rather than developed and running red deficits on county and city ledgers."[43]

By 2009, the Laramie Foothills Group had protected about 100,000 acres of private lands through conservation easements and purchases and connected them to 110,000 acres controlled by local, state, and federal governments.[44] In Knight's words, "Due to several large ranches being placed on the market or their owners seeing the wisdom of placing conservation easements on their land; and some risk-prone individuals working for city and county governments, a progressive land trust, and an international conservation organization; forces converged to protect an east-west swath of land nearly 22 miles wide and 20 miles deep."[45]

One of the Laramie Foothills Project's primary sources of funding for conservation easements are city and county open-space sales taxes. To acquire and maintain such support, the group has to straddle the rural-urban divide. One of the ways it does so is to encourage a local food movement and to establish a place-based education program in local schools.[46] Tellingly, the vision and initiative for the Laramie Foothills Project have all come from the private sector and local government: ranchers, the Colorado chapter of the Nature Conservancy, Larimer County, and the city of Fort Collins—"a new way of doing conservation," according to Knight.[47] "For, truth be told, the U.S. Forest Service (USFS) was not involved in this immense conservation effort," Knight goes on to say. "Indeed, the only entity from the federal government that did play a vital role was the Natural Resources Conservation Service (NRCS). This agency has always worked with private landowners; its employees have learned that to be successful one must focus on listening and show due respect. There is little room for a top-down approach when working with private landowners who control approximately two-thirds of our country."

## OPPORTUNITIES AND CHALLENGES

To be fair to the Forest Service, it is hardly a monolithic organization. The U.S. Forest Service has been a key partner in the Malpai Borderlands Group's efforts to restore fire to their immense, dry landscape. But top-down, command-and-control models of land management rarely allow for the flexibility and innovation necessary to meet the conservation challenges of the twenty-first century. That is why CBCCs are evolving across the West. "Today, watersheds across the country are increasingly self-organizing; with citizen-based groups coming together and finding ways to make where they live, work, play, and worship healthier, both conserving land and water but also strengthening economies," Knight concludes.[48]

Partnerships based on trust are perhaps the greatest accomplishments of CBCCs. If those partnerships continue to develop and grow, they may provide an enduring new model for land management in the West, as Knight and others hope. But trust takes a long time to mature, especially when it emerges from conflict.[49] One of the major challenges CBCCs face is how to bring rural producers, environmentalists, scientists, sportsmen, and agency personnel together when legitimate differences, and the paranoia that often erupts from polarization, threaten to drive them apart. Margerum provides a good overview and literature review of the many issues facing ongoing collaborative efforts.[50] Describing that literature as simultaneously both "vast" and "sparse," Margerum identifies four critical challenges to sustaining collaboratives: collaborative leadership, board governance, organizational change, and external pressures.[51] The first three concern the internal dynamics of collaborative conservation groups. The last—a catch-all that encompasses everything from economics to government policy—is largely outside the control of CBCCs.

One of the perennial internal problems facing collaborative conservation organizations is how to define themselves. Should membership be open or restricted? What are the principles of inclusion and exclusion? Some organizations, like the Madison Valley Ranchlands Group in Montana (see spotlight 11.1), have dues-paying members who participate in various events and projects. Others, like the Altar Valley Conservation Alliance, have made conscious decisions not to open their groups to individuals who are not landowners and rural producers. The Malpai Borderlands Group has no formal membership but restricts the large majority of the seats on its board to local landowners. Both AVCA and Malpai feel that their missions might be diluted or compromised without strong local control.

Another challenge facing foresters and ranchers is a lack of formal say-so in decisions that affect their ability to make a living on federal lands. For example, neither foresters nor ranchers have a legal "seat at the table" when environmental groups sue federal agencies over noncompliance with the Endangered Species Act or the National Environmental Policy Act—unless they themselves file a lawsuit, which most cannot afford to do.

At present, producer-led CBCCs struggle to make their voices heard through informal rather than formal means. Their only trump cards are the private land of their members and the legitimacy given them by their longstanding ties to local landscapes. If rural producers can no longer stay in business, the only value their private lands hold is as real estate. The real-estate subdivision of ranch and forestlands, however, fragments ecosystems and removes fire from the tool kits of land managers. Large-scale ecosystem management becomes difficult, if not impossible. John Wesley Powell's nineteenth-century concerns are today as real as drought, and just as far-reaching in their potential effects. For effective ecosystem management to take place, agency personnel and conservationists should recognize that it is much easier to deal with a few ranchers or foresters than with hundreds of subdivision residents. This has enabled groups like the Malpai and AVCA to take leadership roles in the reintroduction of fire on desert grasslands. But those partnerships are informal, rather than statutory or contractual. Moreover, they depend on agency personnel acting as problem solvers rather than gatekeepers. Unfortunately, such partnerships could easily dissolve if the political climate changes.

The power of CBCCs lies in their ability to present a unified voice in natural resource management and decision making. A potentially fatal weakness is dissension within the group. Just as collaborative conservation demands compromise and a rigorous search for common ground, which often means allowing parties to agree to disagree about peripheral issues, CBCCs need to respect differences among their members. One strategy is to eschew taking positions on controversial issues and focus on conservation on the ground. The Altar Valley Conservation Alliance concentrates primarily on watershed restoration and the reintroduction of fire. As mentioned earlier, it spearheaded the development of a fire management plan for the entire 600,000-acre watershed and has also sponsored workshops by Bill Zeedyk to teach people how to restore gullies and design rural roads that reduce, rather than induce, erosion. The alliance treads lightly, in contrast, around endangered species issues or state trust land reform.

One final internal problem is time. Collaborative conservation demands endless meetings where producers, environmentalists, scientists, sportsmen, and agency land managers sit down together, identify problems, and work toward solutions. Agency personnel and paid staff of environmental nongovernment organizations do this as part of their jobs. Rural producers have to spend their own time and money to go to these meetings, diverting themselves from necessary agricultural tasks. Typically, a handful of volunteers do most of the work in CBCCs. When those volunteers have families to raise and family businesses to run, they often cannot respond in a timely fashion to grant deadlines, meeting schedules, or political crises. They get frustrated and burn out; unless other volunteers step up to take their place, their CBCCs may wither and die.

Being able to hire people to do some of the work is perhaps the most critical transition that CBCCs, like all grassroots organizations, have to make. Of the eight groups surveyed at the Quivira Coalition's annual meeting in 2009, only three have paid staff, including the Madison Valley Ranchlands Group and the Malpai Borderlands Group. The Altar Valley Conservation Alliance is just beginning to make that transition, with a half-time program director supported through grants from sources ranging from the National Fish and Wildlife Foundation to the giant mining company Freeport-McMoRan Copper and Gold and a half-time program coordinator who is paid from grant and AVCA funds. It is very difficult to find money to hire permanent staff, however. Foundations and the federal government provide funds for specific projects that may include salaries, but once those projects are over, the salaries evaporate as well. One of the greatest needs of CBCCs, then, is to secure funds for capacity building, including a bare-bones permanent staff that can administer existing projects and carry on the never-ending search for new grants.

## CONCLUSIONS

The term "working landscape" implies an embodied sense of place, one that often reflects knowledge of local ecological processes that can only be accumulated through generations of daily experience with climate, water, plants, animals, and soil. There are significant differences between those who remain in place—whose workday is defined by seasonal rhythms—and those who return to homes in metropolitan areas at the end of a workday defined by clocks and the need to return vehicles to motor pools. Agency personnel, although they are frequently members of the same rural communities of which ranchers and foresters are a part, often turn over rapidly; those who have gained knowledge of the local social and natural environments may be transferred; or they may not stay in one place long enough to acquire such knowledge. Those individuals are certainly positioned to accomplish a great deal, both from the field and from the office, and many of them share a strong passion for successful natural resource management. But a critical difference emerges between those individual agency personnel and the ranchers who venture out after a rain to determine exactly where and how it fell and who know what it meant to multiple species, including their livestock. It shows up in the fact that weekends, for rural producers, are workdays. People who engage with places in such ways extend their practices, their days, and their bodies beyond the usual confines and comforts afforded by paid wage labor.

It is necessary, at the same time, to avoid romanticizing rural producers. They can overexploit, and have overexploited, the land for profit. Left to their own devices, they may also perpetuate old ways of producing beef or timber that reduce, rather than maintain, biodiversity and ecosystem services. The pendulum swings both ways. What is needed is a perpetual dialectic between conservation and production, between new knowledge and experience, between regulation and experimentation. Adaptive management has to be flexible and inclusive. Above all, it has to be attentive to local variations in the social and natural environments that can never really be separated on particular landscapes.

The promise of collaborative conservation is the possibility of realizing some of the most powerful potentials of a democratic system of governance for people and places. This possibility hinges on developing frameworks that institutionalize accountability, transparency, equity, and flexibility. If natural resource practitioners can define forests and rangelands as dynamic, open systems, they can also conceive of collaboratively managed landscapes as comprising meeting rooms, pastures, offices, the cabs of trucks—all of the places where social and ecological networks are lived and produced in a constantly evolving series of relationships.

In complex social and natural environments, it is logical for management to be considered and conducted as a process, not a final product, requiring a wide repertoire of skills to succeed. These skills range from negotiation and mediation abilities to technical knowledge that includes ecology, law, and finance. Some of these skills exist locally within CBCCs, but others require the coordination of agencies, universities, nongovernmental organizations, and other research groups in creative ways that can respond to changing needs. Communication is vital. It is the medium of social exchange that forges bonds between individuals, facilitates the sharing of knowledge and experience, and enables different interests to negotiate and find common ground. CBCCs encourage everyone to attend to the *lived* part of working landscapes as a tried and true way to move land management beyond the stakeholder, zero-sum game paradigm.

## NOTES

1. R. M. Cawley and J. Freemuth, "A Critique of the Multiple Use Framework on Public Lands Decision Making," in *Western Public Lands and Environmental Politics*, ed. C. Davis (Boulder, CO: Westview Press, 1997), 32–44; P. Brick, D. Snow, and S. Van de Wetering, eds., *Across the Great Divide: Explorations in Collaborative Conservation and the American West* (Washington, DC: Island Press, 2001); C. Klyza and D. Sousa, *American Environmental Policy, 1990–2006: Beyond Gridlock* (Cambridge, MA: MIT Press, 2008); M. Nie, *The Governance of Western Public Lands: Mapping Its Present and Future* (Lawrence: University Press of Kansas, 2008).

2. C. Sunstein, *Infotopia: How Many Minds Produce Knowledge* (New York: Oxford University Press, 2008), *On Rumors: How Falsehoods Spread, Why We Believe Them, What Can*

*Be Done* (New York: Farrar, Straus and Giroux, 2009), and *Going to Extremes: How Like Minds Unite and Divide* (New York: Oxford University Press, 2009); Brick et al., *Across the Great Divide*; C. White, "No Ordinary Burger: Trust Fosters Economic Diversity in Support of Ranches," *ACRES: The Voice of Eco-Agriculture* 41, no. 1 (2011), reprinted at http://www.awestthatworks.com/2Essays/No_Ordinary_Burger/Diablo_Burger.pdf; R. D. Margerum, *Beyond Consensus: Improving Collaboration to Solve Complex Public Problems* (Cambridge, MA: MIT Press, 2011).

3. T. Koontz, T. Steelman, J. Carmin, K. S. Korfmacher, C. Moseley, and C. Thomas, *Collaborative Environmental Management: What Roles for Government?* (Washington, DC: Resources for the Future, 2004).

4. Margerum, *Beyond Consensus*; B. Gray, *Collaborating: Finding Common Ground for Multiparty Problems* (San Francisco: Jossey-Bass, 1989); E. Moore and T. Koontz, "A Typology of Collaborative Watershed Groups: Citizen-Based, Agency-Based, and Mixed Partnerships," *Society and Natural Resources* 16, no. 5 (2003): 451–60; A. Cheng and S. Daniels, "Getting to the 'We': Examining the Relationship between Geographic Scale and Ingroup Emergence in Collaborative Watershed Planning," *Human Ecology Review* 12, no. 1 (2005): 30–43.

5. For other case studies of collaborative conservation, see P. Brick, D. Snow, and S. Van de Wetering, eds., *Across the Great Divide: Explorations in Collaborative Conservation and the American West* (Washington, DC: Island Press, 2001); R. Brunner, C. Colburn, C. Cromley, R. Klein, and E. Olson, *Finding Common Ground: Governance and Natural Resources in the American West* (New Haven, CT: Yale University Press, 2002); E. Weber, *Bringing Society Back In: Grassroots Ecosystem Management, Accountability, and Sustainable Communities* (Cambridge, MA: MIT Press, 2003); C. Wilmsen, W. Elmendorf, L. Fisher, J. Ross, B. Sarathy, and G. Wells, *Partnerships for Empowerment: Participatory Research for Community-Based Natural Resource Management* (London: Earthscan, 2008); and Margerum, *Beyond Consensus*.

6. M. McKinney and W. Harmon, *The Western Confluence: A Guide to Governing Natural Resources* (Washington, DC: Island Press, 2004); Klyza and Sousa, *American Environmental Policy*; C. Wilmsen et al., *Partnerships for Empowerment*; Margerum, *Beyond Consensus*.

7. D. Snow, "Coming Home: An Introduction to Collaborative Conservation," in *Across the Great Divide*, ed. Brick et al., 3.

8. Brick et al., eds., *Across the Great Divide*; Wilmsen et al., *Partnerships for Empowerment*; Margerum, *Beyond Consensus*.

9. J. W. Powell, "Institutions for the Arid Lands," *Century Magazine* 40 (1890): 115.

10. Ibid., 113–14.

11. See www.altarvalleyconservation.org for more about the Altar Valley Conservation Alliance. For examples of other collaborative efforts, especially those jump-started by endangered species issues, see chap. 13 in this volume as well as discussions of the northern spotted owl, salmon and grizzly bear in Brick et al., *Across the Great Divide*; Koontz et al., *Collaborative Environmental Management*; and Klyza and Sousa, *American Environmental Policy*.

12. S. Fenske, "The Rancher's Revenge," *Phoenix New Times*, May 26, 2005; T. Sheridan, "Working Landscapes vs. Wilderness: The Lessons from the Altar Valley Conservation Alliance," in *Five Ways to Value Working Landscapes of the West*, ed. S. Silber, M. G. Chanler, and G. P. Nabhan (Flagstaff: Center for Sustainable Environments, Northern Arizona University, 2007), 33–42.

13. N. F. Sayre, *Ranching, Endangered Species, and Urbanization in the Southwest: Species of Capital* (Tucson: University of Arizona Press. 2002).

14. S. Pyne, *The Story of the Great Fires of 1910* (New York: Penguin Books, 2002); T. Egan, *The Big Burn: Teddy Roosevelt and the Fire That Saved America* (Boston: Houghton Mifflin Harcourt, 2009).

15. O. Stewart, *Forgotten Fires: Native Americans and the Transient Wilderness* (Norman: University of Oklahoma Press, 2007); Pyne, *The Story of the Great Fires of 1910*; C. Bahre, *A Legacy of Change: Historic Human Impact on Vegetation of the Arizona Borderlands* (Tucson: University of Arizona Press, 1991).

16. D. F. Gori, and C. A. Enquist, *An Assessment of the Spatial Extent and Condition of Grasslands in Central and Southern Arizona, Southwestern New Mexico, and Northern Mexico* (Tucson: Nature Conservancy, 2003).

17. R. Humphrey, *The Desert Grassland: A History of Vegetational Change and an Analysis of Causes* (Tucson: University of Arizona Press, 1958).

18. See www.malpaiborderlandsgroup.org. Restoring fire to desert grasslands is also a major goal of AVCA, which, along with the Tucson office of the federal Natural Resources Conservation Service, spearheaded the development of the watershed-wide Fire Management Plan approved by the U.S. Fish and Wildlife Service in 2008 (see ftp://ftp-fc.sc.egov.usda.gov/AZ/AltarValley_Fire_Management_plan.doc).

19. K. Cash, "Malpai Borderlands: The Search for Common Ground," in *Across the Great Divide*, ed. Brick et al.; N. Sayre, *Working Wilderness: The Malpai Borderlands Group and the Future of the Western Range* (Tucson: Rio Nuevo, 2005).

20. Quoted in Sayre, *Working Wilderness*, 41.

21. Quoted in ibid., 42.

22. Cash, "Malpai Borderlands"; Sayre, *Working Wilderness*.

23. An incidental take permit, issued under section 10(a) of the U.S. Endangered Species Act (50 CFR 17.31 and 17.32), allows private, nonfederal entities to undertake otherwise lawful activities that may result in "take" of a listed species. "Take" is defined in the act as "to harass, harm, pursue, hunt, shoot, wound, kill, trap, capture, or collect or to attempt to engage in any such conduct."

24. Sayre, *Working Wilderness*.

25. Ibid., 109.

26. B. Zeedyk, *Water Harvesting from Low-Standard Rural Roads* (Santa Fe, NM: Quivira Coalition, 2006); B. Zeedyk and V. Clothier, *Let the Water Do the Work: Induced Meandering, an Evolving Method for Restoring Incised Channels* (Santa Fe, NM: Quivira Coalition, 2009).

27. R. Cooke and R. Reeves, *Arroyos and Environmental Change in the American South-West* (Oxford: Clarendon Press, 1976).

28. Sayre, *Ranching, Endangered Species, and Urbanization*.

29. Westland Resources, "Altar Wash Grade Control Structure Environmental Assessment," Report Submitted to the Arizona Water Protection Fund Commission, 2000. The Arizona Water Protection Fund Commission administers the Arizona Water Protection Fund, which was created by the Arizona State Legislature in 1994. The Altar Valley Conservation Alliance received a grant from the Arizona Water Protection Fund to carry out a Watershed Resource Assessment of the Altar Valley. Westland Resources was contracted to do portions of the assessment.

30. Ibid., 4.

31. P. Starrs, *Let the Cowboy Ride: Cattle Ranching in the American West* (Baltimore: Johns Hopkins University Press, 1998); R. Knight, W. Gilgert, and E. Marston, eds., *Ranching West of the 100th Meridian: Culture, Ecology, and Economics* (Washington, DC: Island Press, 2002); A. Pearce, "Uncommon Properties: Ranching, Recreation and Cooperation in a Mountain Valley" (PhD thesis, Department of Anthropological Sciences, Stanford University, 2004).

32. Pearce, "Uncommon Properties," 166.

33. R. Liffmann, L. Huntsinger, and L. Forero, "To Ranch or Not to Ranch: Home on the Urban Range," *Journal of Range Management* 53 (2000): 362–70; Sayre, *Ranching, Endangered Species, and Urbanization*; Sheridan, "Working Landscapes."

34. N. F. Sayre, L. Carlisle, G. Fisher, L. Huntsinger, and A. Shattuck, "The Role of Rangelands in Diversified Farming Systems: Innovations, Obstacles, and Opportunities in the USA," *Ecology and Society* 17, no. 4 (2012): 43, http://dx.doi.org/10.5751/ES-04790-170443.

35. According to the Humane Society of the United States, the number of cattle slaughtered in the United States between 2000 and 2010 ranged from a low of 32,539,000 million in 2005 to a high of 36,416,000 in 2000 ("Farm Animal Statistics: Slaughter Totals," Humane Society of the United States, June 27, 2013, www.humanesociety.org/news/resources/research/stats_slaughter_totals.html).

36. E. Ostlind, "The Big Four Meatpackers," *High Country News*, March 21, 2011.

37. B. Fussell, *Raising Steaks: The Life and Times of American Beef* (Orlando: Harcourt, Inc. 2008), 202.

38. Diablo Burger established its first restaurant in downtown Flagstaff in 2009. It opened a second restaurant in downtown Tucson in 2013.

39. White, "No Ordinary Burger" (n. 2 above, this chap.).

40. Robert E. Lang, Andrea Sarzynski, and Mark Muro, *Mountain Megas: America's Newest Metropolitan Places and a Federal Partnership to Help Them Prosper* (Washington DC: Brookings Institution, 2008), 3.

41. R. Knight, "The Laramie Foothills Group: Conservation at the Scale of a Watershed," *Ground Truth: Quarterly Publication of the Diablo Trust* (Spring 2011), 8, http://www.diablotrust.org/newsletters/Spring_2011.pdf.

42. Laramie Foothills Survey, Grassroots Collaborative Conservation Organizations Survey: Laramie Foothills Mountains to Plains Conservation Project (2009). Unpublished survey in possession of the author created and passed out by T. Sheridan at the Quivira Coalition's Annual Meeting in 2009.

43. Knight, "The Laramie Foothills Group," 8.

44. Laramie Foothills Survey, Grassroots Collaborative Conservation.

45. Knight, "The Laramie Foothills Group," 8.

46. Laramie Foothills Survey, Grassroots Collaborative Conservation.

47. Knight, "The Laramie Foothills Group," 9.

48. Ibid.

49. R. Knight and C. White, eds., *Conservation for a New Generation: Redefining Natural Resource Management* (Washington, DC: Island Press, 2008).

50. Margerum, *Beyond Consensus*.

51. Ibid., 149.

# Spotlight 4.1

# HISTORIC PRECEDENTS TO COLLABORATIVE CONSERVATION IN WORKING LANDSCAPES

The Coon Valley "Cooperative Conservation" Initiative, 1934

*Curt Meine and Gary P. Nabhan*

IN BRIEF

- One of the earliest formal efforts of collaborative conservation—called "cooperative conservation" at the time—began in the 1930s in Coon Valley, Wisconsin.
- As many as 418 private farming families worked with researchers and agency personnel to restore soils, watercourses, forest cover, wildlife habitat, and recreational value to 40,000 acres of land degraded by poor farming practices.
- The cooperative effort continues today; the region has become a hub for sustainably produced and organic products, including the highly successful business, Organic Valley.

Intentional community-based efforts toward the collaborative conservation of working landscapes in the United States are not a recent phenomenon. Although watershed councils in the Pacific Northwest and rancher-environmentalist alliances in the Southwest both emerged in the early 1990s, perhaps the earliest formally organized effort to pursue integrated, landscape-level conservation and restoration of a food-producing watershed began in Coon Valley, Wisconsin in 1934.[1] The same year that Aldo Leopold began to offer the first game management courses in any American university, he also became involved as an adviser to the first watershed-scale soil conservation demonstration area designated by the U.S. Soil Erosion Service (now the USDA Natural Resources Conservation Service).[2]

The demonstration area was located in the Coon Creek watershed in the erosion-prone Driftless Area of southwestern Wisconsin. The farmers, soil scientists, foresters, and wildlife biologists involved in Coon Valley did not use the term "collaborative conservation" at that time. However, Leopold's first publication on the project in May 1935 did use the phrase "cooperative conservation" to encompass many of the same principles and processes that the Malpai Borderlands Group and Diablo Trust (see chap. 4) would pioneer in the Southwest in 1993 six decades later.

At the time the Coon Valley initiative began in the 1930s, Leopold described this agrarian landscape as one where residents not only had damaged their own natural resource base but had also generated problems that had impacts downstream: "Coon Valley is one of a thousand farm communities which, through the abuse of its original soil, has not only filled the national dinner pail, but has created the Mississippi flood problem, the navigation problem, the overproduction problem, and the problem of its own future continuity."[3] With technical support from the University of Wisconsin, the U.S. Soil Erosion Service, the Civilian Conservation Corps, and other agencies, Coon Valley's farmers coordinated their efforts to rescue, conserve, and restore soils, watercourses, forest cover, wildlife, and aesthetic and recreational values on their privately owned farms.[4] Through regular meetings on farmsteads, in makeshift offices, and even at the local bar and café, food producers and natural resource professionals came together to "show that integrated use is possible on private farms, and that such integration is mutually advantageous to both the owner and the public."[5]

Just as western ranchers and foresters have done over the last two decades, the farmers of Coon Valley voluntarily participated in farm- and watershed-scale planning to repair incised gullies, reseed barren areas, reforest vulnerable slopes, modify livestock grazing practices, and slow water flows (and soil loss) through contour plowing and the construction of a variety of retention structures.[6] At its peak of activity in the late 1930s, Coon Valley's soil conservation demonstration area engaged 418 farming families and 200 additional employees in working to restore and sustain production on 40,000 acres of enrolled land. The agencies and universities that collaborated with the Coon Valley farmers provided technical assistance, seeds, nursery plants, materials for erosion control structures, and in some cases, equipment. Coon Valley's bankers were involved in the effort, working with farmers and conservationists to provide the financial resources needed to adopt innovative farming practices.

Over the next several decades, contour farming, strip cropping, and fish and wildlife habitat enhancement became the norm in Coon Valley, clearly regenerating rather than depleting the ecological and economic wealth of the watershed. Although intense flood events, urban encroachment, and current high prices for corn and soy today pose threats to the watershed's land stewardship legacy, Coon Valley has amply demonstrated that "cooperative conservation" can indeed work over the long haul for the local economy, for the maintenance of the rural cultural community, and for the resilience of the biotic community.

The Coon Valley watershed continues to attract attention from scientists and land stewardship advocates.[7] The improved streams of the area now host a thriving trout fishery and a flourishing recreational fishing economy.[8] Interestingly, the area around Coon Valley has also become one of the more important hubs for direct marketing of sustainably produced dairy products, eggs, and organic crops through the business Organic Valley, which in 2011 produced revenues to its farmers and shareholders in excess of $70 million.[9] Begun in 1988 in the Driftless Area as the CROPP (Coulee Region Organic Produce Pool) cooperative, Organic Valley has expanded well beyond its home in the Kickapoo River watershed to include more than 1,600 farmer-owners in 33 states and four Canadian provinces.

The Coon Valley experience suggests that, over time, community-based collaborative conservation can enhance ecosystem services, economic productivity, and social cohesion in a landscape. Agencies can support such endeavors, but their personnel, their resources, and even their names and missions shift over time. It is the internalization of a land ethic among local land stewards and food producers that accounts for the persistence and successes of the Coon Valley Watershed Project over eight decades.

## NOTES

1. A. Leopold, "Coon Valley: An Adventure in Cooperative Conservation," *American Forests* 5 (May 1935): 205–8; see also R. Anderson, "Coon Valley Days: A Short History of the Coon Creek Watershed," *Wisconsin Academy Review* 48, no. 2 (2002): 42–48.
2. C. Meine, *Aldo Leopold: His Life and Work*, 2nd ed. (Madison: University of Wisconsin Press, 2010), 313–314. Also see the website of the Aldo Leopold Foundation, www.aldoleopold.org, for more information.
3. Leopold, "Coon Valley," 206–7.
4. Meine, *Aldo Leopold*, 313.
5. Leopold, "Coon Valley," 205.
6. Anderson, "Coon Valley Days," 43–44.
7. For example, S. W. Trimble, "Fluvial Processes, Morphology and Sediment Budgets in the Coon Creek Basin, WI, USA, 1975–1993," *Geomorphology* 108 (2009): 8–23;

A. S. (Tex) Hawkins, "Return to Coon Valley," in *The Farm as Natural Habitat: Reconnecting Food Systems with Ecosystems*, ed. D. L. Jackson and L. L. Jackson (Covelo, CA: Island Press, 2002), 57–70.

8. L. Gaumnitz, "Restoring Life to a Watershed," *Wisconsin Natural Resources* (February 2002), http://dnr.wi.gov/wnrmag/html/stories/2002/feb02/coonval.htm; see also J. Luhning, "Birth of a Conservation Movement," *Edible Madison* 14 (Fall 2013): 34–42, http://www.ediblemadison.com.

9. C. Pine and C. Spengler, eds. *CROPP Cooperative Roots: The First Twenty-Five Years* (Blue Jay, CA: Hawking Books, 2013). For more information on CROPP and Organic Valley, also see www.organicvalley.coop.

# 5

# THE QUIVIRA EXPERIENCE

Reflections from a "Do" Tank

*Courtney White*

IN BRIEF

· Improving and maintaining the health of degraded riparian areas and
rangelands can be done fairly easily; however, viable conservation tools will
only work if they are economically feasible and not reliant on grants and
other subsidies.
· Although collaborative conservation is now widely accepted, implementing
large-scale collaborative conservation projects remains difficult.
· Overworked and underfunded federal agencies have begun asking their
partners to pay for ecological restoration activities on public lands—activities that should be the government's responsibility; this poses a major
threat to collaborative conservation efforts.
· Two things still needed to conserve working landscapes and their biodiversity are (1) an economic model that values ecological restoration; and (2)
strong leadership at the county, state, and federal levels to break through
business-as-usual paradigms and policies, and promote innovation and
entrepreneurship.

INTRODUCTION

Recently, an acquaintance asked me what I did for a living. After explain-
ing that I run a nonprofit that works with ranchers and conservationists in
the Southwest on land health and sustainability issues, he said summarily

"Oh, you run a think tank." Without pausing, I replied, "No, Quivira is a 'do' tank," which elicited a nod and smile from my friend. Afterward, I thought about this brief exchange. What did I mean? Partly, I was being provocative. I wanted him to understand that we are an organization that implements new ideas and does not merely promote them. But we are also, of course, a think tank in the sense that we are committed to finding these new ideas in the first place and sharing the results of our work—both triumphs and challenges—with others.

A rancher, another conservationist, and I founded the Quivira Coalition in 1997 to build bridges among ranchers, conservationists, scientists, and public land managers around concepts of progressive cattle management, innovative stewardship, and improved land health.[1] We vowed not to file lawsuits or advocate legislation. Instead, we concentrated on expanding the "radical center"—to literally bring people together at the grassroots level and to serve as a vehicle for information as well as a catalyst for change. During our first five years, we promoted the concept of the "new ranch," which we defined as an emerging progressive ranching movement that operates on the principle that the natural processes that sustain wildlife habitat, biological diversity, and functioning watersheds are the same processes that make land productive for livestock.

By 2002, however, we embraced an even more holistic vision of landscape health, one that included major restoration efforts as well. Our current mission is to build resilience by fostering ecological, economic, and social health on western landscapes through education, innovation, collaboration, and progressive public and private land stewardship. "Resilience" has been defined as "the ability to recover from or adjust easily to misfortune or change." In ecology, it refers to the capacity of plant and animal populations to respond to the effects of fire, flood, drought, insect infestation, or other disturbance. Socially, resilience describes a community's ability to adjust to changes in economic or environmental conditions. In order to build resilience, we focus on three areas of concern: reversing ecosystem service decline, creating sustainable prosperity, and relocalizing food. We do so through three program areas: education and outreach; land and water; and capacity building and mentoring. Since Quivira's founding, at least one million acres of rangeland, 25 linear miles of riparian drainages, and over 12,000 people have directly benefited from Quivira's collaborative efforts.

What follows is a reflection on Quivira's experience—what has worked, so far, and what has not. It is important to note that most of the ideas and practices that follow came originally from innovators working largely in isolation from one another—where innovation often starts—and were developed by Quivira primarily in an effort to break through paradigmatic logjams in the mainstream. Quivira didn't invent these ideas, but it gave voice to them, provided an ongoing forum to disseminate them, and was among the first organizations to test them on the ground.

## THE RADICAL CENTER

The term, "radical center" was coined by rancher Bill McDonald of the nonprofit Malpai Borderlands Group in the mid-1990s to describe an emerging consensus-based approach to land management challenges in the U.S. West. At the time, the conflict between ranchers and environmentalists over who would and would not have access to grazing rights on public lands had reached a fever pitch, with federal agencies caught in the middle of these decisions. This conflict balkanized the West and led to gridlock. Very little progress was made on projects that would provide long-term environmental or social benefits to the region, including prescribed fires, improvements to the habitat of endangered species on private land, and efforts to help ranchers find viable economic alternatives to selling their private land to real-estate developers. The radical center movement challenged various orthodoxies of the mid-1990s, including the belief that conservation and ranching were part of a zero-sum game where one's gains equaled another's losses.

In 1997, two Sierra Club activists—Barbara Johnson and me—and rancher Jim Winder decided to put the radical center concept to the test in New Mexico. Jim had the idea of starting a nonprofit organization that would sidestep disputes between ranchers and environmentalists by taking a new approach—one that emphasized land management practices that were both economically and ecologically sustainable. We chose the word "Quivira" because on old Spanish colonial maps of the region it was used to designate "unknown territory"—which is exactly where we felt we were headed. Our small coalition became the Quivira Coalition, which invited to its discussions any rancher, conservationist, agency official, scientist, or member of the public who was interested in developing solutions to rangeland conflict. We took a collective vow of "no legislation and no litigation," committing ourselves to finding grassroots solutions. Quivira was different from other radical centrist groups in existence at the time principally because we weren't confined to a watershed or other bounded region. We conducted workshops, tours, conferences, clinics, and public-speaking engagements across the Southwest. In the process, we helped to define the terms, contours, and significance of the radical center within the so-called grazing debate.

In 2003, 20 ranchers, conservationists, and others signed the "Invitation to Join the Radical Center," a pact that we hoped would signal the end of conflict and the beginning of an era of collaboration to heal the land itself (see the introduction to this chapter). The following is an excerpt from this pact:

> We . . . reject the acrimony of past decades that has dominated debate over live-stock grazing on public lands, for it has yielded little but hard feelings among people who are united by their common love of land and who should be natural allies.
>
> And we pledge our efforts to form the "Radical Center," where:
>
> · The ranching community accepts and aspires to a progressively higher standard of environmental performance;
> · The environmental community resolves to work constructively with the people who occupy and use the lands it would protect;
> · The personnel of federal and state land management agencies focus not on the defense of procedure but on the production of tangible results;
> · The research community strives to make their work more relevant to broader constituencies;
> · The land grant colleges return to their original charters, conducting and disseminating information in ways that benefit local landscapes and the communities that depend on them;
> · The consumer buys food that strengthens the bond between their own health and the health of the land;
> · The public recognizes and rewards those who maintain and improve the health of all land;
> · And that all participants learn better how to share both authority and responsibility.[2]

Was the pact, including its concept of the "radical center," successful? Yes and no. Significant shifts in popular opinion have occurred since the 1990s, including a more positive attitude toward ranchers and livestock by lawmakers, newspaper letter-writers, conservationists, and other opinion leaders. As a result, the "grazing wars" have largely faded from view. More important, some of the radical center objectives have been successfully addressed. Across the West, progress has been made as agencies, private landowners, scientists, and conservationists work together toward common goals. Es-

pecially encouraging has been the explosion of watershed-based collaboratives. The concept of collaborative conservation has become mainstream to
the point of institutionalization by universities, national nongovernmental
organizations, and agencies.

At the same time, the overall scale of these efforts remains small. Many
of the objectives defined in the pact have not been addressed at a magnitude
significant enough to effect fundamental reform. Several other challenges
have also become evident: higher standards of environmental performance
by ranchers remain elusive on many rangelands; constructive engagement
by leaders of environmental groups in working landscapes remains sparse;
agencies are still too hidebound by standard procedures to focus on tangible
results; and a gulf exists between scientists and practitioners. There has been
progress on all of these fronts, but its pace has been too slow to keep up with
the rapidly evolving challenges of the twenty-first century. These challenges,
including climate change, threaten to fragment the West even more completely than did the issue of livestock grazing on public lands.

## THE NEW RANCH

The Quivira Coalition's work to foster progressive cattle and land management practices, through the principle of the radical center, was embodied in
an approach that we refer to as the "New Ranch." But the New Ranch isn't
just a philosophy; it is a philosophy accompanied by a set of land management practices. Before land can sustainably support a value, such as livestock
grazing, hunting, recreation, or wildlife protection, it must be functioning
properly at a basic ecological level.

Application of the New Ranch philosophy includes the following: utilizing herding or other rotational grazing strategies on ranches that control the
timing, intensity, and frequency of livestock impacts on the land (sometimes
called planned grazing); documenting the success of land management practices with scientifically credible monitoring protocols, and articulating their
results to diverse audiences; building a common vocabulary for ranchers,
scientists, agency officials, and conservationists to use in addressing rangeland health issues; educating diverse audiences about the complexity and
difficulty of effectively managing rangelands; and engaging in collaborative
conservation and restoration projects.

Initially, the number of ranches across the West adopting the New Ranch
approach was small. Many New Ranch ranchers considered themselves outli-

ers—neither part of the orthodox model of livestock management nor the anti-cattle mindset of conventional urban environmentalists. Over time, however—especially as success stories began to circulate—the New Ranch model began to gain support among ranchers, agencies, and the public. Especially important was a cross-fertilization of ideas and practices—such as ranchers doing riparian restoration work, consumers requesting grass-fed beef (see spotlight 5.1), and agencies willing to use livestock to curtail invasive weed species. Today, although the number of New Ranchers is still comparatively small, they are indisputably no longer viewed as outliers. And, as innovation and cross-fertilization continue, their numbers will grow.

The range of activities of the Quivira Coalition is best demonstrated by looking at three on-the-ground efforts: grassbank development, riparian restoration, and conservation efforts in the Carson National Forest, New Mexico.

### CHALLENGES TO THE NEW RANCH APPROACH:
### THE VALLE GRANDE GRASSBANK

In 2004, Quivira took over management of the Valle Grande Grassbank, a project in northern New Mexico created by author and conservationist Bill deBuys. This created an opportunity to pull many of the New Ranch ideas together and put them into action. A grassbank is a voluntary collaborative process in which forage on the grassbank is exchanged for one or more tangible conservation benefits on the lands of neighboring stock raisers. It is also the physical place where such efforts occur. Stock raisers on the neighboring ranches may remove stock from one or more of their pastures and graze them on the grassbank to allow their pasture(s) to recover from a drought or to increase the fuel load before a prescribed burn, for example. DeBuys started the Valle Grande Grassbank in 1997 on a 36,000-acre allotment of Forest Service land on Rowe Mesa, 25 miles east of Santa Fe. The allotment was associated with a piece of private property that had been purchased by the Conservation Fund and came with a year-round federal grazing permit but no cattle. He had three goals:

- Improve the ecological health of public-grazing lands for the benefit of all creatures dependent on them;
- Strengthen the environmental and economic foundation of northern New Mexico's ranching tradition, arguably the oldest in the nation;
- Demonstrate that ranchers, conservationists, and agency personnel can work together for the good of the land and the people who depend on it.

The grassbank idea originated among ranchers of the Malpai Borderlands Group in southwestern New Mexico (see chap. 4), who were granted access to forage on the vast, privately owned Gray Ranch in New Mexico in exchange for placing conservation easements on their private land. On the Valle Grande allotment, instead of buying and grazing cattle, deBuys proposed to offer grass as a "bank" to other national forest permittees around the region in exchange for restoration work such as forest thinning and prescribed fire projects on their Forest Service grazing allotments. The permittees could move their cattle to the grassbank for two or three years while they worked with Forest Service staff to undertake restoration work on their home allotments. The project worked well for a while, with a variety of restoration projects being accomplished on national forest lands and with nine preexisting grazing associations signing onto the grassbank. When the Quivira Coalition took over its management in 2004, it attempted to build on the successes of the project, principally by adding additional New Ranch elements. These elements included the creation of a land-health map of the entire grassbank allotment; new monitoring procedures; an approach to livestock handling that is low-stress on the landscape; and an entrepreneurial approach to the operation.

Unfortunately, by 2007, the Valle Grande Grassbank had ceased functioning for a number of reasons. First, the modest environmental gains on the home allotments of the permittees ended during the final three grazing seasons (2004–6), when no restoration work was completed by the Forest Service. Restoration work was not completed for a variety of reasons, including drought, National Environmental Policy Act hurdles, and budgetary tensions within the agency. This lack of progress exposed a weakness in the grassbank model: relying on an overworked and understaffed federal agency to perform the restoration work. Second was the decline in participation, over time, by ranchers in the region. With this decline in participation came a decline in their enthusiasm and support for the project. Partly, this occurred for financial reasons—ranchers had to pay transportation costs to the allotment, and, as diesel prices rose, growing numbers of ranchers dropped out. Also, the slow progress of restoration work on ranchers' home allotments discouraged participation.

A third major reason the endeavor failed was that the grassbank's budget was entirely funded by grants, and when the grants dried up in 2006, so did the project. Quivira had been warned of this risk by a rancher on its board of directors, who had said bluntly, "This place has all the costs of a ranch and no income!" This observation raised an important question: How can grassbanks pay for their operation without grants or other subsidies? If the model is to become a viable conservation tool, it needs to be economically feasible to do so.

## INNOVATIVE RIPARIAN RESTORATION PROJECTS FOR LAND HEALTH

The term "land health" was coined in the 1930s by the conservationist Aldo Leopold. He was referring to the ecological processes that perpetuate life, including the proper cycling of water and nutrients in the soil. Leopold considered "land health" to be a self-perpetuating system in which all parts—soil, water, plants, animals, and other elements of the ecosystem–would endlessly renew themselves when unimpaired. Leopold frequently employed words such as "stability," "integrity," and "order" to describe this "land mechanism," utilizing metaphors of engines and bodies to describe the smooth functioning of an integrated system. By contrast, land became "sick" when its components fell into disorder or broke down.[3] After World War II, the rapidly emerging science of ecology refined Leopold's ideas. The engine and body metaphors were replaced by a dynamic, even chaotic, conceptualization of nature as ceaselessly changing rather than consisting of homeostatic, self-regulating systems. This revised idea of ecological health still focused on self-renewal and self-organization, but now scientists saw nature not as static but as fluid. This view also employed a new set of terms and concepts, including "resilience," "historic range of variability," "sustainability," "diversity," and "perturbation." Those activities that encouraged resilience, for example, could be seen as promoting land health, while those activities that reduced an ecosystem's ability to recover from a disturbance could be considered deleterious to land health.

A further refinement of the land health idea began in 1994, when the National Research Council launched a collaborative effort to address persistent disagreement among range scientists, environmentalists, ranchers, and public agency personnel about the health of the nation's public rangelands.[4] Not only was there a substantial lack of data on the condition of the land itself, but there was also a lack of agreement among range experts on how, and what, to monitor. These voids contributed significantly to the acrimonious debate raging at the time about livestock grazing on the nation's public lands. Were rangeland conditions improving or degrading? An interagency team of government scientists was organized to develop both qualitative and quantitative criteria for assessing and measuring the health of the land. This effort reached fruition in 2000 when the team settled on 17 indicators of land health.[5] These indicators of land health were grouped into three categories:

1. *Soil stability*. The capacity of a site to limit redistribution and loss of soil resources (including nutrients and organic matter) by wind and water. This is a measurement of soil movement.
2. *Watershed function*. The capacity of the site to capture, store, and safely release water from rainfall and snowmelt; to resist reduction in this capacity; and to recover this capacity following degradation. This is a measurement of plant-soil-water relationships.
3. *Biotic integrity*. The capacity of a site to support characteristic functional and structural plant communities in the context of normal variability; to resist the loss of this function and structure due to a disturbance; and to recover from such disturbance. This is a measurement of vegetative health.[6]

This important work laid the foundation for a variety of land management practices aimed at both restoring and maintaining land health. Quivira now had clear methods and indicators for measuring success.

Quivira's opportunity to implement an on-the-ground land health restoration program utilizing these measurement methods began in 2000, when it met riparian specialist Bill Zeedyk, a retired Forest Service biologist who had developed a suite of low-cost, low-impact ways to heal eroded gullies and build roads that capture water rather than causing erosion. Soon, Quivira was working with Zeedyk on a creek project at the Williams Ranch in western Catron County, New Mexico, that employed Zeedyk's innovative restoration methodology, termed "induced meandering."[7] Within a few years, Quivira had been awarded two substantial grants from a U.S. Environmental Protection Agency Clean Water Act program to conduct riparian restoration work on the Dry Cimarron River, in northeastern New Mexico, and on Comanche Creek, within the Valle Vidal unit of the Carson National Forest. Both grants also contained funding for educational workshops, publications, and symposia on land health and restoration topics. Eventually, Quivira expanded its restoration work to a variety of public and private landscapes across the Southwest.

From its work on riparian restoration, Quivira learned two principal lessons: (1) land health can be improved and maintained relatively easily, and at low cost, and (2) almost anyone can do it. The key is understanding natural processes, including how water naturally flows across the land ("Think like a creek and let nature do the work," says Zeedyk); the role of riparian vegetation in soil stability; and how grazing animals use the land. Quivira's restoration projects have been highly successful, particularly in their goal of improving and maintaining land health (see chap. 4 for a discussion of Zeedyk's work with the Altar Valley Conservation Alliance in southern Arizona). In case after case, Quivira has documented the recovery of riparian health that has resulted from induced meandering and other techniques, including the repair of low-grade ranch roads. Quivira's role in this work included not only organizing the restoration work itself but also fostering collaboration—providing workshops, symposia, training seminars, and other educational opportunities—that helped change the mentality around restoration work in the region.

The components of land health and of the restoration toolbox to improve and maintain it are now well-developed, thanks to many people and a lot of hard work. What remains to be accomplished, however, is to make this work economically feasible by figuring out a way to compensate landowners and others for their efforts. Quivira can now confront the West's legacy of degraded riparian areas and rangelands proactively; it is hoped that soon we and others will be able to do so profitably as well.

## A MULTIPARTY PROJECT IN THE CARSON NATIONAL FOREST

The idea of collaborative conservation has now become widely accepted among many landowners, agencies, researchers, ranchers, and conservationists. What remains a challenge, however, is implementing large-scale collaborative conservation projects. Since 2001, the Quivira Coalition has led a multiparty habitat restoration project on Comanche Creek in the Valle Vidal unit of the Carson National Forest. Comanche Creek is typical of many areas in the West that have experienced adverse human impacts, including poor timber management, overgrazing by livestock, and toxic waste caused by mineral extraction. These activities led to numerous inadequately constructed and maintained roads, overgrazed grasslands, depleted vegetation in riparian zones, and unprotected stream banks, which in turn increased the erosive tendency of the land and amplified the fine-sediment load within watersheds. The goal of the Comanche Creek project is to restore degraded portions of the 27,000-acre watershed, which includes improving the survival chances of the Rio Grande cutthroat trout (*Oncorhynchus clarkii virginalis*), New Mexico's state fish.

By 2000, populations of native Rio Grande cutthroat trout had been reduced to 10 percent of their historic range in the American Southwest due to a variety of factors, including competition from nonnative trout species, habitat degradation and loss, surface-water diversion and depletion, stream fragmentation, and isolation. Today, the few remaining populations face a significant new challenge: climate change. This challenge includes a likely reduction in the quantity of clear, cold water that trout require for survival; rising water temperatures; increased incidence of diseases and parasites; decreased abundance of insect food sources; decreased dissolved oxygen levels; increased demand for water by human populations; increased potential for flooding; and increased fragmentation of habitat. Taken together, all of these stressors point to the possibility that the Rio Grande cutthroat trout will be pushed to the brink of extinction unless immediate action is taken to restore habitats and increase populations.

Quivira's Comanche Creek project addresses these challenges by implementing effective, cost efficient, and low-impact riverine restoration techniques that feature the use of native raw materials that are locally available. Over the past eight years, 200 in-stream structures and stream exclosures for elk and livestock have been constructed with the aim of reducing erosion, improving water quality, and restoring riparian vigor to the creek. Our previous experience had taught us that viable restoration solutions include: (1) in-stream structures that stabilize stream-bank erosion, increase stream-bank water storage capacity, and improve riparian zone vegetative cover and diversity; (2) side-stream restoration activities that reduce erosion, stabilize headcuts, rewet meadows, and improve hydrological cycles; (3) mitigation or elimination of "bad" roads and road-related features (such as poorly placed culverts) that increase sediment erosion into creeks; (4) encouraging the growth of bank-side native plants (to shade the water for the fish); (5) managing the impacts of grazing; (6) annual maintenance and modification of structures, as needed; and (7) annual monitoring and assessment of progress.

The project is ongoing, but already Quivira has learned some new lessons about collaborative conservation (for a summary of lessons from the broader collaborative conservation literature, see chap. 4):

- It's hard. The technical challenges of creek and habitat restoration pale in comparison to juggling the varied and strong interests of the parties involved. The project manager must be equal parts diplomat, agitator, ringmaster, and delegator and must have persistence, patience, and a good sense of humor.
- Diversity is critical. The power of collaborative conservation comes from the ability to look at one problem, or one landscape, from multiple perspec-

tives. That means having a variety of perspectives represented, not just those of specialists. Collaborators must respect each perspective and learn from other people's ideas.

· Organizations and agencies must keep innovating and remain open to new ideas. It is important not to get stuck in a restoration rut.

· Monitor, monitor, monitor. Collect qualitative as well as quantitative data at every opportunity.

One sobering lesson from our experience is that it is becoming increasingly difficult to do collaborative conservation work on public lands because of government budget cuts. The level of complexity involved in dealing with federal agencies has steadily increased over the eight years of our work on Comanche Creek, nearly becoming a disincentive to collaborative work. For example, the Forest Service recently imposed a requirement that organizations like Quivira pay for National Environmental Policy Act costs associated with new work, such as Quivira's work on the creek. The rationale for passing these substantial costs on to Forest Service partners includes reduced Forest Service staffing, increased workloads for staff, internal priority shifts, and a general trend toward outsourcing certain government functions in order to reduce costs. The practical effect of this policy change, if widely implemented, will be devastating to conservation efforts. Where will organizations find this money? Grant-making foundations are very reluctant to pay for work they consider to be the government's responsibility. Nonprofits operate on very tight budgets. Unless Quivira can soon come up with the money to pay for the mandatory National Environmental Policy Act costs to continue work on the Comanche Creek, the project will languish.

At the same time, federal agencies say they recognize the need for more partnerships, flexibility, and innovation in order to meet rising management challenges on federal lands. However, the view from the trenches is not encouraging. Added to the administrative complexity of working with the Forest Service are a bewildering array of congressionally mandated laws and regulations, plus a diverse constituency, many of whom have conflicting expectations of federal agencies. This is a recipe for gridlock on public land. Some of the gridlock could be remediated through policy changes, but much of the problem is social, bound up with the personalities of individual agency personnel, who have a lot of on-the-ground power. Unless there is wholesale policy reform (which is highly unlikely), innovation and entrepreneurship will become increasingly rare on public lands. The government is unable, or unwilling, to provide these elements, and the private sector is discouraged—

or even prohibited—from trying. Ultimately, it is likely that some sort of co-management model will have to emerge on public lands in order to meet new and rising challenges.

## CONCLUSIONS

In the 17 years since its founding, the Quivira Coalition has explored many resilience-building strategies, enduring its share of failures along with successes. Initially, Quivira focused on land health, collaboration, and progressive livestock management. Over time, this work expanded to incorporate riparian restoration. Quivira has also worked to disseminate both the lessons learned from our own experience, and the innovative ideas of others through a vigorous outreach program. In addition to an annual conference, Quivira has organized over 100 educational events on topics as diverse as drought management, riparian restoration, fixing ranch roads, reading the landscape, water harvesting, low-stress livestock handling, grassbanks, and grass-fed beef (also see spotlight 5.1). Quivira's members have published numerous newsletters, journals, bulletins, field guides, and books, including a rangeland health monitoring protocol and a riparian restoration manual, *Let the Water Do the Work*, by Bill Zeedyk and Van Clothier.[8]

The Quivira experience to date demonstrates that building resilience on private and public lands is possible, practical, and potentially scalable to landscapes of varying size, including watersheds. Many of the elements needed to stitch the West back together have been developed and field-tested by individuals and organizations across the region. But two important elements are lacking: first, an economic model that values regeneration and restoration and, second, strong leadership at the county, state, and federal levels to break through business-as-usual paradigms and policies. Both have proven elusive in recent years, but those of us in the Quivira Coalition are hopeful that, as more and more organizations take the lead by doing (i.e., collaborative conservation work) and telling (i.e., telling others of their challenges and accomplishments), others will follow and contribute their own innovations and entrepreneurial energy. As Quivira's work continues, the coalition's members plan to integrate all of these ideas into mitigation and adaptation strategies for climate change and resource depletion, which together may be the greatest conservation challenges of the twenty-first century. Meeting these twin challenges will mean not only stitching the West back together as fast as possible—socially, economically, and ecologically—but doing so in a way that creates a resilient fabric that can be flexible without rending apart.

## NOTES

1. See www.quiviracoalition.org for additional information on the Quivira Coalition and its history.
2. Excerpt from "Invitation to Join the Radical Center," A West That Works: A Mural of Writing Images and Ideas (website), by Courtney White, 2003, http://www.awest thatworks.com/2Essays/Radical_Center/Radical_Center.pdf.
3. A. Leopold, *A Sand County Almanac and Sketches Here and There* (Oxford: Oxford University Press, 1948).
4. National Research Council, *Rangeland Health: New Methods to Classify, Inventory, and Monitor Rangelands* (Washington, DC: National Academy Press, 1994).
5. M. Pellant, P. Shaver, D. A. Pyke, and J. E. Herrick, *Interpreting Indicators of Rangeland Health*, Version 3, Interagency Technical Reference 1734–6 (Denver, CO: U.S. Department of the Interior, Bureau of Land Management, National Science and Technology Center, Information and Communications Group, 2000).
6. Ibid.
7. B. Zeedyk, and V. Clothier, *Let the Water Do the Work: Induced Meandering, an Evolving Method for Restoring Incised Channels* (Santa Fe, NM: Quivira Coalition, 2009).
8. Ibid.

# Spotlight 5.1

## GRASS-FED AND GRASS-FINISHED LIVESTOCK PRODUCTION

### Helping to Keep Working Landscapes Intact

*Gary P. Nabhan, Carrie Balkcom, and Amanda D. Webb*

#### IN BRIEF

- Some ranchers have responded to declining incomes by targeting niche markets for their meat—such as grass-fed and grass-finished meats—and by marketing their products directly.
- Grass-fed and grass-finished meat production and direct marketing may help conserve working ranchlands by increasing profit margins and incomes for ranchers, reducing the pressure to sell their ranches, providing more money to invest in land-stewardship activities, and requiring a certification process for grass-fed meats based on protocols that require environmentally sound grazing and land management practices.

It is a dilemma that many ranchers pondered near the end of the last millennium: from 1970 to 2000, essentially no new substantive income was added to the ranching economies of the Rocky Mountains, while income streams for recreational and second-home development were rapidly changing the landscape.[1] In rural Custer County, Colorado, the net annual business income of ranchers and farmers peaked in 1976 at $6.3 million, and by 2000, it had dwindled to $2.8 million, less than half the income generated a quarter century before.[2] This forced some ranchers to break off part of their private property for wildcat home developments or to sell their entire ranches to get out from under debt. A similar situation exists in other rural areas throughout the West.

Ironically, even though food production is one of the ranching community's primary goals, mainstream commodity markets of the food system are, in a sense, working against good land stewardship investments, which require the economic viability of working ranches. In 1910, the average farmer or rancher earned roughly 60 cents of every U.S. consumer dollar spent on food. By 2000, the proportion of food dollars returning to food producers had diminished by at least a factor of four and, in some commodity markets, by a factor of 10. This meant that farmers and ranchers were only receiving six to 15 cents per consumer dollar spent on food at the beginning of the twenty-first century.[3] It also meant that ranchers had fewer liquid assets to invest in sustaining or restoring the environmental health and heterogeneity of rangeland resources than ever before.

This shift in the rancher's return on investment may have aggravated the rate of ranch and farmland loss, with rural lands being converted to urban and exurban uses in the 1960s and 1970s at a rate three times the historic norm, in part because the return on investments in working landscapes did not "pencil out" (i.e., weren't expected to be profitable). By the 1990s, farm and ranch lands were being lost at a rate 51 percent higher than that of the 1980s, and by 2000, that loss was consistently reaching one million acres a year.[4] In California in particular, in the years leading up to the economic downturn that began in 2007, rates of rangeland conversion increased every year, and these rates were higher than the rates of farm or forestland conversion.[5] In short, the value of ranchlands that produced beef for the commodity market was not enough to save them from conversion; the standard "utilitarian" reasons that farmland preservationists offered the public for preserving working landscapes were insufficient to ensure their long-term protection.[6]

As a result of these interrelated factors, around the turn of the twenty-first century, a number of western ranchers began to transition from exclusive dependence on the feedlot-finished, commodity beef market, into a variety of niche markets not thoroughly explored by them previously. Rather than waiting for quasi-philanthropic strategies like carbon credits to kick in so that they could increase their revenue streams, some ranchers took a purely market-driven approach to enhancing their income through the direct marketing of their meat products. One of those niche markets used the "eco labels" of grass-fed and grass-finished meats, including beef, mutton, goat, and bison.

The term "grass-finished" has no legal definition and is subject to wide-ranging interpretation but involves supplemental feeding of forage other than corn, sorghum, or other grains during the finishing period prior to slaughter. There is a legal definition of "grass-fed" used by the U.S. Department of Agriculture, but it fails to take into account all of the parameters that most ranchers need to manage a healthy herd and a healthy business. The

American Grassfed Association has therefore devised a more comprehensive definition and grass-fed labeling system based on production practices that meet certain minimum standards.[7] American Grassfed Association standards recommend that grass and natural forage be the feed source consumed for the lifetime of the ruminant animal, with the exception of milk consumed prior to weaning. The cattle's diet should be derived solely from forage consisting of grasses (annual and perennial), forbs (including legumes and mustards), or browse (such as mesquite foliage and pods). Animals cannot be fed grain or grain by-products (starch and protein sources) and must have continuous access to pasture. In contrast, grass finishing implies that animals are moved from natural range conditions for finishing in irrigated pastures where a sequential set of forage crops are planted for their consumption until the animals are slaughtered. The assumption—in both grass-fed and grass-finished scenarios—is that there are both nutritional benefits (such as omega-3 fatty-acid composition) and reductions of greenhouse-gas emissions (methane, fossil fuel use) gained through range production relative to consumption in (contained) feedlots during the life cycle of the livestock in question. While this assumption remains debated within scholarly and policy-making circles, it has been widely accepted by consumers.

The quality control of products and veracity of such claims depend on rigorously followed protocols to meet well-defined standards. Most grass-fed ranchers agree to an on-ranch audit by the nonprofit third-party certifier Animal Welfare Approved, through which ranchers can become certified as grass-fed or grass-finished livestock producers. They can then use this certification in their niche marketing to command higher prices than the commodity market would allow. Price points vary widely, but natural, grass-fed, antibiotic-free beef typically sells for at least 25 percent more than commodity beef. This entire premium will not always be additional profit, since there are both one-time certification costs and costs associated with maintaining the certification protocols over time. Nevertheless, the prospect of receiving $11–15 per pound for ground beef in wholesale orders from high-end restaurants has attracted a number of ranchers to begin transitioning to grass-fed production.

These trends promise to offer ranchers a far greater proportion of the consumer dollar spent on meat than has been witnessed in the last half-century. Furthermore, because there are now fewer cattle on the range and higher beef prices per pound than at any point since World War II, ranchers will have additional economic incentives to transition to grass-fed or grass-finished production and the related strategy of direct marketing over the next three to five years.[8] Nevertheless, as more ranchers adopt grass-fed and grass-finished

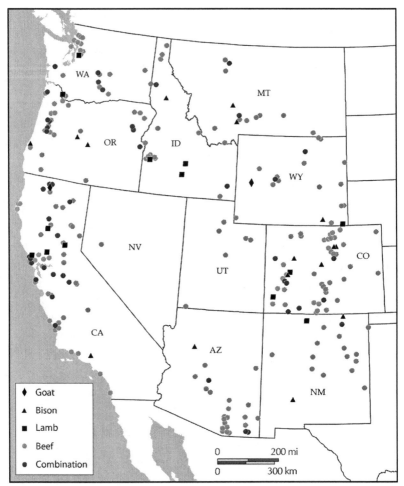

**Figure 5.1.1.** Grass-fed labeling "hotspots" west of the 100th meridian. Map by Darin Jensen and Syd Wayman, 2012.

production, labeling, and marketing, there will inevitably be more competition, occasional "greenwashing," attempts at corporate consolidation of some operations, and mismatches between supply and demand. As with any innovation or start-up business market, the path to success isn't necessarily lined with flowers the entire way.

The accompanying map (fig. 5.1.1) is the most complete to date to highlight the "hotspots" west of the 100th meridian where grass-fed labeling is already being used by ranchers having wild rangeland forage resources at their disposal. It does not include free-range poultry or small livestock on farms. Of course, not all of these ranchers are currently placing 100 percent of their stock into grass-fed markets, nor should they until they sense that the return on investment appears adequate. All those who directly market even a portion of their grass-fed or grass-finished meats are aware that there are additional time, energy, and financial costs in doing so. Nevertheless, well over 370 ranch families west of the 100th meridian are now engaged in this experiment, one that may or may not provide sufficient returns to keep families employed in the working landscapes where they now reside. The growing acceptability of grass-fed production among even the most conservative and risk-averse ranchers suggests that this market-driven solution to declining incomes and declining financial resources for land stewardship may be far more socially acceptable in the West than other market-driven strategies, such as direct payment for ecosystem services.

In short, the key link between grass-fed and grass-finished production and working landscape conservation is not simply that ranchers gain more income during an era when production costs are rising and profit margins thinning; it is that grass-fed production also has land stewardship protocols consistent with the ethics of valuing the integrity of working landscapes in a holistic manner. Only time will tell the results of this experiment in terms of income returns on investment, land stewardship quality, and reduced fragmentation rates in working landscapes.

## NOTES

1. H. Wolfson, "The West: Ranchers Turn to Non-traditional Practices," *Las Vegas Review-Journal Associated Press*, March 12, 2001.

2. T. Wilkinson, *A Pilgrimage to Community: The Story of Custer County's Journey to Find Its Future* (Tucson, AZ: Sonoran Institute, 2002).

3. B. Alexander, *The New Frontier of Ranching: Business Diversification and Land Stewardship* (Bozeman, MT: Sonoran Institute, 2001).

4. This data come from the American Farmland Trust, Farmland Information Center's analysis of National Resources Inventory data ("Fact Sheet: Why Save Farmland?" January 2003, p. 1, http://farmland.org/programs/states/wa/documents/APPENDIXC-c-WhySaveFarmland.pdf).

5. M. W. Brunson and L. Huntsinger, "Ranching as a Conservation Strategy: Can Old Ranchers Save the New West?" *Rangeland Ecology and Management* 61 (2008): 137–47.

6. M. J. Mariola, "Losing Ground: Farmland Preservation, Economic Utilitarianism and the Erosion of the Agrarian ideal," *Agriculture and Human Values* 22 (2005): 209–33.

7. See Standards," American Grassfed Association, http://www.americangrassfed.org/about-us/our-standards/.

8. A. Nation, "World Beef Prices Continue to Rise," *Allan's Blog*, Stockman Grass Farmer, June 22, 2011, http://wincustomersusa.com/stockman/index.php?option=com_content&task=view&id=449&Itemid=9.

# 6

# PLACE-BASED CONSERVATION
# FINDS ITS VOICE

A Case Study of the Rural Voices for Conservation Coalition

*Maia Enzer and Martin Goebel*

### IN BRIEF

- Developing and promoting policy solutions to the economic and ecological challenges that western rural communities face is more successful when done through a shared organizational structure than through individual advocacy by single organizations.
- A diverse, broad-based, prepared, and well-organized coalition gains more attention and better access to political leaders at the national level than groups that work and advocate alone, that focus on a single issue or limited purpose, or that are conflict-driven.
- Local organizations and rural community leaders can more effectively address complex ecological and economic issues at the national policy level by collaborating with other groups that face similar issues, building their capacity, and developing workable, broadly supported solutions to natural resource management problems.
- Substantive and reliable federal investment in place-based, collaborative conservation efforts—in the form of money or staff resources—makes sense for federal agencies and taxpayers and should underpin the policy initiatives of government administrations that aim to support these efforts.

For many rural communities located near public lands in the West, federal forests have historically been working forests in which timber harvesting and processing provide forest-based jobs. In the early 1990s, conflict over what constituted best practices for restoration, conservation, and use of federal forests reached fever pitch. Conflict produced winners and losers and generated deeply held resentment between competing interests.

Conditions have changed considerably since then. A less shrill, more effective approach to forest management emerged, one that embodied a new vision for federal, working forests and forest-based jobs in public land communities. Place-based and collaborative approaches to natural resource management began to take hold, even as traditional urban-based interests remained mired in debates over science and a desire to maintain traditional political influence in the executive and legislative branches of government.

Place-based initiatives grew as local community leaders, interested citizens, and county commissioners struggled to reorient their towns toward a new focus on ecosystem management and protection of biodiversity. These efforts sought to focus on solutions that would work within the local ecological context and the realities of local economies. Essential to their success was the recognition that progress cannot be made without collaboration between local and distant stakeholders possessing a variety of perspectives and expertise.

The new focus on place-based strategies that fit local conditions, culture, and capacities coupled with the use of collaboration dramatically changed many communities across the West. As previously antagonistic stakeholder groups engaged in a new dynamic, ripples spread from the local level to regional and even national discussions. Incremental yet fundamental changes have begun to unfold in the processes, relationships, and laws that govern how western rural communities work with land management agencies, especially at the federal level.

Republican and Democratic members of Congress saw that place-based collaboration could offer solutions that would allow members of both parties to satisfy the competing interests within their states and districts, providing a possible win for everyone. The federal land management agencies were slower to see these potential benefits. Some leaders viewed collaboration as a possible threat to their own decision-making authority (and some still hold this view), but other officials saw that collaboration offered agency personnel the best chance of meeting the public's multiple demands on national forests. These officials increasingly embraced collaboration as a better pathway to public-private land interactions.

The win-lose era thus appears to be waning; replacing it are a growing number of place-based, collaborative efforts that have dedicated themselves to finding solutions that recognize an inextricable link between environmental and community resiliency. Disparate place-based efforts quickly discovered shared barriers and common solutions that depended on changes in federal lands policy. With the groundswell of local innovation came a need to connect and organize these efforts in order to ensure that the lessons from individual efforts could be aggregated and, collectively, have an impact on federal land management policies.

## RURAL VOICES COME TOGETHER

In 2001, Sustainable Northwest, a nonprofit organization based in Portland, Oregon, and dedicated to working with rural communities to achieve environmental and economic sustainability, convened 30 individuals representing many of the earliest adopters of place-based collaborative efforts across the West. They discussed how to work together in their shared struggles with the federal legislative, administrative, and regulatory obstacles that stood in the way of local goals. Some of the organizations participating at this initial meeting included the Watershed Research and Training Center from a tiny mountain town in northern California; Wallowa Resources nestled in the mountains of northeast Oregon; the Ecosystem Workforce Program based out of Eugene, Oregon; Framing Our Community from the tiny end-of-road community of Elk City, Idaho; Lake County Resources Initiative located in south-central Oregon adjacent to the Fremont-Winema National Forest; and Northwest Connections from the Swan Valley of Montana. The participants shared a desire to create environmental and economic changes to an existing policy framework designed during an era of industrial forest management that no longer fit new mandates for ecosystem management and biodiversity protection and did not easily accommodate the new approach of place-based, collaborative conservation. Two years after the first gathering of this small group, the Rural Voices for Conservation Coalition (RVCC) was formally created.

The coalition, an initiative of Sustainable Northwest, is policy focused and advocates for reform of federal rules and laws to support place-based, collaborative practices to natural resource management across the West. It serves as a regional network of rural organizations that share values and believe that by working together to solve problems they can accelerate learning, dissemination, and adoption of best practices. Over the past decade, RVCC has developed a cadre of rural leaders across 11 western states that serve as spokespeople for this new approach to natural resource management. For the most part these new leaders speak with voices aligned with neither the current environmental movement nor resource extraction industries. As local people whose livelihoods once depended on stewardship of public and private lands, they largely represent the rural middle class that disproportionately suffered the impact of changes in federal lands policies enacted in the early 1990s.[1]

Collaboration regarding natural resource management became a new pathway for rural communities to gain a seat at the table long dominated by traditional interests representing environmental groups, the timber industry, and the Forest Service. Collaboration provided a meaningful way for local people to share concerns and knowledge and offer insight into approaches to solve ecologically critical forest health problems. For the environmental community, collaboration offered a new route to conservation objectives. Appeals and litigation had largely succeeded at stopping the most egregious forest management practices but proved ineffective at facilitating environmental improvement. The environmental community lacked the tools to restore degraded public lands; collaboration offered a way to do that. For the timber industry, collaboration offered an opportunity to show that business approaches, and wood utilization, could contribute economic viability to these conservation strategies. The Rural Voices for Conservation Coalition filled a growing need for regional organizing as the multifaceted collaborative movement matured.

Existing administrative and legislative constructs frequently prevented collaborative efforts on public lands. Traditional timber sales and service contracts, for example, were deemed too limiting to achieve new goals of forest restoration and local community benefit.[2] Discussions among community leaders, loggers, and concerned citizens led to the realization that agencies like the Bureau of Land Management and the Forest Service would need new policies to ensure the use of collaborative processes to advance quality restoration work, improve administrative efficiencies in contracting complex restoration projects (e.g., thinning dense stands of trees, restoration of roads to reduce sedimentation in adjacent streams, prescriptions designed to enhance wildlife habitat, culvert removal, etc.), provide local job opportunities, ensure local utilization of the material removed, and improve monitoring of the impacts of all these activities.

Community leaders proposed the adoption of stewardship contracting, first authorized as a pilot program in Section 347 (c)(1) of the Omnibus Consolidated Appropriations Act of fiscal year 1999 and subsequently authorized for a 10-year period in 2003.[3] Stewardship contracting is one example of how collaboration between community leaders, environmental groups, land management agency personnel, the forest products industry, and concerned citizens (representation by whom is required in stewardship contracts) can achieve better conservation outcomes and create local economic benefit. The enabling legislation also required the use of best-value criteria to score and award bids to ensure better attainment of desired ecological outcomes and community benefit.[4] The legislation further stipulated multiparty monitoring to build trust and accountability in the collaborative process.[5] (See also chap. 9 for a discussion of stewardship contracting in the Siuslaw National Forest.)

Armed with early policy successes like legislative authorization of stewardship contracting, community leaders realized that developing and promoting policy remedies to public land management problems through a shared organizational structure could yield much better and, perhaps, faster results than could individual advocacy. As Sustainable Northwest convened its rural partners to discuss policy barriers to working on federal lands, remedies were proposed for communities in similar situations. Policy challenges included poorly structured agency budgets and performance measures, inadequately funded programs, outdated forest management priorities, and a lack of attention to innovative uses of small-diameter trees and traditionally underutilized tree species along with the development of markets for these products.

## RVCC OBJECTIVES AND FOCUS AREAS

The Rural Voices for Conservation Coalition set out with a mission to find policy solutions to the ecological and economic challenges facing the rural West. Overarching issues of concern included public and private lands management, forest and rangeland health, integrated biomass utilization, and climate change. Critical to these themes was RVCC's focus on ensuring that policies support and enable collaboration, community capacity building, local wealth retention, intergenerational land transfers, and multiparty monitoring. The coalition structure supports and encourages participants to reexamine regularly the challenges facing their individual communities and the region as a whole. For example, concerns about private inholdings in the vast landscape dominated by federal ownership were an early priority for the coalition, particularly with respect to the impact of degraded federal lands adjacent to private lands. Coalition members sought ways to fund work on private lands that could ensure watershed or landscape-scale ecological resiliency.

In the early days, RVCC members mostly focused on convincing Congress to fund Forest Service programs that provided financial and technical assistance to private landowners and on salmon recovery funds. Another early focus was the so-called Wyden Amendment, first authorized in 1998 and since reauthorized (Public Law 105–277, sec. 323, as amended by Public Law 109–54, sec. 434; see also discussion in chap. 9). This amendment allows the Forest Service to enter into cooperative agreements to improve resources within watersheds containing National Forest System lands. Here, agreements may be with willing federal, tribal, state, and local governments, private and nonprofit entities, and landowners to conduct activities on public or private lands to accomplish a variety of land management objectives.

As the participants in RVCC became more involved in landscape-scale restoration and came to recognize that additional natural resource management was vital to community health and resiliency, interest in working on rangeland and ranching issues emerged. Today concern about the protection and viability of rangelands and the ranching families that are so vital to the western landscape and rural economies runs strong among RVCC members. Historically, RVCC's advocacy related to the Farm Bill focused on the Forestry Title. More recently the coalition is committed to shaping not only the Forestry Title of the 2012 Farm Bill, but also the conservation, rural development, and energy sections of the bill. Private lands conservation in the West happens in a public lands–dominated context. If landscape-level restoration and stewardship are to be successful, federal policies and investments must provide the enabling conditions necessary for long-term conservation success.

Through collaboration and persistence, RVCC has grown and sustained itself for nearly a decade. Although coordinated and managed by Sustainable Northwest, RVCC's strength lies in its growing volunteer membership and its commitment to operating in a manner that supports member-driven working groups. Working groups focus on specific topics and may change over time as priority issues change. Examples of topics around which working groups have formed are woody biomass and energy; climate change; private lands; public lands management and economic development; appropriations and legislative strategies; workforce, labor, and contracting; and ranching and grasslands. One of the unique characteristics of RVCC is the lack of formal membership requirements. Organizations are simply asked as a condition of participation that they commit to a spirit of collaboration with the aim of promoting solutions that reflect the values of the coalition as a whole. This voluntary opt-in structure allows organizations to participate as they are able, support policy proposals they agree with, and opt out of the policy proposals they oppose or don't consider relevant to their local constituency. The term "member" is used very loosely. The membership of RVCC gathers yearly near Portland, Oregon, at an annual policy meeting to learn about new issues and developments, to discuss and determine priorities for policy action in the coming year, and to assign duties to working group members.

During the first quarter of each year, RVCC members develop policy-issue papers. Dozens of issue papers have been prepared on topics related to forest restoration and producing community benefit from federal forestlands. Topics have included community capacity building, collaboration, performance measures, biomass utilization, stewardship and best-value contracting, workforce and labor issues, climate change, green jobs, private lands, wildfire cost containment, sustainable ranching, working landscape conservation, and the role of community-based organizations in western conservation.[6] Member organizations of RVCC are given copies of the policy papers and asked voluntarily to sign on, designating their agreement to support the priorities and solutions expressed in that paper and to work to advance the solutions proposed in that document.

When papers are final and RVCC members have completed their sign-on process, Sustainable Northwest organizes an education and outreach week in Washington, DC, called the Western Week in Washington. Its aim is to connect RVCC members with agency, administration, congressional, and interest group leaders to discuss the issues and solutions of highest priority to the coalition. The week entails small meetings with and briefings and presentations to decision makers, as well as peer-to-peer learning.

More than 100 community-based, collaborative organizations currently participate in RVCC. Most are from the West, but national organizations and local entities have joined from as far away as Vermont and Tennessee. Ten organizations form RVCC's steering committee. This core group advises Sustainable Northwest on all matters relevant to membership, governance, process, annual work plan, evaluation, and monitoring. The steering committee is also instrumental in helping to reach out to new organizations and securing in-kind assistance and necessary financial resources.

## RVCC'S APPROACH TO POLICY CHANGE

Unlike many traditional conservation efforts that view policy change through the singular lens of an advocacy campaign, RVCC has employed a nontraditional approach to changing national policy. The coalition was created to achieve multiple objectives, among them, to change laws and administrative procedures that block the ability of communities to participate in the restoration and long-term stewardship of public and private lands across the West; to showcase and promote local efforts that are successfully achieving land stewardship goals; to allow networking among rural leaders to accelerate learning and adoption of best practices; to build a cadre of new rural leaders who can speak for themselves about the impact of policies on their communities and adjacent landscapes; and to contribute to a national movement of place-based conservation.

The leadership of RVCC has long recognized that the single-issue approaches to conservation typical of past campaigns by major environmental groups and the timber industry contributed to legal and administrative constructs that, coupled with agency culture, stood in the way of good stewardship. The coalition saw that these constructs, although they were sometimes effective at preventing harm to the environment, failed to create conditions necessary to support restoration and stewardship of degraded public and private land.

From the beginning, RVCC has sought to change the culture and dialogue at the national policy level as well as to change specific laws and administrative procedures. As a coalition of rural people and their supporting organizations, it simply did not have the political clout or financial resources that mainstream interest groups had; it would be a long time, if ever, before RVCC could garner comparable resources and access. Instead, RVCC's leadership believed that if solutions that resonated in their constituent communities could be integrated into the agendas of established national interests, they might have a chance to promote change. In many cases, RVCC has been successful at making "our" issues "their" issues.

The coalition advances several core principles that enable its vision for working landscapes to be achieved. These core principles are aimed at ensuring:

- collaboration among diverse stakeholders and community leaders as a primary strategy embraced by the land management agencies;
- investments in community capacity to perform needed restoration and stewardship across the landscape and in local infrastructure to make the fullest use of wood material removed from public and private forests, including locally manufactured valued-added products and energy production;
- multiparty monitoring to support adaptive management, ensuring that diverse stakeholders can learn together and regain the trust lost through the "timber wars";
- the evaluation of land management agencies' activities based on the ecological, social, and economic outcomes achieved, rather than on commodity outputs such as timber harvest;

- federal budgets that recognize the role of rural communities in achieving landscape-level resiliency and health;
- limited federal dollars are targeted to low-income and vulnerable communities; and
- projects on public lands are designed at a scale appropriate to the ecological and economic constraints of the landscape and the heritage of public land communities.

Two examples of how RVCC has successfully implemented its approach to policy change are described below.

## SAVING FORESTS FROM FIRE

At the time RVCC was founded, the driving concern of rural public land communities was the increasing intensity of wildfire on national forests. The nation had witnessed some of the worst wildfire seasons in decades. There was deep concern about community safety, local infrastructure and capacity to respond to wildfire, and the potential loss of critical habitat, soil fertility, and ecological resiliency.

In August 2000, President Clinton directed the secretaries of agriculture and interior to develop a response to severe wildland fires, reduce fire impacts on rural communities, and ensure sufficient firefighting capacity in the future. In turn, Congress provided direction and funding to the agencies to work with governors on a long-term strategy to deal with wildland fire and the hazardous fuels situation, as well as the need for habitat restoration and rehabilitation. In response to these concerns, the Western Governors' Association (WGA) led the creation of the National Fire Plan, codified under the Clinton Administration and supported by members of both parties in Congress, which appropriated millions of dollars to implement the plan. After George W. Bush was elected president, the National Fire Plan framework led to the Healthy Forest Restoration Act in 2003. The Western Governors' Association continued to take a leadership role in moving the framework to implementation and ensuring that federal, state, tribal, and private landowners coordinated efforts to manage the impacts of wildfire. The National Fire Plan was a first attempt to manage wildfire across a working landscape made up of many different landowners and managers, in essence an all-lands approach.

Members of RVCC were very engaged with the Western Governors' Association and several secured seats on its Federal Forest Health Advisory Council, positioning RVCC members to integrate their concerns through this collaborative process. The coalition released two issue papers in 2004 that codified members' top concerns regarding both the National Fire Plan and the Healthy Forest Restoration Act.[7] Coalition members worked successfully with other interests in the WGA, including the Wilderness Society and the Center for Biological Diversity, to secure adoption of a landscape definition of restoration fully supported across interest groups, changes in the performance outputs and outcomes required of the land management agencies, and recognition of the need to support communities in the development of community wildfire protection planning. However, RVCC members were unable to get the WGA or other interest groups to spend political capital to ensure that the most vulnerable and low-income communities would receive priority for grant funds. In 2010, WGA released a policy resolution, *Large Scale Forest Restoration*, calling on federal agencies to establish a community capacity and collaborative support grant program to ensure that rural communities and businesses were engaged in landscape-level forest restoration and contributing to forest health solutions where they live. The policy resolution is an example of the way in which RVCC has been able to integrate its priority issues into the agenda of other groups.

### INFLUENCING CONSERVATION LEGISLATION

Early in its history, the RVCC sought to promote comprehensive, landscape-scale restoration in a manner that would support rural communities and businesses. In 2001, members of RVCC worked on a legislative concept that embodied much of the current thinking about how to create the right policy environment to support place-based, collaborative conservation and to achieve landscape-level objectives. In 2002 the U.S. Senate took up this concept and drafted and passed Senate bill 2672, the Community-Based Forest and Public Lands Restoration Act. The bill had bipartisan support from leaders including senators Jeff Bingaman (D-NM) and Larry Craig (R-ID), who served as chairman and ranking member, respectively, of the Senate Energy and Natural Resources Subcommittee on Forests and Public Lands. The purposes of the legislation were:

1. to create a coordinated, consistent, community-based program to restore and maintain the ecological integrity of degraded National Forest System and public lands watersheds;

2. to ensure that restoration of degraded National Forest System and public lands recognizes variation in forest type and fire regimes, incorporates principles of community forestry, local and traditional knowledge, and conservation biology; and, where possible, uses the least intrusive methods practicable;

3. to enable the Secretaries to assist small, rural communities to increase their capacity to restore and maintain the ecological integrity of surrounding National Forest System and public lands, and to use the by-products of such restoration in value-added processing;

4. to require the Secretaries to monitor ecological, social, and economic conditions based on explicit mechanisms for accountability;

5. to authorize the Secretaries to expand partnerships and to contract with non-profit organizations, conservation groups, small and micro-businesses, cooperatives, nonfederal conservation corps, and other parties to encourage them to provide services or products that facilitate the restoration of damaged lands; and

6. to improve communication and joint problem solving, consistent with Federal and State environmental laws, among individuals and groups who are interested in restoring the diversity and productivity of watersheds.

Despite bipartisan support, bill 2672 failed to pass the Senate. However, the tenets of this legislation framed and informed much of the future work of RVCC. When a national environmental group hired a member of the coalition, she was able to integrate many, but not all, of the concepts embedded in this early legislative attempt into the group's work. Now with the political influence wielded by a national environmental group, RVCC participated in development of legislative concepts that resulted in passage of the Forest Landscape Restoration Act as part of the PL 111–11 Omnibus Public Land Management Act of 2009. This legislation created a priority landscape program to encourage collaborative, science-based ecosystem restoration of priority forest landscapes, supporting restoration that would ensure healthy working lands across ownerships at an ecologically significant scale. The program, administered by the U.S. Forest Service, is called the Collaborative Forest Landscape Restoration (CFLR) program. This program provides a means to encourage ecological, economic, and social sustainability; leverage local resources with national and private resources; facilitate the reduction of wildfire management costs, including through reestablishing natural fire regimes and reducing the risk of uncharacteristic wildfire; demonstrate the degree to which various ecological restoration techniques achieve ecological and watershed health objectives; and encourage sale of forest restoration by-products to offset treatment costs, to benefit local rural economies, and to improve forest health.[8]

This legislation meets many of the objectives of RVCC but is missing several core principles. The largest omissions include a lack of funding to support collaboration, community capacity building, and mechanisms to secure local access to contracts offered on public lands. The legislation is authorized for 10 years, but the authorities for stewardship contracting embedded in the legislation expire in 2013, threatening the ability of the act to secure local benefit and employment from the increased financial investment in land treatments on public lands. However, passage of this law has created a new financial incentive for collaborative groups and their National Forest partners to work together on landscape-scale planning and project design. Ten landscape projects were selected in the first round of competitively submitted proposals. Members of RVCC are represented by one or more organizations in five of these projects.

### LESSONS LEARNED

Perhaps the most important lesson of the RVCC experience to date is that small organizations and community leaders can feel comfortable and confident when addressing complex topics that affect their communities and in communicating with their elected representatives in Washington, DC. On countless occasions we have heard RVCC members make statements such as, "I didn't know I could really do this," "I could never do alone what we are doing together," and "It's empowering to be advocating for solutions rather than just whining about my community's problems." On numerous occasions we have heard congressional aides and committee staffers say, "We are consistently surprised to see such a diverse, well-prepared group that so effectively knows what it wants and how to ask for it."

While the majority of RVCC participants have indicated in evaluations and surveys that they feel included and empowered by their participation in the coalition, some report challenges. We often hear from the newest participants that they feel overwhelmed by the information or intimidated by the knowledge of other coalition participants. We have heard from a few participants that it can be difficult to get new issues or topics prioritized and to receive the level of attention earned by issues that have long-standing support among coalition participants. We have worked to address these concerns with tactics that include assigning mentors to new participants to help guide them through their first meeting, conducting an orientation to RVCC at the beginning of each annual policy meeting, publishing on the web and distributing a print version of our vision statement and operations manual to explain our processes and identify group members, and conducting facilitator trainings to improve our own group processes. With each tactic we have made RVCC more inclusive and approachable, but as a practical matter we have learned that the most effective and engaged members are those that are committed to learning the process and seeing how their involvement can help them resolve local issues over the long term.

We have also learned that we cannot accommodate everyone and that the coalition may not meet the needs of each individual or interest group. At a certain point we believe it is most effective for us to focus on those who invest in their own participation. That said, there are certain issues that the leaders of the coalition take care to keep front and center, even in the absence of a vocal constituency in support (such as workers rights and protections) because coalition members recognize that our collaborative work style may not be easily accommodated by those constituencies. In such instances we have strived to secure the involvement of associations or intermediaries that can serve as proxies for those voices; while imperfect, this has been relatively successful at keeping certain issues at the center of the RVCC platform.

We have learned that a diverse, broad-based, well-prepared, and organized collaborative group garners far more attention, and therefore better access to key leaders in Washington, than individuals or groups who advocate alone or belong to a traditional, limited-purpose lobby or interest group. For example, in 2010, RVCC members held 120 meetings with legislators, legislative staffers, and agency leaders. While meetings alone do not guarantee change, the "face time" that RVCC is able to secure with key policy makers does make such change more likely.

By bringing together a diverse membership of individuals and organizations that have a breadth and depth of practical knowledge, groups like RVCC can become a forum for learning and networking. Meetings of RVCC provide members with the opportunity to discover how to engage with each other, exchange ideas, and anticipate challenges and opportunities. Participation in issue-paper development and joint decision making helps to sharpen members' skills in discussion, debate, and collaborative decision making. Seminars and lectures at the annual policy meeting on issues such as biomass energy, climate change, and ecosystem services markets prepare members to be informed leaders and advocates in their own communities, state capitols, and Washington, DC. Moreover, bi- and multilateral collaborations have emerged from members' participation in RVCC. These exchanges have resulted in publications, handbooks, and research projects. Such collaborations bring healthy challenges such as working through leadership issues, deciding who gets credit for accomplishments, and finding sufficient financial support to sustain collaborations.

Although aligning place-based work with a variety of regional and national groups to achieve specific and shared goals is a powerful way to influence agency, administration, congressional, and interest group decision makers, durable policy changes come from sustained effort over time. The group must stay organized, focused, and adaptive. Challenges faced by the RVCC have included insufficient funding to support the staff capacity that place-based and collaborative groups need to sustain their involvement in policy, the number one factor groups cite to explain why they are able or unable to engage in coalition activities. Limited understanding of legislative and administrative processes has required that we invest in building rural leaders' capacity to engage effectively in policy development. At times, members have raised concerns about becoming too similar to conventional interest groups and have expressed a desire to maintain a focus on integrated approaches and solutions, rather than simply advocating against various proposals.

Sustainable Northwest itself struggles to support RVCC participants adequately, augment their expertise, and help them access the financial and media resources needed to convey the urgency of RVCC's issues effectively. The RVCC's most difficult challenge stems from the reality that conflict and single-issue interests drive political processes. A multi-issue coalition, RVCC comprises people and organizations interested in solutions, civil dialogue, and compromise. In a political system that normally rewards those who use the harshest and most blunt political and communications tactics, it has proved, at times, difficult to get traction on core issues and recognition for the leadership role that rural communities play.

Members of RVCC have not let this discouraging political dynamic deter them. However, it has slowed the coalition's progress, frustrated participants, and at times even triggered despair. But one of the most powerful social dynamics within the coalition is the willingness of members to share hope, determination, resources, and vision when their colleagues are struggling. This is the benefit most often articulated by members. Experience has shown that diverse groups can effectively work together on both broad and specific policy initiatives.

Today, RVCC is becoming a voice for a much larger constituency, not all of whom are coalition members—for example, federal agency personnel cannot formally participate in the RVCC but often report that the coalition aligns with their own values. Other coalitions, including the National Rural Assembly, have aligned with RVCC although they are not formal members, and many organizations participate in the deliberations of the RVCC but may not ultimately sign on to issue papers for internal reasons or larger political considerations.

The coalition's members typically represent small place-based and collaborative groups that perform restoration work on public and private lands. Many rural, place-based organizations have tended to blame the environmental community for the economic decline of their communities and for inadequate management of public land. However, groups that once resented and feared urban environmental organizations have learned that they can, sometimes, rely on these groups to be fair advocates and interlocutors for their interests. Moreover, the RVCC now includes a growing number of traditional environmental organizations such as the Wilderness Society, the Center for Biological Diversity, Western Environmental Law Center, and Oregon Wild that were once reluctant to trust rural people, seeing rural communities and workers as synonymous with the timber companies that they held responsible for the poor management of public lands. Many environmental organizations have learned both from direct engagement in place-based projects and from their involvement in RVCC that rural communities can be honest stewards of the landscape. Alignment born from trust developed through the collaborative process has enabled these diverse organizations to work together in the RVCC to develop and promote public policies that support shared values and goals.

## THE PATH AHEAD

As RVCC grows, it must address both operational and issue-specific challenges. Continued effectiveness will depend on an organizational model that supports a nimble and inclusive work style, while adding structure sufficient to support growth in the numbers and types of organizations that participate. Groups from across the country, eager to join the RVCC, often bring a narrow-issue agenda with them. The coalition must be both creative and organized to balance the interests of old and new members successfully.

Significant opportunities lie ahead. The Clinton, Bush, and now Obama administrations have each proposed some form of initiative to support the growing movement for collaborative, place-based conservation. During the Clinton years, it was called collaborative stewardship; the Bush administration called it cooperative conservation; and the Obama administration has codified it as a core strategy for America's Great Outdoors initiative. Yet, to date, none of these initiatives have directed money or staff resources to support collaborative efforts. These policy initiatives have counted on rural communities and interest groups to raise money from private foundations and donors to make their collaborations successful.

The challenge for all involved in place-based conservation is to make the case that federal investment in these efforts makes sense for both federal agencies and taxpayers. The Rural Voices for Conservation Coalition is one example of how western communities are organizing to establish a credible voice for federal investment in place-based conservation and related collaborative and partnership efforts. But without a national movement uniting rural America, a movement in which communities that affect all lands—public and private—are engaged and articulate, our efforts will remain limited. We will face boom-and-bust levels of support driven by the whims of Congress, a pattern likely to make meaningful environmental, economic, and social outcomes on the lands that sustain us difficult to achieve.

## NOTES

1. For a larger discussion of the policy changes that catalyzed the emergence of place-based conservation, see S. L. Yaffee, *The Wisdom of the Spotted Owl: Policy Lesson for a New Century* (Washington, DC: Island Press, 1994).
2. When the Forest Service wants to sell timber, they follow the rules of the National Forest Management Act and offer a timber sale that is awarded to the contractor who will pay the most for the timber. When the Forest Service wants to hire an entity to provide a service such as planting trees, repairing a stream, conducting wildlife surveys, or other similar activities they offer a service contract. Service contracts, in practice, are often to the entity that offers to do the work for the lowest price. While both timber sales and service contracts do have a lot of complexity to them, the general system of highest or lowest bid deemphasizes quality work and local benefit.
3. Stewardship End-Results Contracting was authorized by law under 16 U.S.C. 2104 when it passed as part of the Omnibus Consolidated Appropriations Act of FY 1999 and amended by Sec. 323 of P.L. 108-7, 2003.
4. When the federal government awards a contract on a best-value basis it must consider three factors: technical approach, past performance, and price. The agency retains discretion over how much weight to assign each factor. For more informa-

tion about Best Value and Stewardship Contracting, see C. Daly, *Best Value and Stewardship Contracting Guidebook: Meeting Ecological and Community Objectives* (Portland, OR: Sustainable Northwest, 2006), http://www.sustainablenorthwest.org/uploads/resources/Best_Value_Report.pdf.

5. Multiparty monitoring is an approach to building trust and practices-adaptive management throughout the life cycle of a natural resource management project. It is a process that involves discussion and mutual learning among a diverse group of individuals representing different groups and interests. For more in-depth discussion of multiparty monitoring, see A. Moote, *Multiparty Monitoring and Stewardship Contracting: A Tool for Adaptive Management* (Portland, OR: Sustainable Northwest, 2011), http://www.sustainablenorthwest.org/uploads/resources/Multiparty_Monitoring_Guidebook_2011_finalV2_links.pdf.

6. For specific examples, see the following policy papers: "RVCC Vision and Policy Priorities"; "2008 RVCC Transition Memo: A Rural Agenda for Stewardship of Natural Resources in the American West," November 26, 2008, http://www.sustainable northwest.org/uploads/resources/FINALRVCCTransitionMemo.pdf; "RVCC Recommendations for Increasing Community Capacity to Deliver Farm Bill Conservation Programs," May 14, 2012, http://www.sustainablenorthwest.org/blog/posts/rvcc-2012-farm-bill-issue-paper; "Rural Capacity for Conservation and Job Creation," May 2011, http://www.sustainablenorthwest.org/uploads/resources/2011_CBO_Capacity_Issue_Paper_color_final.pdf; "Sustainable Biomass Energy: The Need for a Comprehensive Energy Policy that Includes Thermal Energy (Joint RVCC/CEFC issue paper)," May 1, 2011, http://www.sustainablenorthwest.org/blog/posts/sustainable-biomass-energy-2011-issue-paper; "Monitoring: An Essential Tool for Achieving Environmental, Social, and Economic Goals," May 1, 2011, http://www.sustainablenorthwest.org/blog/posts/monitoring-an-essential-tool-for-achieving-environmental-social-and-economi; "Stewardship Contracting: Successes in the Field," May 1, 2011, http://www.sustainablenorthwest.org/blog/posts/stewardship-contracting-successes-in-the-field—all available at http://sustainablenorthwest.org.

7. See 2004 RVCC issue papers: "Healthy Forests Restoration Act: Implementing New Legislation for Forest and Rural Community Health," http://www.sustainable northwest.org/resources/rvcc-issue-papers/2004%20HFRA.pdf (no longer available online); and "Wildfire, Poverty, and the National Fire Plan," http://www.sustainable northwest.org/resources/rvcc-issue-papers/Wildfire%20and%20Poverty_March2004 .pdf (no longer available online).

8. "Collaborative Forest Landscape Restoration Program," U.S. Department of Agriculture Forest Service, http://www.fs.fed.us/restoration/CFLR/index.shtml.

# PART THREE

## CASE STUDIES OF WORKING FORESTS

In chapter 2 we described transformations in the forestry sector that have been occurring in the West since the 1980s, changes that have affected harvests from both federal and private timberlands as well as employment in the forest products industries. These transformations were triggered by concerns over the environmental impacts of timber harvesting and changing market conditions, technologies, and public values relating to how forests are managed. Today, timber harvesting on federal lands most often occurs in the context of ecosystem restoration, while timber harvesting on private corporate lands occurs under a changing set of management goals and objectives. Both can be subject to heavy public scrutiny. The key question that public and private managers of working forestlands face today is, How can these forests be managed to produce forest products in a manner that meets the "triple bottom line" of social, economic, and environmental sustainability?

The three chapters and three spotlights in this section each provide a different answer to this question, illustrating some of the approaches that have emerged to keep working forests working across ownerships. The Swan Valley case (chap. 7) captures an effort to achieve landscape-scale conservation in a place characterized by checkerboard ownership patterns and skyrocketing land values. There, the Plum Creek Timber Company had large private industrial timberland holdings interspersed with federal lands and decided to sell them for real-estate development, threatening to cause habitat fragmentation and wildlife declines. One message of that chapter is that working landscape conservation sometimes happens through a series of experiments with different strategies that don't always work out; nevertheless, commitment and persistence can be rewarded. The conservation strategies tried in this case included land exchanges, conservation easements, community forests, land-use planning, and land acquisition. In the end, a combination of strategies prevailed, the main one being land acquisition because a broad group of local stakeholders in partnership with regional, state, and national interest groups rallied behind it. What emerged was the Montana Legacy Project—the largest private land conservation deal in the United States—encompassing 310,000 acres of corporate timberlands around Missoula, Montana, including the Swan Valley. With federal and state funding, these lands were purchased and the majority transferred to the U.S. Forest Service and state agencies for management, thereby erasing the checkerboard ownership pattern in the Swan Valley and the threat of real-estate development there. The deal included provisions for supporting local mills and ensuring a continued supply of wood fiber to support local economies.

New models are being tried on private timberlands as well. Two examples in this section come from the Redwood Region of California, where a long history of unsustainable timber harvesting led to extreme political conflict and social divisiveness over the issue. The Mendocino Redwood Company—a family-owned business engaged in industrial-scale forestry—is attempting to overcome this history of mistrust among stakeholders by operating in a transparent manner, adhering to high environmental standards, and taking a long-term, patient approach to seeing returns on its investment (chap. 8). The company sought Forest Stewardship Council certification, and developed an 80-year combined Habitat Conservation Plan and Natural Community Conservation Plan for their property to protect over 40 threatened and endangered species and meet federal and state regulatory requirements simultaneously as part of their strategy. The family has since acquired more lands in the Redwood Region and created the Humboldt Redwood Company, drawing on lessons learned through the Mendocino Redwood Company experience and expanding what appears to be a successful model for running a working forest from a business standpoint based on an ethic of environmental stewardship. The nearby Garcia River Forest (spotlight 8.1) presents another approach to sustainable forestry in the Redwood Region on what was the first large working forest in California owned by a nonprofit organization. The Conservation Fund purchased the forest with a combination of private and public funding to prevent its conversion to developed or agricultural land uses. Sustainable timber harvests (certified by the Forest Stewardship Council and the Sustainable Forestry Initiative) and the sale of carbon-offset credits provide revenue to finance the cost of maintaining, managing, and restoring the forest.

In the case of federal lands, stewardship contracts and stewardship agreements have emerged in the past decade as one innovative tool for accomplishing timber harvesting as part of broader ecosystem restoration work on Forest Service and Bureau of Land Management lands, while promoting collaborative conservation and local job creation. The case of the Siuslaw National Forest in Oregon (chap. 9) illustrates the use of this tool to accomplish much-needed thinning to improve forest health, while creating local jobs and supplying local mills with wood to help maintain forestry infrastructure around the forest. Stewardship contracting has also generated revenue to fund forest restoration projects on the Siuslaw and nearby private lands that share watersheds. A local collaborative group—the Siuslaw Stewardship Group—was formed to help identify, plan, implement, and monitor these projects. Their success has inspired the creation of other stewardship groups across the Siuslaw National Forest and extensive use of the stewardship-contracting authorities there. The Siuslaw's highly productive forests, which contain commercially valuable wood—together with local business capacity for undertaking restoration work, a local market for wood harvested, and the presence of a strong local collaborative that agrees on how to approach the shared stewardship and comanagement of public and private lands and is willing to innovate—make this model work.

Community forests—where local community residents have some degree of responsibility and authority for management—represent another approach to working forest conservation. Two types of community forests from California are described in spotlights 7.1 and 9.1. The Arcata Community Forest (spotlight 7.1) is owned by residents of the city of Arcata, California, and managed jointly by the city and an advisory committee composed of local volunteers, with significant opportunities for input by local residents. Today the Arcata Community Forest is an example of a successful, sustainably managed, multiple-use forest that generates enough revenue from timber harvests to cover the costs of management and help fund the acquisition of additional conservation lands. The Weaverville Community Forest (spotlight 9.1) is located on federal lands and is managed in collaboration with the local Trinity County Resource Conservation District in accordance with community objectives and priorities using stewardship agreements (similar to stewardship contracts) with the Bureau of Land Management and the U.S. Forest Service. These agreements make it possible to manage local forests in a manner that leverages resources, creates local jobs, accomplishes forest management projects, and meets community goals.

What all of these examples tell us is that different approaches to working forest conservation are appropriate in different places, depending on the local context, and that collaborative conservation is key to mobilizing successful efforts. They also show that working forest conservation on a single-ownership parcel has benefits for working forest conservation on other nearby parcels. In addition, they provide inspiration by demonstrating innovative models that show that it is not always easy, but it is possible, to manage working forestlands successfully today in a manner that meets the triple bottom line.

# 7

# SWAN STORY

*Melanie Parker*

IN BRIEF

- A mixture of tools and approaches geared to the social and ecological realities of a particular place are often needed to conserve working landscapes; options may include land exchanges, conservation easements, land-use planning, and land acquisition.
- Investing public dollars in working landscape conservation can save taxpayers money in the long run because the costs of fire suppression, endangered species recovery, service delivery to scattered homes, and other problems can be more expensive than the costs of conservation.
- Building alliances between local, regional, state, and national interest groups that share common goals is a strategic way to achieve large, landscape-scale conservation; such conservation efforts should build on, rather than supplant, local initiatives.
- Government funding to support land acquisition and conservation easement programs is critical for landscape-scale conservation in areas having substantive private landholdings, especially where there is high development pressure.

The Swan Valley of western Montana is a place of deep forests, abundant wildlife, and a relatively small rural population. The land itself is unusually rich with biological resources—unique wetlands, diverse forests, and critical habitats for threatened and endangered species such as grizzly bears (*Ursus arctos horribilis*), lynx (*Lynx canadensis*), wolves (gray wolf, *Canis lupus*), and bull trout (*Salvelinus confluentus*). About 1,000 permanent residents in the Swan Valley live in a loosely associated, unincorporated rural community. This chapter describes how members of the community worked for over a decade to build a coalition to prevent the widespread conversion of working forestlands in the Swan Valley. In the end, these efforts resulted in the public acquisition of 310,000 acres of corporate timber lands in western Montana, made possible by leveraging federal and state funds to purchase the land from the Plum Creek Timber Company. Without this project, many thousands of acres would have been converted, over time, into real-estate developments with their associated clearings, roads, fences, and nonnative species. Resident wildlife species would have slowly been eliminated.

This conservation strategy was only settled on after long consideration. Several key lessons emerged from the Swan Valley experience:

1. landscape-scale conservation works when it builds on, rather than supplants, local initiatives;
2. current tools and funding mechanisms to support working landscape conservation are limited compared to the scale of this need across the West;
3. each place requires its own carefully crafted solution based on site-specific social, political, economic, and ecological realities.

This chapter tells the story of what took place in the Swan Valley, elucidates several of the strategies considered, and concludes by demonstrating the importance of land acquisition as one element of the toolbox for working landscape conservation.

To understand Swan Valley's conservation strategy for working forestlands, it is important to know something about its geography and land ownership. Altogether, the Swan Valley covers roughly 440,000 acres. Residents of the valley live within small patches of private land that are surrounded by a checkerboard of public and corporate lands (see map, fig. 7.1). The checkerboard is situated between two wilderness areas that are part of one of the last large connected landscapes in the world, the Crown of the Continent Ecosystem. The Crown of the Continent is a 10-million-acre mosaic of wild habitat encompassing Glacier National Park, the Bob Marshall Wilderness, and surrounding public and private ranch and forestlands. According to the Nature Conservancy, the Crown of the Continent is one of the most intact landscapes on the planet.

The checkerboard land ownership pattern is the legacy that Swan Valley, and many other communities across the West, inherited from the railroad

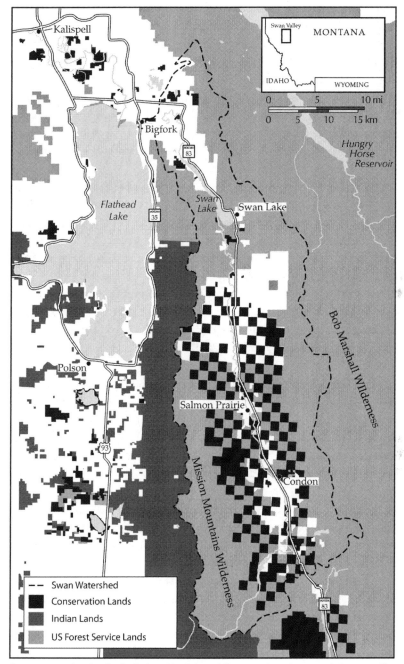

**Figure 7.1.** Map showing the checkerboard land ownership pattern that exists not only in Swan Valley, Montana, but in many places across the American West. Map by Darin Jensen and Syd Wayman, 2013.

land grants of the 1860s. Alternating square miles across the northern tier of the United States were granted to the Northern Pacific Railroad to stimulate the expansion of the transcontinental railroad and the settlement of the West. In the Swan Valley, these Northern Pacific lands eventually became Burlington Northern holdings, converted in recent decades to Plum Creek Timber Company lands. Through mergers and acquisitions, Plum Creek Timber Company is now the largest private landowner in the United States, and owns approximately eight million acres. Much of this acreage is intermingled with federal lands, as is the case in the Swan Valley. In the early 1990s, Plum Creek owned 80,000 acres of land there.

In the mid 1990s, land values skyrocketed in western Montana. Prices for forested lands in the Swan Valley increased from an average of $500/acre to $5,000/acre over a 10-year period. Motivated by this economic climate, Plum Creek Timber Company began to convert some of its land portfolio in western Montana from working forest to real estate. In 1997, Plum Creek identified 2,500 acres along the shores of Lindbergh Lake at the headwaters of the Swan River as "higher and better use" lands. They explained to community members that some lands, in terms developed by real-estate appraisers, have higher and better uses than long-term timber management. These lands were surrounded by National Forest lands and adjacent to federal and tribal wilderness. The higher and better use was determined to be residential home development. In other words, profits received in the short term for selling the land far outstripped all potential future profits that would come from manufacturing wood products from the trees growing there. This was local residents' first lesson in the shifting economic conditions of the region. Working landscapes, whether they were forest, range, or farm, were beginning to have a hard time outperforming the real-estate market and so were increasingly vulnerable to conversion.

In 1997 a group of residents, including myself, who were concerned about Plum Creek's proposed activities met with a representative from the Trust for Public Land. The trust is a national nonprofit organization highly skilled at working with federal acquisition programs; they helped ensure that the first acres divested in the Swan Valley were for public acquisition. After three years and a $13 million investment, the Trust for Public Land was able to purchase all 2,500 acres using federal Land and Water Conservation Funds, transferring the acreage to the Flathead National Forest. The Land and Water Conservation Fund is a mitigation fund created by Congress to offset the impacts of offshore oil drilling. The funds are allocated through a competitive grant program that is dependent on strong community support.

In the following years, Plum Creek sold off more and more acres in the Swan Valley, and the pace and scale of divestment began to outstrip the residents' ability to respond. In 2000, Plum Creek company representatives announced a 10-year plan to sell 10,000 acres in the valley. One year later, they quadrupled the rate and announced the intent to sell off 20,000 acres in five to seven years.

As a local resident of the Swan Valley, I set out with a small group of other dedicated volunteers to figure out what, if any, options existed for a large-scale transaction in the Swan Valley that would essentially remove the development potential from all corporate checkerboard timberland. Our constellation of local actors evolved from a subcommittee of the Swan Valley Ad Hoc Committee, a nascent group working on natural resource issues. We then morphed into a subcommittee of the Swan Ecosystem Center, the formal nongovernmental organization that grew out of the ad hoc committee efforts of the early 1990s.

Our group found its most productive configuration, however, as the Swan Lands Coordinating Committee, which included every public and private entity that considered itself a stakeholder in the Swan Valley. We invited representatives from state and federal agencies like the U.S. Fish and Wildlife Service, the U.S. Department of Agriculture Forest Service, the Montana Department of Natural Resources and Conservation, the Montana Department of Fish, Wildlife and Parks, county commissioners, several environmental groups like the Montana Wilderness Association and the Friends of the Wild Swan, as well as industry groups like the Montana Logging Association, land trusts like the Montana Land Reliance and the Trust for Public Land, and local citizens and landowners. We began meeting quarterly in 2001 to consider each of several potential conservation strategies. Such strategies included land exchanges, conservation easements, community forests, land-use planning, and land acquisition. Each of these tools offered certain opportunities and presented some obstacles that will be addressed in the following sections.

## LAND EXCHANGES

Many in our community, as well as among our state and national conservation partners, believed that the Swan Valley would benefit from erasing the checkerboard land ownership pattern because of the difficulty of managing natural resources and maintaining ecological partnerships across highly fragmented ownerships. In the early 1990s, community members worked on a small-scale project called the Elk Creek Land Exchange. A land exchange is a nonsale real-estate transaction in which a nonfederal party exchanges private land for federal land, causing it to change ownership. The Elk Creek transaction was designed to prevent road building and logging above some extraordinary bull trout spawning beds along Elk Creek, a major tributary of the Swan River. This land exchange succeeded in consolidating some of the lands within Elk Creek into Forest Service ownership in exchange for consolidating a few sections of Plum Creek lands in the northern valley.

However, the project had two significant downsides. First, it was opposed by local environmental groups who did not support the notion of public lands being converted to corporate timber lands through a land exchange. Second, it took five years to work its way through the environmental review process. In the end, Plum Creek told the Swan Valley Ad Hoc Committee that they would not be interested in future land exchanges unless they were enacted by special legislation, thereby avoiding the lengthy public review process and its inherent uncertainty. Typically, when a land exchange is proposed, it goes through a full National Environmental Policy Act process, which—because of the significance of possible impacts to the environment—is long and detailed, and includes consultations with the U.S. Fish and Wildlife Service on impacts to threatened and endangered wildlife species. The time it takes to accomplish such a review, and the uncertainty as to whether the proposed action will be supported by the review, leave the landowner in a difficult position: the owner cannot alter the condition of the land while the project is in the queue, so economic returns from that land are reduced dramatically, even though there is no guarantee that the exchange will occur. Subsequent to the Elk Creek land exchange, the Swan Lands Coordinating Committee deliberated seriously over scenarios that would enact special legislation to exchange all 80,000 acres of Plum Creek Timber Company land in the Swan Valley for public lands elsewhere in the state. But such an exchange never gained traction because strong environmental constituents exist for public lands almost everywhere, and the group could identify no likely tract in Montana where an exchange out of public lands would be supported.

## CONSERVATION EASEMENTS

Although erasing the checkerboard had its allure in terms of simplifying land management, it was also always clear that preventing the land from being converted to real estate was the primary goal. In that vein, the Swan Lands Coordinating Committee sought opportunities to purchase not the land itself but simply the development rights. They took inspiration from a project that the Trust for Public Land, the Rocky Mountain Elk Foundation, and Plum Creek were completing on a large landscape in northwest Montana. This project, the Thompson-Fisher Conservation Easement, prevented development on 120,000 acres of corporate timber lands, while Plum Creek maintained ownership of those lands and management of the timber.

Our local committee asked the Trust for Public Land to approach Plum Creek about a similarly scaled conservation easement across the Swan Valley. But Plum Creek was not motivated to agree to such an easement in the Swan Valley watershed because Swan Valley real-estate values were comparatively high. The Trust for Public Land and Plum Creek did settle on pursuing a conservation easement on a subset of lands within the Swan River State Forest, where real-estate values were somewhat lower. Over the course of three years, 7,200 acres were protected using federal Forest Legacy Program dollars. The Forest Legacy Program, administered by the U.S. Forest Service in cooperation with state partners, supports the conservation of private forestlands by funding the purchase of conservation easements. Plum Creek, however, maintained that a valley-wide conservation easement was still off the table.

## COMMUNITY FOREST

In the late 1990s, the director of the Swan Ecosystem Center visited Arcata, a little town in northern California that had purchased and was managing its own community forest (see spotlight 7.1, Arcata Community Forest). She returned to the Swan Valley with a vision for how its citizens could acquire and manage some of the Plum Creek Timber Company Land. Community leaders representing diverse stakeholders—including loggers, outfitters, forest managers, and conservationists—were enthusiastic about the plan. The community forest concept matched local values. Condon, Montana, a community of around 600 households in the upper part of the Swan Valley watershed, had been pushing for sustainable land management practices on its federal, state, and corporate lands for two decades. Residents understood that if they wanted land management that supported both the land and the local economy, they might have to figure out how to do it themselves. After clamoring for jobs, fire protection, weed suppression, and wildlife conservation on public and corporate lands, it was time to build the strategic partnerships necessary to achieve these goals in a community forest.

In the early 2000s, a representative from the National Wildlife Federation came to the Swan Valley with a slightly different vision. He proposed sponsoring special legislation to purchase all of the Plum Creek Timber Company lands in the Swan Valley and then make the entire watershed a community forest within the federal public land system. The idea was to put a diverse committee of individuals representing local, regional, and national interests in charge of setting the management direction for a chunk of federally owned land in a way that local community benefits were given high priority. The Swan Lands Coordinating Committee spent more than a year working on the details and engaging the broader public before it was told that Plum Creek Timber Company was not interested in supporting the legislation at that time.

Another option presented itself in the potential to sell community forest bonds. A consulting firm, U.S. Forest Capital, had been developing this economic tool in the state of Washington. The idea was that a local government or nonprofit organization could issue bonds to raise acquisition money and, then, service the bonds over time with sustainable timber harvest from the project lands. This tool looked promising for Swan Valley but presented two problems. First, the Internal Revenue Service had not yet approved this particular bonding mechanism. Second, the Swan Valley lands had been harvested to the point where reasonable returns from timber harvest were not likely to occur in the time frame of even a long-term bond. If such a funding mechanism were to be used, it would have to be coupled with other revenue streams.

While the community forest bonds were not utilized in the Swan Valley, a quasi community forest finally was established in 2005. The idea of land owned and managed by local interests has taken hold in the Swan Valley, but the mechanisms by which this could succeed were not clear or simple. The Elk Creek Conservation Area, as the Swan Valley community forest is named, covers 640 acres and is owned by the Swan Ecosystem Center and the Confederated Salish and Kootenai Tribes. Mitigation funds allocated for endangered fish recovery through the Bonneville Power Administration and private philanthropic dollars were used to help initiate the project. The Swan Ecosystem Center plans to add to the footprint of the community forest in the future. Current efforts also revolve around developing new shared ownership and governance models.

## LAND-USE PLANNING

Over the years, community leaders in the Swan Valley have been criticized by some environmental groups for developing strategies like land exchanges and conservation easements to prevent real-estate development on working forestlands. Why spend money or give away public resources, they argued, if you can accomplish the same end with simple land-use planning? Our group decided that while land-use planning is a valuable end goal, it is so contentious, complex, and difficult in rural Montana that our working landscape would be lost long before a plan could be negotiated. In the Swan Valley, this is made more difficult by the fact that our watershed straddles two counties in Montana: Missoula County and Lake County. The Missoula County portion of the watershed had developed a comprehensive growth management plan and updated it twice, in 1987 and 1996. When the last iteration was completed, the Plum Creek Timber Company lands were not subject to analysis, as they were considered working forestlands. Thus, the plan currently in effect offers no clear direction to the county commissioners when considering proposed developments on former Plum Creek Timber Company lands. Updating the plan once the higher and better use sales began would have made some sense, but for the fact that in 2001 the Montana state legislature weakened counties' abilities to use a comprehensive growth plan to make rulings on proposed subdivisions. County commissioners told the Swan Valley communities in no uncertain terms that they were not interested in seeing the comprehensive plan revised unless it was linked with what they called "implementation"—in other words, zoning.

A protracted fight over zoning in the valley was not going to galvanize a coalition to support large-scale conservation on Plum Creek Timber Company lands. Zoning cuts right to the heart of the cantankerous debate in the West over private property rights and public goods, and it was clear that debate could not be resolved on a local level in time to stem the tide of land sales already occurring in the valley. By 2005, Plum Creek had sold 4,000 acres of what had been open space, working forest, and wildlife habitat for private home development. The public had lost access to these lands, the wildland-urban interface had grown, and grizzly bears had lost more acres of core habitat.

At that point, community efforts had helped to acquire roughly 5,800 acres of Plum Creek's "higher and better use" lands for the Flathead National Forest and to purchase the development rights of another 7,200 acres in the northern part of the valley. Each transaction, however, was growing more expensive, as land values continued to rise. More and more money was being spent for less and less land.

## LAND ACQUISITION

By 2005, it had become clear that land acquisition was the best tool for a landscape-scale solution to the problem of corporate land divestment in the

Swan Valley, and the community began to come together on this idea. The new regional director of the Trust for Public Land met with several local conservation and community organizations, and asked, "Which part of the Swan Valley can we acquire and protect, and then call this project a wrap?" The community's response can be summed up pithily: "Go big, or go home"— in other words, protect the entire Swan Valley, or quit throwing increasingly large sums of money through retail transactions to purchase small parcels within the valley. Several key partners agreed on the importance of finding a way to protect the whole valley; going piecemeal was not worth the effort.

Since 2005 a colossal effort has been put forth by a large number of dedicated professionals and volunteers. The magnitude of the project required collaboration among conservation groups that frequently compete with each other for limited resources. The Trust for Public Land and the Nature Conservancy collaborated to negotiate a project with Plum Creek Timber Company. By 2007, Plum Creek Timber Company finally decided it could part with the Swan Valley and all of its development potential if the sale also involved lands across western Montana that were nonstrategic for the company. The result is a project that encompasses 310,000 acres of corporate timber lands centered primarily on Missoula, Montana. The Montana Legacy Project is, as of 2013, the largest private land conservation deal to date in the United States.[1]

The obvious benefits of this project for the Swan Valley were that it simultaneously prevented the conversion of land for residential real estate and erased the checkerboard legacy by putting the majority of lands into public ownership. Since the price tag for 310,000 acres was nearly half a billion dollars, this project required a new and monumental funding source. Montana's Senator Max Baucus successfully championed the creation of a new fund— the Qualified Forestry Bonds program—in the 2007 Farm Bill for the protection of large landscapes of over 40,000 acres that are also under a Habitat Conservation Plan for native fish and wildlife. The Montana Legacy Project was able to apply for funds authorized through the Qualified Forestry Bonds program. This fund, or one like it, would be a valuable tool to be renewed in subsequent Farm Bills for large, valuable working landscapes under threat of conversion in other portions of the country.

In addition to significant federal funding, state funding is also being allocated to this project. In 2010, a strong bipartisan coalition passed paired bills through the state legislature to acquire some of the lands within the footprint of the Montana Legacy Project and create a revolving loan fund for local mills simultaneously. Conservationists and the timber industry within the state realized their overlapping interests in saving both the landscape and the

economic infrastructure that depends on this landscape. Not all lands across
the project area will end up in public ownership. Some will be transferred
to private landowners after the Nature Conservancy and the Trust for Public Land place conservation easements on them. In the Swan Valley, there are
about 2,500 acres still hanging in the balance, and the community is working
to determine the final disposition of those lands.

Because the Montana Legacy Project involved transferring so much land
to the Forest Service (90,000 acres) and state agencies (approximately 110,000
acres), it generated concern from people who do not support public land management. But despite such concern, and despite the strong "no-net-gain of
public lands" rhetoric in the daily papers, this project has enjoyed tremendous
support overall. One reason is that, in the spirit of protecting working forests,
the Trust for Public Land and the Nature Conservancy negotiated a fiber supply agreement whereby wood fiber will continue to be harvested from the acquisition lands and delivered to local Plum Creek mills for at least 10 years.
This will be done even on lands conveyed to the National Forest System by
having the conservation groups hold a "timber reserve" on lands passed to the
Flathead and Lolo National Forests. Another reason for support is that people
knowledgeable about this particular landscape—the Crown of the Continent
Ecosystem—understand the significant benefits of public lands acquisition,
especially for places like the Swan Valley. Plum Creek Timber Company lands
in the Swan Valley were more like inholdings in a public lands setting to begin with, so blocking them with the surrounding land management agencies
made good sense to the majority of stakeholders and local citizens.

## CONCLUSIONS

Large working landscapes with irreplaceable conservation values across the
West are experiencing increasing development pressure. In many cases, it is
possible to stem the tide of private land conversion by increasing the economic returns to private landowners for maintaining or enhancing conventional land uses. But in the Swan Valley, the discrepancy between the value
of the land for timber and the value of the land for real estate was too great
to remedy with these sorts of strategies. The only viable options were to purchase the development rights from the landholding company, so that they
retained timber management, or to purchase the entire fee title for the properties and find new landowners, such as public agencies, that lacked incentive to develop real estate.

Three factors ended up determining the success of the Swan Valley project. First was the willingness of all the partners to work at a large landscape scale. That scale helped convince supporters and partners that their efforts would result in significant ecological benefits to the region. The second factor was the ability to secure broad support among the local constituencies. County commissioners, community leaders, ranchers, loggers, conservationists, and business owners were involved in considering each of the conservation strategies mentioned and provided strong testimony for the project at critical junctures. The third factor was the very strong alliance built between these local constituencies and regional, state, and national interest groups. The project found champions in the state and federal government, within the national land trusts, in the environmental community, and among local residents. By having their "heads in the clouds but their feet firmly on the ground," as I like to say, members of our coalition managed to build a project that has permanently protected 310,000 acres directly, and close to a million acres indirectly.

Moving forward, this same constellation of partners is charged with ensuring that all the benefits of a working landscape are actually realized in the context of a public lands acquisition. Efforts are ongoing to promote stewardship projects that emphasize local employment, train local contractors in restoration techniques, secure restoration funding to support a robust program of work on federal lands, and create new collaborative management agreements between local nonprofits and the federal government. All of the conservation strategies that were investigated in the Swan Valley are important options for rural communities across the West. To continue such conservation efforts, funding for public acquisition programs and conservation easements will need to be increased; the capacity for land-use planning in rural counties will need support; and community forest bonds will need to be fully authorized. It is important for policy makers to evaluate the full range of opportunities for protecting working landscapes in the rural West and to consider where policy and funding changes could increase the chances for success in stemming the tide of land conversion.

The conservation of working landscapes is not a frivolous public expenditure; in many cases, the public dollars for such projects actually constitute prudent investment in long-term cost reduction to tax payers. The costs associated with fire suppression around homes, endangered species recovery in fragmented landscapes, and delivery of services to widely scattered homes far exceed the costs of preventing those problems in the first place. Cost savings are particularly stark when addressing the remaining checkerboard landscapes across the region, where development is poised to occur on alternating square miles within an otherwise public lands setting. To be clear, the people of the Swan Valley do not think that land acquisition is the only—or even the optimal—way to protect working landscapes across the West. Market-based incentives, tax incentives, and several other tools are needed in our region. But the story narrated in this chapter, our story, does illustrate that some places— in particular, checkerboard landscapes where land values have skyrocketed— are good candidates for public lands acquisition strategies.

### NOTES

1. "The Montana Legacy Project," Nature Conservancy, http://www.nature.org/our initiatives/regions/northamerica/unitedstates/montana/placesweprotect/montana -legacy-project.xml.

# Spotlight 7.1

# ARCATA COMMUNITY FOREST

*Mark Andre*

## IN BRIEF

- The Arcata Community Forest (ACF) serves as an example of how a working community-owned and managed forest can provide multiple benefits to local residents.
- The ACF produces a number of ecosystem services; in addition, revenue generated from sustainable timber production is reinvested in conservation by funding the public acquisition of additional lands that have conservation value and by the purchase of conservation easements on private forest and near the city.
- Successful community forests in the United States and Canada share in common the presence of a technical advisory committee or oversight board to help the governing body make sound management decisions.
- Community forests are an appropriate model of working landscape conservation in places where productive timberlands interface with urban areas, and multiple goals can be achieved by maintaining working forest lands.

The Arcata Community Forest serves as an example of how a working community-owned and -managed forest can provide multiple benefits to local residents. The Arcata Community Forest produces a number of ecosystem services; in addition, revenue generated from sustainable timber production is reinvested in conservation by funding the public acquisition of additional lands that have conservation value and by purchasing conservation easements on private forest lands near the city.

Successful community forests in the United States and Canada share in common the presence of a technical advisory committee or oversight board to help the governing body make sound management decisions. Community forests are an appropriate model of working landscape conservation in places where productive timberlands interface with urban areas and where multiple goals can be achieved by maintaining working forestlands.

One of the first community-owned forests in the West, and the first in California, was the Arcata Community Forest (ACF). It is owned by residents of the city of Arcata, in Humboldt County, northern California, and managed jointly by the Natural Resources Division of the city's Environmental Services Department and Arcata residents. The city owns a total of 2,235 acres of forestland: 793 acres belong to the original ACF; 1,442 acres belong to the Jacoby Creek Forest, which was later acquired and added to the ACF; and an additional 850 acres of wetlands and riparian lowlands. The forests are second-growth forests, approximately 100–130 years old, dominated by coast redwood (*Sequoia sempervirens*) and Douglas fir (*Pseudotsuga menziesii*) trees.

Throughout the 1880s, the forest that was to become the ACF was heavily logged. After logging opportunities were exhausted, the land was used for animal grazing and human water supply—the ACF contains the headwaters of many creeks. In the 1930s, the city of Arcata gained title to the land in order to provide water to the town. In 1942, the city purchased the more remote Jacoby Creek Forest tract with the intention of developing a dam on Jacoby Creek; however the dam was never built. In 1955, the city council—recognizing the lands as being the first municipally owned forests in the state of California—declared that the forests must be managed for the benefit of citizens of the city. Special attention would be given to watershed maintenance, recreation, and timber management. From 1955 to 1970, most of the forestland was commercially thinned, with around 16.7 million board feet cut. Sales revenue was used to pay for city services and infrastructure. Harvest plans were developed by forestry faculty members of Humboldt State University and approved by the three-member Forest Advisory Council. This style of management continued until 1979, when voters passed an initiative mandating that the forest be managed according to "the principles of ecological forestry and perpetual sustained yield." The vote prompted the city to adopt the Arcata Community Forest/Jacoby Creek Multiple Use Management Plan, which they revised in 1994 to reflect an adaptive management approach and an emphasis on biodiversity. The plan was developed with the assistance of the Arcata Forest Management Committee and outlines in greater detail the following objectives:

1. maintaining the health of the forest system, specifically the integrity of the watershed, wildlife, fisheries and plant resources, and their relationships and the process through which they interact with their environment;
2. producing marketable forest products and income to the city in perpetuity, balancing timber harvest and growth;
3. managing the forest to provide recreational opportunities for the community;
4. promoting the city's forests as models of managed redwood forests for demonstration purposes.

No tax revenues are used for management purposes; the ACF generates approximately $300–500 thousand in revenue per year, more than is needed to be self-supporting. City staff is responsible for implementing the management plan, while policies and procedures continue to be developed by the city council with input from the Arcata Forest Management Committee. This seven-member committee of volunteers, appointed by the city council, has expertise in the areas of botany, forest ecology, wildlife, fisheries, geology, recreation, and forestry. Monthly committee meetings and frequent field trips are public, and attendance and participation by community members is encouraged and is key to the success of Arcata's community forest management.

In 1998, the ACF became the first municipal forest in the United States to be certified by the Forest Stewardship Council, and it is therefore subject to the council's standards. Thirty percent of the land base is held in reserve, and the remaining 70 percent has a maximum allowable annual harvest of half the annual growth increment. In 2008 the Forest Guild recognized the ACF as a "model forest," acknowledging its success as a sustainably managed, working, multiple-use forest. On average, 400–700 thousand board feet of timber are harvested annually from the forest using uneven-aged silviculture, growing trees in excess of 100 years old to produce high-quality redwood lumber and structurally diverse forest habitat. Carbon sequestration is another management emphasis, and a portion of the forestland base has been registered under the California Climate Reserve. Other benefits provided by the forest include protection of sensitive habitat and wetlands, preservation of open space, and recreation.

Additions to the ACF are planned strategically to prioritize the protection of sensitive habitat and create linkages to other public open-space areas. In most cases, the land is purchased outright, using donated money from private foundations, revenue from timber harvests, or (most commonly) grants from state and federal agencies.

While the forest is owned by the citizens of Arcata, surrounding communities also benefit. Many streams flowing through state, federal, and other local wildlife areas originate in the ACF, and the quality of those areas is therefore influenced by the forest's management policies. Revenue from the forest is used to purchase and protect sensitive wetlands, create creek-side conservation easements, and acquire new open space, all of which protect and improve the extended Humboldt Bay ecosystem. The forest protects biodiversity associated with old second-growth redwood/Douglas fir forests and provides valuable habitat for associated fish and wildlife species. Finally, the forest provides opportunities for local recreationists and serves as a scenic backdrop for the town.

Community Forests are relatively rare in the western United States, but recently there has been increasing interest and efforts to establish community forests in other areas. Arcata's city-owned model is similar to some of the town forests of New England and county forests in the upper Midwest. Common to successful models of community forestry in the United States and Canada is the establishment of a technical advisory committee or oversight board to assist the governing body. The ACF program strives to integrate ecological, economic, and social strategies where local decisions on forest management benefit the local community. Arcata's city-owned community forest and adaptive management approach is a model that can be used wherever productive timberlands interface with urban areas, and multiple goals are realized by maintaining forested open space.

# 8

# TAKING A DIFFERENT APPROACH

Forestland Management in the Redwood Region

*Mike Jani*

## IN BRIEF

- Using tools such as Forest Stewardship Council certification and a joint Habitat Conservation Plan and Natural Community Conservation Plan, the Mendocino Redwood Company created a successful business while implementing environmental stewardship on its working forestlands.
- Sustainable forestry approaches on working forestlands owned by private industry can help encourage and support sustainable forestry practices on private nonindustrial forestlands and can benefit owners economically.
- Collaborative processes are more likely to be successful in finding solutions to forest management conflicts if they focus on meeting the needs, not the wants, of stakeholders, and if participants—including the timber industry, government, and the public—listen to one another.
- Rather than simply policing the industry, the appropriate role for government in relation to private industrial forest management is to set goals rather than mandate step-by-step processes or piecemeal regulations; empower communities to solve their own problems; and be a partner that cooperates in helping forest industry figure out how to meet those goals and solve those problems.

In Mendocino County, California, forests and the timber industry have played a huge cultural role stretching back at least 160 years (see map, fig. 8.1). Jerome Ford staked a claim in 1852 for the first Mendocino sawmill at the headlands of Big River. The next 100 years produced not just dozens but hundreds

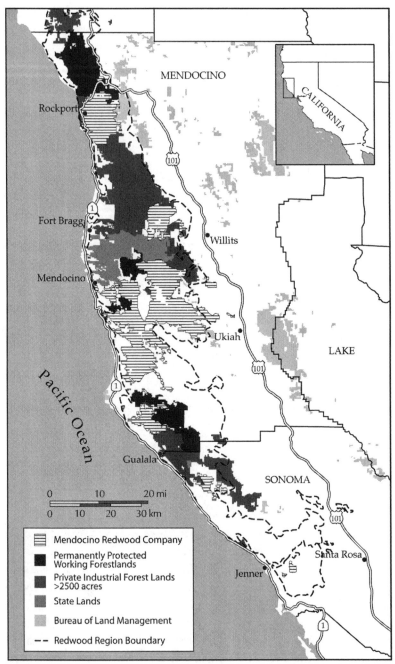

**Figure 8.1.** Mendocino Redwood Company lands in Mendocino County, California. Map by Darin Jensen and Syd Wayman, 2012.

of sawmills, small and large, across Mendocino County, where coastal red-woods and Douglas fir dominated the landscape and several timber compa-nies bought up productive timberlands. A subculture grew up around the loggers and sawmill workers who supplied wood from these industrial tim-berlands. Today, however, only two redwood sawmills remain in the county, and the entire redwood region has become a place where new models of sus-tainable forest management are being tested.

The main threat to working forestlands in Mendocino County during much of the twentieth century was the manner in which forestry was typi-cally carried out—by repeated extensive removals of higher, timber-grade old-growth timber, followed by multiple reentries in the older second-growth stands, which resulted in large expanses of industrial forestlands with very low volume inventories of conifer timber and vastly degraded habitat in both the upland and riparian forest. Simply put, timber was harvested much faster than the forest could naturally recover. Today in Mendocino County, the chal-lenge in conserving industrial forestlands as working landscapes is to figure out how to practice forestry in a manner that enhances the environment but is also economically viable and politically acceptable. This has proven to be challenging in an area where years of conflict have divided the forest com-munity. Creating a business that allows for a great deal of open, transparent dialog regarding the management of privately held timberlands has helped to preserve timber-related jobs, while at the same time put the forests' biologi-cal trajectory back on the path toward recovery.

## A HISTORY OF FOREST CONFLICT

Until about 1960, most residents of Mendocino County were connected in some way to the timber industry or to sheep ranching. The 1960s brought an influx of new homesteaders who wanted to flee the city and "get back to the land," including many from San Francisco. This relocation of people from urban to rural areas injected a completely different view of life into the Mendocino re-gion. During the 1970s and 1980s, the two subcultures—old and new—began to clash. The culmination of their unease came in a well-orchestrated series of protests and demonstrations from June to September of 1990.

Redwood Summer, as it was called, was organized by Earth First!—a radi-cal environmental action group that advocated trespassing, sabotage, and property damage and believed the mainstream environmental movement had sold out. A wide range of environmental and social justice organizations brought an estimated 3,500 activists, including many college students, to northern California. Their strategies mimicked, in some cases, the civil rights protests of the 1960s, which depended on civil disobedience and disruption. Demonstrators went onto private property to surround old-growth trees and halt bulldozers. In the end, loggers were pitted against protesters.

On one side—the side of the loggers—were those who wanted to continue a way of life that extended back through several generations, and on the other were environmentalists who wanted to save the threatened ancient redwoods from what they called Greed, Inc. Logging trucks sported bumper stickers that read "Earth First!—We will log the other planets later." The environmentalists wanted to protect the last few remaining old-growth stands that had not been set aside in parks and to slow the rate of harvest down to a point from which the forest could begin to recover. An event flier (purportedly circulated by Earth First!) shows two shadowy figures, wrenches clenched in their fists as they approach a parked road grader, and the slogan, "Go out and do something for the Earth . . . at night." There were individual extremists among both sides, and any possibility of reasoned compromise that would allow for continued timber harvesting appeared remote.

Even after the Redwood Summer protesters departed, the conflict remained. In Mendocino County, a population of well-organized, vocal, intelligent, environmentally minded individuals who knew how to apply pressure to those in power to get what they wanted turned to the courts and government agencies in Sacramento to put pressure on large timber corporations to become more accountable to the public and to the government. Louisiana-Pacific, the largest timberland owner in Mendocino County at the time and a publically traded company on the New York Stock Exchange, decided to put its land up for sale and pull out of the area. It had simply become too difficult for Louisiana-Pacific to operate in California and the company felt it could gain a higher rate of return for its shareholders by moving operations to the southeastern United States.

## A NEW BEGINNING

Mendocino Redwood Company (Mendocino Redwood) was started by the Fisher family of San Francisco, the same family that founded the GAP, a highly successful retail clothing company known worldwide. They purchased the Louisiana-Pacific timberlands—228,000 acres of primarily coastal redwoods—in July 1998. An additional asset that came with the land purchase from

Louisiana-Pacific was the Ukiah Sawmill. At the same time that the Fisher family set up Mendocino Redwood, they established a sister company—Mendocino Forest Products—to run the mill. The forest products company manufactures decking, fencing, and dimension lumber (lumber cut to standard sizes) at the Ukiah mill from logs grown on Mendocino Redwood lands and lands owned by other private forest owners in the region. Mendocino Redwood's vision for the future was to demonstrate productive forestlands management following a high standard of environmental stewardship that stakeholders in the region could support while maintaining a successful business. But along with the Louisiana-Pacific land and mills came some unwanted, but not unexpected, baggage.

Clearly, at the outset, Mendocino Redwood wished to take a new, more sustainable approach to managing the forest, but the company faced big hurdles in the form of public distrust and lack of credibility stemming from the prior owner's poor forest management. Early on, Mendocino Redwood also made a crucial decision that hurt the public image it was trying to build. California requires timberland owners to file timber harvest plans every time they decide to harvest trees. Shortly after the purchase of the Louisiana-Pacific lands, while still developing the details of a new operating plan, Mendocino Redwood decided to complete existing plans that had been submitted by Louisiana-Pacific and approved by the California Department of Forestry. By finishing these timber harvest plans, which included clear-cutting, Mendocino Redwood was able to keep the mill running and people working. However, this decision haunted the company for several years, as photos of newly clear-cut areas on what had become Mendocino Redwood land appeared in local media, on the Internet, and on banners outside GAP clothing stores in San Francisco and New York City. Needless to say, this bad press was disconcerting to the owners.

## MOVING FORWARD

For a number of reasons, Mendocino Redwood's earliest efforts therefore amounted to "social forestry," meant to address festering social issues surrounding the timber industry in Mendocino County. They recognized that, for most of the twentieth century, large corporations had dominated the scene, including Union Lumber, Boise Cascade, Georgia-Pacific, and Louisiana-Pacific. Publicly traded companies such as these generate additional capital from corporate stocks and bonds; they need to show early and consistent profits for shareholders. The business solution that had proven profitable for these companies—high-volume clear-cuts—had sparked the protests and litigation that had stemmed from the local community and environmental activists in the 1990s. The founders of Mendocino Redwood believed they had a better model. As a family-owned business, Mendocino Redwood would be in the hands of patient investors whose goals were to restore the land and increase its value steadily over time while still generating enough of a cash flow to pay the bills. To do that, the investors had to be willing to stick with the company for the long run. The Mendocino Redwood investors were committing to an effort that required not just a desire to save the redwoods but also high environmental standards dictating that the company "do it right or don't do it at all."

During the due diligence period prior to the Louisiana-Pacific land purchase, the Mendocino Redwood investors, led by company president Sandy Dean, had done their homework. Dean identified influential people locally and statewide and asked them several key questions about the social context for forestry in the region. What are the issues? What can the company expect from the public and government agencies? How will local "noise" affect Mendocino Redwood's ability to keep the business running?

Based on the answers that Dean received, it was clear that the lands Mendocino Redwood proposed to buy had been degraded and that earlier harvest rates would not be sustainable. Mendocino Redwood would need to cut fewer trees and pay attention to the entire infrastructure of a working forest. Furthermore, the company owners and investors would need to be patient. In forestry, the return on investment only comes about slowly. As the trees grow over decades, the forest improves ecologically. Similarly, as money is pumped into upgrading roads, installing new culverts, rebuilding bridges, cleaning up streams, controlling erosion, and conducting other routine maintenance, the value of the land gradually increases.

Mendocino Redwood determined to run an environmentally sound and transparent business. If, by doing so, Mendocino Redwood would have to spend less money litigating, Dean reasoned, it would have more money to invest in the real work of restoration and sustainable forestry. Dean made himself readily available to critics. He saw the stakeholders in Mendocino Redwood's success or failure as the environmental groups, neighboring landowners, loggers, government agencies that regulate forest management, local politicians, local businesses, and certainly his own employees. By 1998, all of these stakeholders had one common interest: to restore peace and eliminate controversy over timber harvesting. People's lives had been absorbed by the "timber wars" for 25 years or more and everyone was tired of fighting. They simply wanted it to stop.

Mendocino Redwood identified six key elements that needed to be addressed immediately: (1) reduce the harvest rate to a level where growth exceeded harvest, (2) eliminate clear-cutting, (3) obtain forest certification by the Forest Stewardship Council (FSC), (4) protect old growth, (5) rethink herbicide use, and (6) keep local people in the loop about what is happening with the forest. As it moved forward, Mendocino Redwood focused on two key strategies: to obtain and maintain FSC certification and to develop and gain acceptance of a Habitat Conservation Plan (HCP) by the U.S. Fish and Wildlife Service and a Natural Community Conservation Plan (NCCP) by the California Department of Fish and Game. An HCP focuses on species under the protection of the U.S. Fish and Wildlife Service and an NCCP on natural communities typically associated with those species under the protection of the California Department of Fish and Game.

## FOREST STEWARDSHIP COUNCIL CERTIFICATION REQUIRES NEW THINKING

Many models exist for building a successful business but very few for implementing successful environmental stewardship in a working forest. Mendocino Redwood wanted a credible, independent, third-party assessment of its operating plans. It turned to the Forest Stewardship Council, a global organization formed in 1993 by a coalition of environmental organizations including the Natural Resources Defense Council, World Wildlife Fund, Greenpeace and the Wilderness Society. Like forestry professionals, council stakeholders have multiple perspectives on what constitutes a well-managed and sustainable forest, but they do agree on standards that require forest management to comply with each nation's laws, respect the use rights of neighboring landowners and indigenous people, reduce the environmental impact of logging activities, maintain the ecological functions and biodiversity of the forest, and, provide operational planning and monitoring.

Certification by FSC is based on annual, systematic, on-the-ground audits and auditor engagement with local stakeholders that identify very explicit areas that the auditors must review. Knowing that the FSC trademark adds value to forest products in the marketplace, Mendocino Redwood first pursued certification in 1999, six months after the company's start-up. Auditors from FSC required that Mendocino Redwood clearly articulate its vision for the land, including management alternatives to clear-cutting, criteria for old growth, and limitations on herbicides used to control tan oak and nonnative invasive plants. After considerable work that challenged long-standing forest management practices for industrial timberlands in the redwood region, Mendocino Redwood received FSC certification in late 2000. Achieving the certification had required the company to reeducate its principals and forest managers, but the certification process ultimately validated that the new management practices would be better for the forest.

Another development that may have a big impact on Mendocino Redwood and its small neighboring landowners is group certification, now in place under the company's FSC certification. Group certification allows multiple forest owners, all managed to FSC standards, to join as a group to share certification costs. All along, Mendocino Redwood has depended in part on logs from small landowners, but for small landowners, the cost of forest certification can be an impediment. Under FSC group certification, Mendocino Redwood can purchase logs, process them, and ship the lumber to market with FSC branding, as long as participating landowners are willing to manage to the FSC standards. This helps Mendocino Redwood and Mendocino Forest Products meet the demands of their primary customer, the home-improvement retailer Home Depot, which continues to maintain a preference for FSC-certified products in its stores. It also helps small landowners, who benefit from the strategic planning involved in the certification process and receive recognition for their responsible management practices.

## DEVELOPING A SINGLE MANAGEMENT PLAN FOR MULTIPLE AGENCIES

Once Mendocino Redwood was up and running, the company began to consider which entities most influenced its operations. While environmental groups made an impact, Mendocino Redwood recognized that its business was largely controlled by government agencies. Almost every aspect of work in the forest requires a permit or compliance with a rule enforced by independent regulatory agencies.

Early on, the National Marine Fisheries Service and U.S. Fish and Wildlife suggested Mendocino Redwood develop an HCP to protect coho salmon (*Oncorhynchus kisutch*) and northern spotted owls (*Strix occidentalis caurina*) in its forestland. The agencies felt that a property-wide, holistic approach to species and habitat protection was much more beneficial to public trust resources than the rather piecemeal timber harvest plan approval process in place in California.

The HCP process was soon coupled with an NCCP. Mendocino Redwood decided to try to address all of these requirements under a single management plan that protected each agency's authority over its portion of public trust resources.

Mendocino Redwood has been developing a joint HCP/NCCP since 2002. The process has taken longer than anticipated, as there was no comprehensive template for the participants to follow. The effort has involved a major financial commitment from the company, as well as contributions from over 50 individuals, including Mendocino Redwood employees, environmental and botanical consultants, staff from the federal and state agencies, and science advisers. If successful, Mendocino Redwood's joint HCP/NCCP will be the first such plan for an industrial timberland in the state of California. Final approval is expected by the end of 2013.

Mendocino Redwood seeks an 80-year period for its HCP/NCCP because it will take that long for most of the timber stands to grow to maturity. Once the HCP/NCCP implementation agreement is signed, the prescriptions of the HCP/NCCP must be met, even if the land passes to another owner. This is as close to a guarantee as one can get that the land will continue to be developed responsibly and its biodiversity protected well into the future.

Mendocino Redwood's HCP/NCCP is comprehensive and complex. In total, 40 species or subspecies of fish, wildlife, and plants are directly affected by it. A multispecies HCP alone must provide for biodiversity; no single management scheme fits each of the 40 species. The forester has to move across the landscape and create habitat mosaics to promote biodiversity (see chap. 13 for a discussion of Pima County's proposed Multispecies Habitat Conservation Plan in Arizona).

The development of a Habitat Conservation Plan requires by law the analysis of several different management alternatives to ensure that the path chosen provides the best alternative to achieve desired outcomes while analyzing all related environmental impacts. This takes shape in an ancillary document called an Environment Impact Report. During both the development and agency approval phase of this planning exercise, the public is invited to participate: first, to provide input on what should be addressed in the development of the plan and then, once the plan is in final draft form, to comment on the adequacy of what is being proposed. Mendocino Redwood maintained respondent/participant lists to keep all interested stakeholders informed as to planning progress and the availability of various drafts for review.

Ultimately Mendocino Redwood hopes to have a publically vetted and agency-approved long-term operating plan that will provide protection for endangered species and, at the same time, act as a very comprehensive Environment Impact Report to which all its Timber Harvesting Plans (required by the state of California to commercially harvest timber) will be tiered. The company believes the programmatic approach (Programmatic Timber Environmental Impact Report) will provide it, the agencies, and members of the public who are interested in its harvesting and management activities a much clearer picture of what is happening on its lands. This should translate into expedited permitting across multiple agencies.

In developing the HCP/NCCP, the company learned five valuable lessons: (1) write in concise, clear, enforceable language; (2) test out conservation measures as they are proposed and before they become set in legal stone; (3) migrate proposed policies to the business plan as quickly as possible so that employees become familiar with them; (4) in future management activities, build on the team approach that brought the company and the agencies through the HCP/NCCP planning process; and (5) keep investors regularly updated, engaged, and supportive of the plan's progress.

## A NEW VENTURE IN SOCIALLY RESPONSIBLE FORESTRY

In 2009, the Mendocino Redwood founders also purchased 210,000 acres of land in Humboldt County from the bankrupted Pacific Lumber Company, lands which for years were the target of protesters and government regulators. Drawing on Mendocino Redwood's earlier experiences, the newly formed Humboldt Redwood Company's owners were determined to engage the passion of the environmentally minded members of the local community rather than react to demonstrations of that passion. At Humboldt Redwood, activists are now working beside foresters to identify old-growth trees to receive protection under Fish and Wildlife and FSC rules. Unlike the early experiences at Mendocino Redwood, effective, intelligible management policies were in place at Humboldt Redwood from day one, based on the guidelines developed over many years at Mendocino Redwood.

## OTHER MANAGEMENT MODELS AND CONSERVATION TOOLS
## IN THE REDWOOD REGION

### Public-Private Partnerships

The Mendocino Redwood model for promoting biodiversity across large landscapes exists alongside others in the Redwood Region. An alternate model is what the Conservation Fund, which partners with government, private foundations, corporate donors, and local communities, calls "the business of conservation." In effect, conservation organizations use public and private money as well as guaranteed loans to purchase and preserve lands. In 2004, the Conservation Fund purchased the Garcia River Forest—24,000 acres of redwood and Douglas fir (*Pseudotsuga menziesii*)—as California's first "large nonprofit-owned working forest" (see description of the Conservation Fund's Garcia River Forest in spotlight 8.1). Then, in 2007, the Bank of America Corporation made it possible for the Redwood Forest Foundation to buy 50,000 acres dominated by young Douglas fir, redwood, and tan oak (*Lithocarpus densiflorus*) in the Usal Redwood Forest. In some respects, the objective of these organizations is social redemption for the irresponsible environmental management actions of previous landowners. They welcome corporate contributions to land conservation as mitigation for the environmental impacts from private development. In addition, they see the preservation of chosen sites as undoing or counterbalancing the social irresponsibility of previous land developers and timber operators.

The question, though, is whether these working forests can remain sustainable without real work and substantial cash flow. A good part of the work of managing a forest is not just harvesting the trees but maintaining roads, replacing culverts, controlling erosion, and monitoring streams and species—all of which takes considerable cash. The steep economic downturn of the late 2000s has affected everyone in the Redwood Region. Only time will tell how these other economic and environmental models will function and survive as their needs to service debt require the harvest of trees to be sold into markets with flat or falling demand.

*Carbon Sequestration*

Another emerging model is that of the forest as first responder to global climate change. The Climate Action Reserve adopted Forest Project Protocol 3.2 in early 2010 to calculate and verify offsets to greenhouse gas emissions. In early 2013, California launched a compliance carbon market and an associated cap-and-trade system, increasing the incentive to produce and sell carbon-offset credits from forestry. The Pacific Gas and Electric Company has already agreed to purchase greenhouse gas reductions from the Garcia River Forest, amounting to 40,000 tons annually for the next five years. Other companies, including Mendocino Redwood and Humboldt Redwood, could sell carbon credits to commercial and individual customers who are interested in reducing their carbon footprint or on California's compliance market to offset the emissions of regulated entities. Certainly in the redwood region, with its large biomass available to capture and sequester carbon dioxide, this model brings new opportunity.

However, for timber companies, there is also reason to be cautious until there is a proven market that can bring a return on investment that is comparable to timber harvesting (see also chap. 14 on payments for ecosystem services). A further reason for caution is that, for forest projects, the crediting period under the Climate Action Reserve is 100 years, which means that purchased carbon in the form of added growth to trees must remain for 100 years. Nevertheless, the potential for additional revenue from forestlands is reason enough for landowners in the region to take a serious look at carbon credits, especially in light of the economic tsunami of 2007–10. Interest in the trading of forest carbon will be directly tied to future government actions at the state, national, or even international level as various mechanisms are developed, including cap-and-trade requirements for carbon dioxide emitters. The Mendocino and Humboldt redwood companies are still waiting to see how California's new carbon market will play out in terms of actually entering the fledgling market. However, they have taken all of the necessary steps toward the verification of their accrued carbon so as to be able to market credits when they feel the time is right.

*Conservation Easements*

In time, both Mendocino Redwood and Humboldt Redwood may look at the expansion of conservation easements as another conservation tool and mechanism for capturing value from their lands (some small easements have been in place since the time the properties were purchased). These require a valid assessment of the sometimes competing values that occur on all of California's timberlands. Once the company has secured the approvals for its long-term planning agreements, it will be able to assess land valuations fairly across its holdings and whether it makes sense to expand existing easements. The benefit of selling easements would be to raise additional funds to invest in forest management and stewardship. Mendocino Redwood has concerns, however, about the way in which easements on private lands are valued in California through the appraisal process, which could be a disincentive for them in pursuing this tool.

*Computer Landscape Modeling*

In the forest management business, computer landscape modeling is an invaluable tool. It allows managers to "grow" trees and habitat in a virtual reality and project what results their decisions might produce after 40 or 80 years. In a sense, this tool is also valuable to the local community and to the government agencies that regulate forest practices. By turning time into space, as it were, managers can present a graphic snapshot of what their land will look like after decades under a particular silviculture. Newer versions of this technology allow the pattern of growth and harvest to be viewed in two dimensions. In effect, managers are able to say, "You don't need to just take our word for it." However, if managers rely solely on a computer model and forget what they have learned from the forest itself, the model can be at best useless and at worst dangerous. The old rules still apply; foresters must regularly make trips to the forest to gather the sampling data on which the model bases its projections. They must use real growth data, and tree lists in the model that derive directly from the information that comes in from their timber cruisers.

## FACING THE CHALLENGES TO FOREST MANAGEMENT

People are always looking for the silver bullet that will make all things right. It rarely exists. Within the timber industry, the timber harvest plan process itself is fragmented, burdensome, and dissatisfying. A more holistic approach is required, through a planning process that focuses on whole landscapes and watersheds, not just on individual projects as timber harvest plans do. Mendocino Redwood hopes that its joint federal-state HCP/NCCP, if successful, will become a template for others to follow or even that the California legislature might adopt and modify the joint HCP/NCCP template and model as a package approach for managing private forests elsewhere in California. If Mendocino Redwood succeeds in implementing its plan, the company will likely begin the same process at Humboldt Redwood with the hope that it will be a smoother road this time.

Most owners of working forest landscapes in Mendocino and Humboldt counties want to do the right thing and probably believe they are doing just that. What often frustrates them is the lengthy and unpredictable process of getting multiple approvals from the various agencies that review timber harvest plans in the state. Twenty years ago, David Osborne and Ted Gaebler wrote *Reinventing Government* (1992), in which they enunciated principles for bringing the "entrepreneurial spirit" to government.[1] Among their suggestions were that government should steer rather than row and should empower communities to solve their own problems. In Mendocino Redwood's experience, that advice is right on the money.

Mendocino County is rich in natural resources. In the economic chain, someone has to make "the first dollar"; in Mendocino County that dollar comes from forestry and agriculture. Over the last 100 years, tax dollars from the timber industry and paychecks from its employees have ignited the economic growth that built the county's infrastructure and services. In recent years, as pressure for urban development has increased, forestry has retained a stable, green environment for Mendocino County and a skilled work force. Jobs are more important than ever in the current economic climate, and they are especially scarce on California's North Coast. A symptom of the lack of jobs in forestry and other formal sectors is the huge marijuana economy. A county-commissioned report indicates that marijuana is now the major cash crop in Mendocino County, accounting for an estimated two-thirds of the local economy and generating perhaps $1 billion annually. Securing and perpetuating forestry jobs depends on the interaction and cooperation of landowners and government. Such jobs are promoted by coordination between the timber industry, government, and the public and are threatened by conflict. In addition, there is a need to change the dynamic in which the government is policeman to one in which it is a partner. Instead of government saying what is to be done step-by-step, it should communicate goals, which are to be worked toward cooperatively.

To manage a commercial forest in California, one cannot be myopic. Everything must be considered—government regulations, environmentalist demands, the political will of legislators, market signals, and the curve balls that nature throws. While it may be extremely difficult for forestry enterprises to juggle these various and competing interests now, the challenge will likely increase as more people move into California. Still, it is important that they understand the dynamic of social engagement and be willing to participate with other stakeholders. One key element of negotiation is identifying what each party at the table needs, not what they want. What do the landowners, government regulators, environmentalists, and local interest groups need to work out a solution? After posing the question, you have to do one more thing. Listen.

## NOTES

1. D. Osborne and T. Gaebler, *Reinventing Government: How the Entrepreneurial Spirit Is Transforming the Public Sector* (Reading, MA: Addison-Wesley Publishing Co., 1992).

# Spotlight 8.1

# THE CONSERVATION FUND'S GARCIA RIVER FOREST, CALIFORNIA

*Chris Kelly*

IN BRIEF

· Restoring degraded timberlands to productivity following a history of overcutting is economically challenging because forest management and restoration are costly and because it takes time for trees to grow larger and improve in quality; this requires patience on the part of investors.

· The sale of carbon offsets from improved forest management on voluntary or compliance carbon markets can be an additional source of income for private forest owners who own large tracts of forestlands having high carbon sequestration potential.

The Garcia River Forest, about 100 miles north of San Francisco in Mendocino County, was the first large working forest in California to be owned by a non-profit organization: the Conservation Fund (TCF). The Conservation Fund is a national organization that engages in land conservation projects that balance ecological and economic goals and address the priorities of local partners. The fund purchased the 24,000-acre Garcia River Forest in 2004 as part of its California North Coast Forest Conservation Initiative, which aims to protect and restore coastal forests in the redwood region of California.[1] The strategy is to purchase large tracts of cutover forestlands that are threatened by conversion to other land uses, such as rural residential subdivision and vineyards, and use sustainable forest management to enhance future forest productivity, generate revenue to cover the cost of management and habitat restoration, and contribute to the local economy. The Garcia River Forest is

a predominantly Douglas fir (*Pseudotsuga menziesii*) and redwood ecosystem
that provides important habitat for species such as coho salmon (*Oncorhynchus kisutch*), steelhead trout (*Oncorhynchus mykiss*), and northern spotted
owls (*Strix occidentalis caurina*). In addition, it has high carbon sequestration
potential: redwood trees store more carbon per acre than any other forest
type in the United States. The forest is managed with frequent input from
local community stakeholders and public agencies.

The Conservation Fund purchased the Garcia River Forest using innovative financing that consisted of a combination of a low-interest loan from the
David and Lucile Packard Foundation, grants from state agencies (the California State Coastal Conservancy and State Wildlife Conservation Board), the
sale of a conservation easement to the Nature Conservancy, and donations
from TCF supporters. The Nature Conservancy is an active partner in the
project, providing conservation planning and restoration expertise, among
other services. Revenue generated through sustainable timber harvests and
the sale of verified carbon offsets covers most of the costs of property taxes,
on-site maintenance, forest management, restoration projects, and loan repayment. The high cost of managing a large tract of timberland combined
with the economic downturn that began in 2007 have posed economic challenges, however.

Timber harvesting has occurred on the Garcia River Forest for more than
100 years. Prior to purchase by TCF, the forest was owned by a series of industrial timber companies that managed it for intensive, commercial timber production. This management significantly reduced the volume of merchantable timber, caused severe soil erosion, and contributed to excessive
stream sedimentation in the Garcia River watershed. As a result, at the time
of purchase by TCF, the forest was dominated by young (30–50-year-old),
densely stocked stands of trees. In contrast to past management, TCF's focus is on increasing stand volumes and forest productivity to benefit wildlife
habitat and increase carbon sequestration, maintaining and enhancing ecological processes, and engaging in sustainable timber production to provide
jobs. A comprehensive management plan developed in collaboration with
community stakeholders and agency partners guides management. This collaborative process has been important for building trust, especially among
members of the local environmental community, who had concerns that
managing the forest for timber production could repeat the problems created by historic industrial forest management.

About 35 percent of the Garcia River Forest has been set aside as an Ecological Reserve, where ecological objectives drive management. Timber harvesting in the reserve takes place only where appropriate to help achieve ecological objectives. "Light-touch" single-tree and small-group selection timber management is used on the remaining 65 percent of the landscape. Silvicultural treatments emphasize uneven aged management to promote a variety of tree sizes and ages within stands. Thinning of the forest's young, dense stands through selective harvesting is intended to promote the growth of larger, high-quality trees. The thinned material is sold as timber to local mills. Eventually, forest management will focus on the selective harvesting of larger, high-value trees for sawtimber. This approach is one of "patient management"; it will take years to restore the forest to desired conditions and generate higher revenues from the harvest of large-diameter, high-quality trees. Forestry operations at the Garcia River Forest have been certified by the Forest Stewardship Council and the Sustainable Forestry Initiative.

The Garcia River Forest is also the first large forest in California to produce third-party verified carbon offsets for the voluntary carbon market to help mitigate global climate change. Carbon offsets are produced through improved forest management techniques that increase forest carbon sequestration—such as thinning, reduced harvesting, and allowing trees to get bigger and older so that they store more carbon. Since the project's commencement in 2004, an average of approximately 190,000 tons of carbon emissions have been offset annually. Carbon offsets from the forest have been verified to Climate Action Reserve (CAR) standards and are registered with the reserve. The Conservation Fund has been selling them over-the-counter to buyers throughout the United States. A main buyer of these offsets for voluntary purposes has been the Pacific Gas and Electric Company, whose customers could elect to offset their energy consumption through its ClimateSmart Program (though the program was discontinued in 2012). In 2011, California adopted final rules to regulate greenhouse-gas emissions at the state level. These rules will allow holders to use Garcia River Forest offsets to meet a portion of their emission reduction obligations. The Conservation Fund estimates that the sale of carbon offsets has more than doubled its net annual revenues from the forest. With the emergence of a compliance market for offsets, demand and prices are expected to rise, further increasing project revenues.

For the next several years, TCF will continue to manage the Garcia River Forest to mitigate climate change, protect water quality, restore wildlife habitat and contribute to the local economy. Eventually, TCF expects to sell the property to a private owner subject to the conservation easement and reinvest the proceeds in other conservation efforts in the region. The Conservation Fund believes that the Garcia River Forest project represents an innovative and cost-effective approach to conserving large tracts of threatened forestland that complements good forest management measures undertaken by other forest landowners in the region.

## NOTES

1. The Garcia River Forest Integrated Resource Management Plan, the North Coast Forest Conservation Program Policy Digest, and other documents detailing TCF's approach to forest conservation and management can be found at www .conservationfund.org/our-conservation-strategy/focus-areas/forestry/north -coast-conservation-initiative/.

# 9

# STEWARDSHIP CONTRACTING IN THE
# SIUSLAW NATIONAL FOREST

*Shiloh Sundstrom and Johnny Sundstrom*

## IN BRIEF

· Stewardship contracting is an innovative and promising administrative
tool that can be used for working landscape conservation on Forest Service
and Bureau of Land Management lands to accomplishing comprehensive,
watershed-scale environmental restoration across land ownership boundaries, allowing for private and public lands comanagement.
· As the mission and management focus of the Forest Service and Bureau
of Land Management shift, new tools may be needed to accomplish forest
management work; these will require internal agency champions and staff
willing to innovate and take risks, along with supportive external partners.
· Stewardship contracting could provide even broader benefits to communities and forests if retained receipts from the sale of forest products could
also be used to fund economic development projects in local communities
that benefit and relate to the needs of federal forest management.
· An important factor in maintaining working forests on federal lands is
maintaining the local forestry infrastructure needed to support them,
including equipment, business capacity, and a skilled workforce that can
respond to the needs and opportunities associated with federal forest management.

The past two decades have brought great changes in the mission and focus of
the U.S. Forest Service and federal land management agencies in general. Not
long ago, the primary activities of Forest Service personnel, and the bulk of

its appropriated budget and income, dealt directly with timber harvest and management, road building and maintenance, and reforestation. National Forest lands in the Pacific Northwest were managed largely as working forests and produced more than one-third of the region's timber.[1] In western Oregon alone, federal forests managed by both the Forest Service and the Bureau of Land Management (BLM) produced, on average, close to 50 percent of the region's timber in the 30 years prior to 1994.[2] Many forest community residents worked as loggers, truck drivers, and mill workers—harvesting, transporting and manufacturing wood and wood products—or directly for the agency.

With the changing global marketplace, the invention of new building materials, and increasing concern for old-growth forests and dependent species, the mission of the Forest Service has shifted.[3] Priorities became the conservation of biodiversity, with a primary focus on species protection in accordance with the Endangered Species Act, as well as the restoration of natural conditions and functions, and providing opportunities for the public to enjoy and access Forest Service lands. Concern over the environmental effects of timber harvesting on the region's forestlands resulted in the listing of the northern spotted owl (*Strix occidentalis caurina*) as "threatened" under the Endangered Species Act and a series of lawsuits that effectively shut down or greatly reduced timber harvesting on federal forestlands in western Washington, western Oregon, and northwestern California.[4]

What emerged in 1994 was the Northwest Forest Plan, a federal forest management plan that was an attempt early in the Clinton administration to resolve the conflict among traditional industrial interests, their community supporters, and environmental advocates or litigious parties. Since 1994, the Northwest Forest Plan has been in effect on 24 million acres of Forest Service and BLM lands in the Pacific Northwest. It was promoted as a compromise that would support local economies by balancing forest protection with a sustained timber harvest, but it has never been fully endorsed by either the timber industry or environmental interests.

These transformations significantly reduced timber harvesting on U.S. Forest Service lands and the associated customary timber income of the Forest Service, causing a rapid downsizing of government capacity. They also resulted in a restructured agency workforce, both through reductions in the number of agency employees and the recruitment of different worker specializations and expertise to replace the previous harvest-dominant labor profile of the Pacific Northwest Region. Additional effects included the substantial loss of forestry-related jobs in nearby rural communities due to declines in timber harvest on both public and adjacent private lands, causing forestry workers to commute farther distances, move away, retire, change professions, or become unemployed.[5]

Communities such as Mapleton-Deadwood in western Oregon's Siuslaw National Forest (SNF) were devastated economically and psychologically when their commercial foundation suddenly seemed to have been pulled out from under them. Although the historic boom and bust cycles of the timber economy were an integral part of this way of life, nothing had ever so directly threatened the existence of the severely impacted timber-dependent towns and populations living throughout the Pacific Northwest's coastal forest-lands. Projected consequences related to the 80 percent decline in harvest and revenues from federal lands produced a variety of responses among these communities, including business closures and severe reductions in school attendance, income, and community support.

On the positive side, there was a gradual emergence of collaborative private and government attempts to deal with the impacts of this nearly catastrophic upheaval and its all-pervasive impacts. In 1994, the newly formed Siuslaw Watershed Council became a forum for these efforts in the Siuslaw River basin, as partners such as the Forest Service, Lane County, and the locally based not-for-profit Siuslaw Institute, struggled to salvage their future with an ambitious agenda of restoration and stewardship designed to involve all sectors of the affected communities and their residents.

Through this collaborative forum, members of the community and their Forest Service counterparts, including then Siuslaw National Forest supervisor Jim Furnish and his staff, developed a broad-based vision focused on transforming their community and National Forest from one that prioritized "getting out the cut" to one that prioritized "restoration with productivity as a result."[6] Among other things, this vision—known as the Siuslaw Option—promoted the revolutionary concept of keeping funds from selected timber sales at the producing forest's level and using those receipts to fund the restoration of aquatic and terrestrial resources, such as habitat for the endangered northern spotted owl and the soon to be listed as threatened coho salmon (*Oncorhynchus kisutch*).[7] The SNF and their partners sought and received designation as one of the first stewardship-contracting pilot project areas in the country.[8]

## STEWARDSHIP CONTRACTING

Stewardship contracting is a management authority that allows national forests and BLM districts to combine several timber sale and service provisions into a single contract. The same authority provides for the use of stewardship agreements, described with an example from the Weaverville Community Forest in spotlight 9.1. Stewardship contracting and its authority originated from concepts advanced by rural forest community leaders, legislators, and Forest Service officials who recognized that processes and procedures held over from the era of industrial timber production limited their efforts to restore forestlands and improve weakened community economies. Stewardship contracting was designed to foster comprehensive forest and rangeland restoration, build closer working relations between federal agencies and communities, and contribute to economic growth and sustainable development in these local and rural communities.

These contracts are significantly different from traditional timber sale contracts (used for commercial logging of federal timber) and service contracts (used to acquire goods and services). Stewardship contracting emphasizes forest restoration over generating federal income by awarding contracts that can be based on "best value," a criterion that allows for the consideration of past performance and benefit to the local community as well as bid price. This provision contrasts with traditional timber sale and service-contracting practices that could only consider the bid price and did not allow preference for utilizing the local workforce and businesses when making awards. Stewardship contracts also allow for the exchange of goods (forest products removed during a project) for services (precommercial thinning, road maintenance, habitat improvements, hazardous fuels reduction, etc.) by using one instrument in which the value of goods offsets the cost of providing services.[9] The offering national forest retains any excess funds produced by the sale of forest products in stewardship projects to be used for the implementation of additional restoration activities, both within forest boundaries and on nearby private lands. These accumulated funds are referred to as "retained receipts" because they remain on the forest where they were generated rather than being deposited in the U.S. Treasury. The now-permanent Wyden Authority (also known as the Watershed Restoration and Enhancement Agreement Authority, first enacted in 1998) makes it possible for the federal land management agencies to spend both federal appropriations and retained receipts on private lands restoration so long as these activities provide tangible benefits to public lands.[10] This is an efficient technique that benefits both the agency and private landowners.

Additionally, stewardship contracts can be designed and awarded for up to 10 years to accomplish long-term restoration goals and to ensure a longer-term supply of forest products. It is hoped that over time these contracts, in turn, will stimulate investment in value-added manufacturing and utilization businesses. Although the legislation authorizing stewardship contracting does not require collaboration for stewardship contracts, the Forest Service and BLM have been directed by the secretaries of agriculture and interior to engage "states, counties, local communities, and interested stakeholders in a public process to provide input on implementation of stewardship contracting projects" and to "make an effort to involve a variety of local interests and engage key stakeholders in collaboration throughout the life of the project, from project design through implementation and monitoring."[11] This direction provides encouragement for the agencies to engage in collaborative conservation, another departure from the traditional way of doing business.

Between 1999 and 2002, Congress authorized the Forest Service to establish 28 stewardship-contracting five-year pilot projects. These projects took an experimental approach to combining new and old contracting authorities to promote more effective ecosystem management and to better meet the needs of rural communities. The pilot authority also required the Forest Service to establish a multiparty monitoring process to evaluate the pilots' stewardship projects and to report on their accomplishments. In 2003, this pilot authority was reviewed by Congress, and the Forest Service and BLM were given a revised statutory 10-year "permanent" authority to contract with private or other public entities for services that would achieve land management goals and ideally meet local community needs (Public Law 108-7).[12] These land management goals include, but are not limited to:

1. road and trail maintenance, or obliteration, to restore or maintain water quality by reducing sediment run-off;
2. soil productivity, habitat preservation for wildlife and fisheries, or other resource enhancement values, such as forest health and species recovery;
3. setting of prescribed fires to improve the composition, structure, condition, and health of forest stands or to improve wildlife habitat;
4. vegetation removal by harvest or other activities to promote healthy forest stands, reduce fire hazards, or achieve other land management objectives;
5. watershed-scale restoration and maintenance;
6. wildlife and fish habitat restoration and maintenance; and control of noxious and invasive species, and reestablishment of native plant species.

Between fiscal years 1999 and 2011, 1,082 stewardship contracts were awarded on National Forest lands nationwide, for a total of nearly 659,000 acres re-

ceiving land management treatments using this instrument.[13] Although the
use of stewardship contracts has been steadily increasing, their use has not
yet become widespread, and many forests have been slow to adopt this tool as
part of their management strategy.

## APPLICATIONS OF STEWARDSHIP CONTRACTING
## ON THE SIUSLAW NATIONAL FOREST

Federal lands in the Pacific Northwest now play a greatly reduced role in timber production, and public lands communities have diversified their economies as the timber industry and wood products infrastructure has declined
and shifted toward private lands harvests.[14] According to the original authors
of the Northwest Forest Plan, the plan "has been more successful in stopping actions thought to be harmful to conservation of late-successional and
old-growth forests and aquatic systems than it has been in promoting active
restoration and adaptive management and in implementing economic and
social policies set out under the plan."[15] It is within this context that the story
of stewardship contracting and the SNF and its adjacent and surrounding
communities in Oregon takes place (see map, fig. 9.1).

The Siuslaw watershed is centrally located in Oregon's Coast Range and
comprises a large part of the SNF. It covers an approximate area of 505,000
acres (773 square miles) and includes valleys and gentle slopes in the eastern
part of the watershed, steep and sharply bisected terrain in the Coast Range
Mountains, and the dunes, broader floodplain, and wetlands of the coast and
its estuary. The climate is mild and rainy, with up to 130 inches of annual pre-
cipitation west of the Coast Range crest, creating some of the most productive timber-growing lands in the world. Fast growing conifers, such as Douglas fir (*Pseudotsuga menziesii*), western red cedar (*Thuja plicata*), and western
hemlock (*Tsuga heterophylla*), have historically covered much of the basin.
Clear-cut timber harvesting significantly reduced the population of older
trees in the watershed on both government and private land, and much of the
remaining forest consists of younger stands of Douglas fir created by regeneration plantings; shifts in forest age and makeup are also a result of severe
wildfires that occurred in the mid-nineteenth and early twentieth centuries.[16]

Over half of the land in this basin is publicly owned, including about 25 percent owned by the Forest Service (a large portion of that in the Coast Range
Mountains), 26 percent by BLM, and 5 percent by the state of Oregon. This
leaves a little over 40 percent of the land in private ownership, consisting

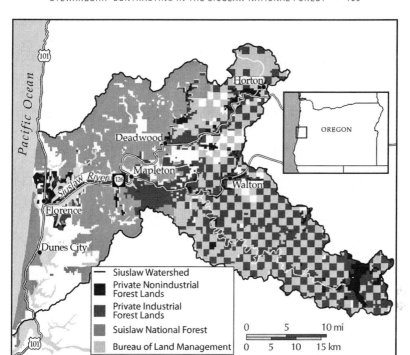

**Figure 9.1.** Suislaw Watershed including land ownership. Map by Darin Jensen and Syd Wayman, 2013.

mostly of a mix of industrial timberlands (31 percent) and smaller holdings (10 percent) of nonindustrial forestland, homes, farms, and small towns scattered throughout its valleys and floodplains.[17] The Siuslaw National Forest once harvested the largest volume of forest products per acre on federal lands in the nation.[18] Dimensional lumber and plywood veneer production supplied a national demand for the finest in straight-grain softwood products during the relatively strong housing and construction market, post–World War II through 1990. Between 1960 and 1990, the Mapleton Ranger District of the SNF alone generated two billion board feet of harvested timber worth millions of dollars, helping to feed the nation's rapidly increasing appetite for such prime materials. Very little of those profits benefited the local area as it was primarily income earned by national corporate entities.[19]

In 1988, an injunction was handed down to stop much of the harvest in the SNF due to landslides related to harvesting in the headwaters. This was followed by an owl-related lawsuit and resulting injunction against further timber harvesting in 1991, and then came the implementation of the Northwest Forest Plan in 1994. Since the plan was implemented, the majority of the SNF has been allocated for the protection and long-term restoration of old-growth habitat for threatened and endangered species in the category of "late-successional reserves" (i.e., long-term development of spotted owl habitat based on improving old-growth forest characteristics, and watershed and aquatic health). However, regeneration harvest (or clear-cutting) and replanting from the 1950s through the 1980s had left a legacy of overstocked and dense young plantations in serious need of both precommercial and commercial thinning treatments. Planting up to 400 trees per acre—in anticipation of rather high mortality that did not occur and with the expectation that these sites would be clear-cut again at age 70—actually created plantations that were fire and disease-prone liabilities with significantly suppressed growth rates. This phenomenon exacerbated the suite of emerging forest health problems from loss of old-growth habitat and excessive road networks. Any hope the SNF had of meeting the goals of the Northwest Forest Plan called for immediate and sustained management attention. The legacy of 40 years of clear-cutting had occurred over about 30 percent of the forest landscape. Although the Northwest Forest Plan allowed for thinning of forest stands under 80 years old and within late-successional reserves in forests with historical fire regimes of infrequent, stand-replacing fires, this active management had been slow to occur across the entire plan area.[20]

In 2000, the Siuslaw National Forest was selected by the Forest Service's Washington Office to participate in the pilot program to test stewardship contracting. The forest was in an advantageous position to utilize the new authority because of its fast-growing characteristics and commercially desirable size of the trees on its overstocked plantations and because of the absence of litigation—evidence of social and agency unanimity on the need for thinning. Above-cost income from thinning projects could be counted on as "retained receipts" for the SNF (see previous section). Cooperating entities already existed to assist the SNF in determining how best to spend these "new moneys" to accomplish extensive restoration actions. For the pilot study, the SNF focused its stewardship efforts in the Siuslaw watershed, utilizing a whole-watershed approach that included both federal and private forest lands and an overall restoration strategy.

This transformation was instituted at the turn of the twenty-first century, following nearly a decade of greatly reduced timber harvests in the SNF. It is important to point out that the successful transition to the stewardship process adopted by the SNF was due in large part to the enthusiasm and expertise of agency personnel at all levels, from the supervisor's office to field personnel. Although turnover in some of these positions sometimes slowed the adoption and application of the authority, there was a nearly universal acceptance of its potential among SNF staff. This enthusiasm for innovation and the willingness of local personnel to take risks gave this effort the impetus that led to its rapid inclusion in the business practices and implementation toolbox of the SNF.

In addition, over a decade of previous collaborative restoration efforts between the Forest Service and local partners like the Siuslaw Soil and Water Conservation District, the Siuslaw Institute, and the local watershed council led to formation of the Siuslaw Stewardship Group (SSG) in 2002. The SSG was specifically associated with this new stewardship authority. Initial representation came from local government, nonprofit organizations, commercial timber interests, private landowners, and regional environmental organizations, with a professional facilitator hired by the Forest Service. The SSG's role was to assist the agency in complying with the authority's mandate for collaboration in stewardship-contracting activities by advising the Forest Service on its stewardship projects and recommending other restoration projects in the watershed that could be funded through a process utilizing the SNF's retained receipts. The group also functioned as the local multiparty monitoring team for the pilot project and has evolved to become a vital partner for mutual education, resource pooling, and overall cooperative work toward watershed and community health and vigor. In 2004, the Siuslaw Basin Partnership was the winner of the annual International Thiess Riverprize given out each year by the Australia-based International River Foundation. The award highlighted the partnership's whole-watershed restoration strategy and its innovative and successful partnering in implementing that strategy.

Since stewardship contracting allows the local forest to retain the receipts from the sale of timber and associated forest products, the Forest Service and the SSG created the Siuslaw Stewardship Fund to oversee the use of these receipts. An agreement between the Forest Service and the SSG was forged that allocates up to 60 percent of the funds generated by stewardship activities for reinvestment in public lands restoration, while 40 percent are dedicated to pay for restoration on private lands through the Wyden Authority. (Note that the Siuslaw effort was the first to test the application of this authority to stewardship-contracting activities and that this required a favorable judgment by the solicitor generals of both the U.S. Department of Agriculture and the U.S. Department of the Interior.) Although the Forest Service is not bound to follow the SSG's recommendations, most participants feel that their input has been highly valued and that the effort is built on trust and commitment to a shared vision within a mutually beneficial process.[21]

Working collaboratively with the SSG, the SNF conducted silvicultural treatments to improve forest health on approximately 2,000 acres of forestland and sold almost 50 million board feet of timber between 2002 and 2007.[22] Initial applications of the Wyden Authority from 2003 to 2005 resulted in approximately $216,000 of retained receipts being awarded to 13 Wyden Amendment projects on adjacent private and nonfederal lands.[23]

In 2005 and 2006, building on the success of the SSG, two new stewardship groups were formed in other parts of the SNF: the Alsea Stewardship Group and the Mary's Peak Stewardship Group (in 2012, a similar group was formed in the Hebo District and now 90 percent of the SNF has citizen-based collaboration in place). From 2006 through 2009, over 1,500 acres of late-successional reserve was commercially thinned, resulting in a harvest of over 19 million board feet of timber. During this time period, $2.9 million of retained receipts was also spent on restoration projects on Forest Service lands and adjacent private and nonfederal lands.[24] Accomplishments for 2007 are presented here as an example of the diverse types of projects completed using this authority across the SNF and adjacent private lands:

### Forest Service Stewardship Contracts

Accomplishments of four ongoing and completed stewardship contracts as of September 2007.[25]

· Thinning treatments on 670 acres of late-successional reserve lands
· Noxious weed control over 13 acres
· Brushing and maintenance work along three miles of trail
· Closure, storm proofing, and maintenance along 27 miles of road
· Sidecast pullback, as needed, on nearly 1,200 miles of road
· Noncommercial thinning on 140 upland and 41 riparian acres
· Upland planting on 75 acres
· Riparian conifer planting and maintenance on 25 acres
· Structure placement by helicopter on three miles of stream
· Value of commercial timber: $967,000
*Total retained receipts: $726,000*

### Retained Receipts Projects on Forest Service Land

A total of 10 restoration projects implemented.

· Planting and maintenance of trees and native grasses on 33 acres
· Noxious weed control over approximately 100 acres

- Placement of large woody debris along one mile of stream
- Reconstruction or decommissioning of approximately nine miles of road
*Stewardship receipts contribution:* $436,000

## Wyden Amendment Projects on Private Land and Other Nonfederal Lands

Accomplishments of 12 Wyden Amendment projects as of September 2006.

- Planting and maintenance of 17,000 conifers, 5,500 deciduous trees, and 10,000 other native plants on private lands
- Placement of large woody debris along more than three miles of stream
- Construction of more than 3,000 feet of fence to exclude cattle from riparian areas
*Stewardship fund contribution:* $291,000

Another benefit of these retained receipts projects is that much of the project funding on private lands is matched by leveraged sources of additional funds. For example, in 2007 nearly $800,000 of retained receipts was spent on more than 30 projects on Forest Service lands and on adjacent private, nonfederal lands using the Wyden Authority. When matched with other funding sources associated with these projects, a total of $1.7 million was invested in the watershed for aquatic, riparian, and terrestrial restoration. In total, up through 2011, 19 stewardship contracts had been awarded for the entire SNF, totaling over 7,000 acres of commercial thinning.[26]

### BENEFITS AND CHALLENGES OF STEWARDSHIP CONTRACTING

As part of the facilitation and monitoring process paid for by the Forest Service appropriated funds, participants in the Siuslaw stewardship projects—including Forest Service employees, community members, environmental organizations, and contractors—were interviewed by researchers about the outcomes, successes, and challenges associated with stewardship-contracting projects generally and the SSG specifically.[27] All participants valued not only the completion of important thinning projects on the SNF but also the new flexibility in directing restoration work to where it was most needed across the watershed, including aquatic areas not associated with traditional timber-harvest-related restoration work and upland areas not previously addressed in traditional watershed restoration projects.

Contractors appreciate the flexibility of stewardship contracts that allows them to obtain desirable timber, make a profit, and keep workers employed while also protecting and enhancing wildlife habitat. Forest Service employees and their local collaborative partners value the transition to a more cooperative land management approach. Local community members feel that the Forest Service respects their opinions, including those on the subject of how funds are used to conduct restoration in the watershed. Through collaboration, partners learn from each other and have developed new approaches to achieving objectives, while building a stronger, more trusting relationship with the Forest Service.

The transition, however, was not without obstacles that had to be overcome in the quest for solutions. Community and commercial engagement in the stewardship-contracting process needed to "grow" by demonstrating results, generating local income, and utilizing the full extent of the powers accorded by the enabling legislation. One of the early tests of what was possible with these powers came in the application of the Wyden Authority. As mentioned before, this issue was resolved at the level of BLM's and the Forest Service's national legal apparatus and culminated in a successful request for the Wyden Authority to be interpreted as giving the authority to use retained receipts (not just appropriated dollars) on restoration. This restoration would occur within certain prescribed limitations on private lands near and related to the same forest where the receipts were generated and would address species and ecological problems encountered within the "green line" as well as on nearby private lands. Stewardship contracting was thereby proven, as illustrated here, to have the ability to create restoration benefits for local private land ownerships as well as public lands.

Additional tests of stewardship contracting, and ones that developed into significant opportunities to utilize this new authority, resulted from further interpretation of the Northwest Forest Plan and the endangered and threatened species listings of the northern spotted owl, coastal coho salmon, and marbled murrelet (*Brachyramphus marmoratus*). The embedded requirement that stewardship contracting focus on forest health and restoration, the need to "bundle" various restoration practices in a single geographic and temporal process, and the urgency for keeping up with the need to provide mill-ready materials gave the SNF the motivation to proceed as quickly as possible to resolve questions regarding this new business model. Significant amounts of time and effort were invested in an ongoing dialogue with the regional U.S. Forest Service office, as well as with other experts in this field, that was designed to improve the uses and the results of the authority.

The new stewardship-contracting authority was readily embraced by local communities anxious to receive retained forest receipts for application to the backlog of much-needed restoration projects, and it was further supported by the existing climate of cooperation, rather than litigation. The environmental group Oregon Natural Resources Council (now Oregon Wild) began cooperating with the SNF in designating stewardship thinning sales, and the Siuslaw Institute provided leadership in representing the local communities and their interests. Willing contractors learned to bid for and manage projects governed by the new procedures. Georgia-Pacific Corporation, in particular, was significant in starting the process, as they had mills and the business appetite for timber products but no timberlands of their own. The need to supply the remaining mills in the area, and the industry's recent shift from processing larger logs to smaller-diameter timber, made this new process a great fit and fostered the successful partnerships among the Forest Service, its contractors, and the SSG.

Stewardship contracting is not uniformly viable across the western landscape, however, nor is it always possible to generate retained receipts to the offering national forest; in many instances, the sale of forest products does not cover the cost of financing restoration-related service work. Fuels reduction projects without a market for the associated product (e.g., small-diameter trees being used by a local biomass power plant or mill) are a good example of this imbalance between cost and income. Nevertheless, the benefits of bundling several restoration activities in a single project, minimizing the number and frequency of entry impacts, utilizing the local workforce, using authorities like best-value contracting, and enlisting the participation of stakeholders in the multiparty advisory and monitoring processes has shown that stewardship contracting is often a useful and desirable alternative to traditional timber sale and service-contracting models. The cost of using the agency's appropriated funds to accomplish work using stewardship contracts will not normally exceed the costs of doing the work using other instruments.

## TWENTY-FIRST CENTURY MOVE TOWARD COOPERATIVE COLLABORATION

As the twentieth century ended and the Forest Service looked forward to entering its own second century, pockets of innovation, spurred by a sense of the economic and environmental urgency of forest restoration, appeared in various places across the national landscape—and even within the U.S. Congress. Public lands communities adjacent to or interspersed with federal lands began to view their role as more than just supplying a workforce for the Forest Service's commercial offerings. A refreshing vision of management that moved beyond jurisdictional boundaries has motivated new exercises in what was being called collaborative engagement, cooperative conservation, or shared stewardship, among other labels. Landscapes are being looked at in a far more holistic manner.

The values and value of stewardship and restoration came to define this new approach. Individuals within both the Forest Service and communities adjacent to national forests began to understand that no one party or agency has all the answers to the problems of landscape-scale management, responsibility, and oversight, and they have begun to address these challenges collaboratively. New business approaches have been developed as conceptual solutions to emerging challenges, and the spirit of innovation has taken hold wherever willing partners and agency personnel have found and engaged in formats for working together. Community-based organizations in many areas have stepped up to lead in this transformation.

The health and vitality of rural public lands communities is essential for cooperative management, such as that exemplified by stewardship contracting. The use of retained receipts depends on restoration project implementation capacity that functions across jurisdictional boundaries. Therefore, an important factor in maintaining federal working forests, sometimes overlooked, is the need to maintain local forestry infrastructure—the equipment, business capacity, and trained workforce capable of responding to the needs and opportunities of federal land management.

As the application and interpretation of the stewardship authority evolved with input both from above (U.S. Forest Service and Congress) and below (on-the-ground discoveries about best practices for land restoration and maintenance), the Siuslaw National Forest, its communities, and local government partners took leadership roles in both utilizing and improving the authority. Yet there remains a great deal of potential for evolving increased usefulness. One part of the stewardship-contracting authority that was originally allowed, but quickly withdrawn due to political pressure, was the use of funds in local communities for economic development and other projects related to the national forest's needs and products. This kind of authority, were it to be reinstated, could be applied to everything from firefighting training and equipment to biomass utilization facilities and natural resources education for workers and students.

The federal Rural Revitalization Act of 1990 grants permanent authority to the Forest Service to provide assistance to rural communities that are located in or near National Forest lands and are likely to be economically disadvantaged by federal or private-sector land management practices (such as Endangered Species Act procedures and the Northwest Forest Plan). The purposes of such assistance include providing aid to diversify communities' economic bases and to improve the economic, social, and environmental well-being of rural America. There is no systemic reason why stewardship contracting could not be adapted to this expanded use, and it would truly aid in building a more unified and comprehensive approach to both watershed health and community well-being. That said, it is still important to emphasize that cooperative conservation and productivity have both grown with increased use of the stewardship-contracting tool, and such organizations as the Siuslaw Institute have taken on more responsibility and have become more widely recognized for their role in advancing the beneficial management of the federal lands and nearby private landholdings.

During the past decade, many tours and workshops have been held in the Siuslaw, both on-site and at regional venues, to explain and inspire other national forests, BLM districts, communities, and contractors to utilize the stewardship-contracting authority and to mobilize greater support for its evolution and application. Attention is now being paid to the 2013 expiration date of the "permanent authority" and the need to ensure its needed reauthorization in Congress. While cooperative stewardship and the whole-watershed approach to planning, management, and monitoring are still in their early stages, the story of stewardship contracting in the Siuslaw National Forest is an instructive example in the new reality and potential of public and private lands comanagement. This example has been notably successful at modeling this new approach to working forestry on federal lands, protecting and utilizing some of our nation's most valuable resources, and contributing to the vitality of our nation's rural communities.

## NOTES

1. R. W. Haynes and E. Grinspoon, "The Socioeconomic Implications of the Northwest Forest Plan," in *Northwest Forest Plan—the First Ten Years (1994–2003): Synthesis of Monitoring and Research Results*, General Technical Report PNW-GTR-651, technical eds. R. W. Haynes, B. T. Bormann, D. C. Lee, and J. R. Martin (Portland, OR: USDA Forest Service, Pacific Northwest Research Station, 2006), 59–82.
2. J. W. Thomas, J. F. Franklin, J. Gordon, and K. N. Johnson, "The Northwest Forest Plan: Origins, Components, Implementation Experience, and Suggestions for Change" *Conservation Biology* 20, no. 2 (2006): 227–87.
3. S. Charnley, E. M. Donaghue, S. Stuart, C. Dillingham, L. P. Buttolph, W. Kay, R. J. McLain, C. Moseley, R. H. Phillips, and L. Tobe, *Northwest Forest Plan—the First*

*Ten Years (1994–2003): Socioeconomic Monitoring Results*, vol. 1, *Key Findings*, General Technical Report PNW-GTR-649 (Portland, OR: USDA Forest Service, Pacific Northwest Research Station, 2006).

4. Thomas et al., "Northwest Forest Plan."

5. Charnley et al., *Northwest Forest Plan*.

6. Ecotrust and Resource Innovations, *Redefining Stewardship: Public Lands and Rural Communities in the Pacific Northwest* (Portland, OR: Economic and Resource Innovations, 2008).

7. Ibid.

8. Jim Furnish, phone conversation with authors, 2013.

9. U.S. Department of Agriculture, "Stewardship Contracting," chap. 60 of *Renewable Resources Handbook*, Forest Service Resources Handbook 2409.19 (Washington, DC: USDA, Forest Service, 2008), sec. 60.5.

10. Public Law 105–277, Section 323, first amended and reauthorized by Public Law 109–54, Section 434, and made permanent by Public Law 111–11, Section 3001. Public Law 111–11, Title III—Forest Service Authorizations, Subtitle A—Watershed Restoration and Enhancement, Section 3001. Amends the Department of the Interior and Related Agencies Appropriations Act, 1999 to make permanent the authorization for the secretary of agriculture to use Forest Service appropriations for the purpose of entering into cooperative agreements with government, private, and nonprofit entities and landowners for the protection, restoration, and enhancement of fish and wildlife habitat and other resources on public or private land and/or the reduction of risk from natural disaster. Makes provisions regarding executive agency use of procurement contracts, grants, and cooperative agreements inapplicable to: (1) watershed restoration and enhancement agreements entered into under such Act; and (2) cooperative agreements between the Secretary of Agriculture and public or private entities for Forest Service programs.

11. USDA, "Stewardship Contracting," sec. 61.12, p. 24.

12. Conservation Title of the 2003 Final Appropriations Bill, Title 16; 16 U.S.C. 2104.

13. From U.S. Forest Service, unpublished data for 1999–2011 from National Stewardship Contracting Coordinator at the Forest Service National Headquarters (USDA Forest Service, 201 14th Street, SW, Washington, DC 20024). Also see "Stewardship End Result Contracting," USDA Forest Service, http://www.fs.fed.us/restoration/Stewardship_Contracting/index.shtml.

14. Charnley et al., *Northwest Forest Plan*.

15. Thomas et al., "Northwest Forest Plan," 283.

16. Ecotrust, *A Watershed Assessment for the Siuslaw Basin* (Portland, OR: Ecotrust, 2002), http://www.inforain.org/siuslaw.

17. Ibid.

18. Dan Karnes, Mapleton Ranger District Silviculturist, in-person conversation with authors, 1995.

19. Ecotrust and Resource Innovations, *Redefining Stewardship*.

20. Thomas et al., "Northwest Forest Plan."

21. M. Kauffman, N. Toth, and J. Sundstrom, *Voices from the Siuslaw* (Deadwood, OR: Siuslaw Institute, Inc., 2005).

22. Ecotrust and Resource Innovations, *Redefining Stewardship*.

23. Kauffman et al., *Voices from the Siuslaw*.

24.  Resource Innovations, "Siuslaw National Forest Stewardship Contracting Multiparty Monitoring Report Fiscal Year 2007"; Integrated Resource Management, "Siuslaw National Forest Stewardship Contracting Multiparty Monitoring Report Fiscal Year 2008," and "Siuslaw National Forest Stewardship Contracting Multiparty Monitoring Report Fiscal Year 2009." These annual reports are no longer available online, but copies may be obtained by contacting Siuslaw National Forest, (3200 SW Jefferson Way, Corvallis, Oregon 97331).

25.  Data for this list comes from Ecotrust and Resource Innovations, *Redefining Stewardship: Public Lands and Rural Communities in the Pacific Northwest* (Portland, OR: Economic and Resource Innovations, 2008), chap. 10.

26.  Tim Dabney, U.S. Forest Service, personal communication, 2010; unpublished Forest Service data.

27.  Kauffman et al., *Voices from the Siuslaw*.

# Spotlight 9.1

## STEWARDSHIP AGREEMENTS

The Weaverville Community Forest, California

*Pat Frost*

IN BRIEF

- The Weaverville Community Forest is an example of a community-managed forest in a federal lands context; a stewardship agreement provides for the joint accomplishment of work by a federal agency and a partner organization for the benefit of both.
- The benefits of community forestry on federal lands include (1) helping federal agencies achieve their management objectives associated with environmental restoration and forest health improvement; (2) helping communities meet their goals associated with federal land management, including creation of local jobs; (3) the ability of local partners to leverage funding and resources to accomplish management activities that agencies could not otherwise pay for; and (4) strengthening relations and partnerships between federal agencies and local communities.
- The presence of a well-functioning, high-capacity local community organization that can engage with public land management agencies in forest management activities is an important asset that helps make working forests work in a public lands context.

The Weaverville Community Forest encompasses 13,000 acres around the town of Weaverville in Trinity County, northern California. It is not a community-owned forest but, rather, a community-managed forest located on federal lands administered by the Bureau of Land Management (BLM) and U.S. Forest Service (USFS) in a county where more than 70 percent of the land is federally owned. The Weaverville Community Forest is managed through the Trinity County Resource Conservation District (TCRCD) through two 10-year cooperative stewardship agreements (one each with the BLM and USFS). A stewardship agreement, like a stewardship contract, is an instrument that can be used by BLM and USFS to accomplish land management goals through stewardship projects that meet local and rural community needs, with the ability to exchange goods for services in order to meet management objectives. Stewardship agreements have the same authorities as stewardship contracts and are used for the same purposes. However, unlike a contract—which is for the purchase of services for the direct benefit of the federal government—an agreement is used for the joint accomplishment of work by an agency and a partner organization for the benefit of both. Stewardship contracts and agreements have been gaining traction within federal agencies in recent years as viable ways to improve forest health, create local jobs, and form strong local partnerships.

The BLM stewardship agreement grew out of citizen concern over a 1,000-acre land exchange proposed in 1999 between BLM and Sierra Pacific Industries, a private timber company. The BLM's interest in the land exchange primarily concerned the fire risk these acres posed to the community. Local citizens were particularly concerned about the visual impacts that intensive logging by Sierra Pacific might have on the "view-shed" of their community should the exchange proceed. Community members banded together as the Weaverville 1,000 and met regularly to discuss the possibility of purchasing the property instead. The Trinity County Resource Conservation District, a county-wide agency with a board of local volunteers and experience dealing with land management issues, was particularly vocal in expressing the need for more local control over the land. In response to community outcry, the agency agreed to halt the land exchange and examine alternative options for reducing fire risk.

In 2001, a wildfire threatened the town of Weaverville and the health of forests in Trinity County, destroying seven homes and highlighting the need for better management. In response, BLM proposed a stewardship agreement with TCRCD. The agreement, signed in 2005, stipulates that while BLM retains ownership of the land, it works cooperatively with TCRCD to manage it in accordance with the community's objectives and priorities. Land management decisions—including those related to fire risk reduction—as well as some costs associated with upkeep, are shared with TCRCD. The stewardship agreement fulfills the agency's goal of reducing fire risk on the landscape and the community desire to manage local forestland as a community forest. In

2008, USFS entered into a similar agreement with TCRCD, adding an additional 12,000 acres to what is now the Weaverville Community Forest. The BLM and USFS community forestlands are largely contiguous and encompass all of the federally managed lands within the view-shed of the town of Weaverville.

The TCRCD is tasked with developing and overseeing a comprehensive management plan for the Weaverville Community Forest that reflects the community's vision by integrating local objectives with broader concerns for forest health. Management is provided by the TCRCD board of directors, which sets policy and budgets and establishes timelines for projects. The board conducts monthly public meetings, during which the community forest is a standing agenda item. The board also gives direction to TCRCD staff tasked with implementing the stewardship agreements. A steering committee composed of local citizens and members of the public provide input and feedback on project implementation, management policy, and strategic planning, with the TCRCD district manager acting as the point of contact among all involved parties. Meanwhile, BLM and USFS ensure that individual and cumulative environmental consequences are analyzed and disclosed to the public prior to authorizing management actions under the strategic plan.

The TCRCD strategic plan emphasizes managing for watershed health, recreation, forest health, education and outreach, economic opportunities for the community, and environmental monitoring. More detailed objectives are laid out within this plan, which is posted on the TCRCD website.[1] While the focus of projects in the forest is not to generate profit, they do sometimes draw revenue beyond what is necessary to cover operational costs. When this happens, the funds are held in an account designated for sole use in future Weaverville Community Forest projects.

The accomplishments of TCRCD to date have included several thinning projects (providing logs to the local mill and firewood to the community), trail construction, invasive-plant removal, water-quality monitoring, and environmental-education programs. Projects are performed by trained TCRCD staff in consultation with resource professionals such as a registered professional forester. Crew members, consultants, and contractors are local residents of Trinity County. The key challenge has been some confusion in the local community concerning the different approaches that BLM and USFS have adopted for implementing the stewardship agreements—even though both agencies work under the same authorities.

Stewardship agreements benefit the federal agencies by providing a local connection between program-level policy development on public land and the community at large. Another advantage of the stewardship relationship for BLM and USFS is that TCRCD has been able to obtain grants from other federal and state partners to help fund new projects in the community forest, making it possible to accomplish additional work by drawing on outside resources.

Stewardship agreements—like stewardship contracts—provide a mechanism for communities to help manage local working forests in a public lands context. Forests are managed in a manner that meets community goals and needs, including creating local economic opportunities, while helping federal agencies accomplish forest management objectives on the ground.

#### NOTES

1. "Stewardship: What Is a Community Forest?" Weaverville Community Forest, accessed October 11, 2013, www.tcrcd.net/wcf/stewardship.htm.

# PART FOUR

## CASE STUDIES OF WORKING RANCHES

The political conflict over grazing on public lands in the West became almost as acrimonious as the battle over logging on national forests, but it did not take the same toll. "Cattle-Free in '93"—the rallying cry of many environmentalists at the beginning of the Clinton administration, when Bruce Babbitt became secretary of the interior—never gained much traction, and no endangered species issues related to grazing resonated with the same power as the northern spotted owl in the Pacific Northwest forestlands.

Nonetheless, ranchers faced daunting challenges in the early twenty-first century: escalating land prices, stagnating beef prices, a West that seemed to be crowding them out in many places. In spite of these challenges, the rise of the collaborative conservation movement gave ranchers an increasingly unified voice in many watersheds, as the spotlight on the Madison Valley Ranchlands Group (spotlight 11.1) in this section demonstrates. Some ranchers even joined together to market their products through innovative ventures like Country Natural Beef (spotlight 10.1) or the Mountain States Lamb Cooperative, of which Lava Lake Land & Livestock (chap. 10) was a founding member.

Conservation-minded ranchers experimented with a diverse mix of innovations, as the next five case studies make clear. Not all of the experiments worked, especially after the housing boom went bust in 2008 and plunged the United States into its worst recession since the Great Depression of the 1930s. But each experiment exhibited a deep commitment on the part of these ranchers to making the ranching landscapes of the West more sustainable in both ecological and economic terms.

As chapter 10's case study of Lava Lake Land & Livestock reveals, private ranches can be important catalysts for large landscape conservation across public and private boundaries. Beginning in 1999, Kathleen and Brian Bean purchased five working sheep and cattle ranches in the Pioneers-Craters region of south-central Idaho, consolidating private deeded land and public leased land into a single operational unit. Because private lands in the region tend to be located along watersheds in middle and lower elevations, Lava Lake was able to conserve critical wildlife corridors.

The Lava Lake example demonstrates a series of conservation and business initiatives aimed at achieving the company's twin goals of conservation and profitability: incorporation of science-based grazing and conservation management plans to balance agricultural productivity with improvements in rangeland conditions and restoration of key habitats; innovations in how Lava Lake approached the marketing of its lamb; and participation in the organization of the Pioneers Alliance of ranchers, environmentalists, agency personnel, and elected officials to achieve long-term conservation of the Pioneers-Crater landscape in the face of real-estate speculation. Whether it was market forces or political pressure, each new challenge stimulated Lava Lake to take a leadership role in launching new initiatives and forming broader coalitions—which, after all, is what collaborative conservation on working landscapes is all about.

Sun Ranch in Montana (chap. 11) was another experiment in conservation ranching launched by outside investors. As both that chapter and the spotlight on the Madison Valley Ranchlands Group (spotlight 11.1) make clear, the Madison Valley is one of those iconic western landscapes whose beauty contains the seeds of its own possible destruction. Beginning in the 1970s, land speculation fueled the conversion of ranchlands into "ranchettes"; by the 1990s, the average price per acre in the Madison Valley rose to triple the agricultural value of the land. The Lang family purchased Sun Ranch in 1998, working closely with their neighbors in the Madison Valley Ranchlands Group to pioneer collaborative conservation in the region. They also quickly diversified, investing several million dollars in a sustainable eco-resort. In 2007, the Langs identified up to 13 potential home sites that protected critical wildlife habitat on their deeded lands while placing the rest of their private lands—98 percent of it—into conservation easements.

By 2008, it looked like Sun Ranch's innovative mix of cattle ranching, eco-tourism, and carefully limited residential development was poised to become a model for profitable conservation ranching in the West. But by the end of that year, innovation and creativity collided with the recession. Nonetheless, the combination of conservation and economic diversification exemplified by Sun Ranch may turn out to be a more successful model as the economic climate improves.

Other innovators have weathered the storm. Perhaps the most remarkable story in this section is that of the Moroney family's 47 Ranch in southeastern Arizona (chap. 12). It illustrates how a ranching family solved two major financial challenges that cripple many ranchers in the West. Through the sale of development rights that established conservation easements on their private lands, the Moroneys paid off their mortgage and reduced the taxable value of their estate, which will make it easier for their heirs to continue ranching when they retire.

The Arizona Land and Water Trust, which assumed responsibility for holding the primary conservation easement and monitoring compliance on the Moroney's ranch, considers itself to be "part of a growing movement of land trusts across the nation, which are in general responsive to an entrepreneurial approach to protecting the land."[1] In the case of the 47 Ranch, that "entrepreneurial approach" helped preserve a working landscape by eliminating debt and reducing taxes. That allowed the Moroneys to experiment with an ever-expanding toolkit of innovations to make their ranch more profitable and more sustainable, including renewable energy and heritage livestock breeds. Their example demonstrates that you do not need deep pockets to survive in ranching if you find the right combination of strategies for your particular outfit.

## NOTES

1. "How We Work," Arizona Land and Water Trust, accessed October 11, 2013, http://www.aolt.org/aboutus/howwework.shtml.

# 10

# LAVA LAKE LAND & LIVESTOCK

The Role of Private Landowners in Landscape-Scale Conservation

*Michael S. Stevens*

## IN BRIEF

- Lava Lake Land & Livestock is a landscape-scale conservation and business enterprise that incorporates a combination of tools and approaches to maintain a viable business while conserving working ranchlands across private and public lands.
- Lava Lake's innovations are highlighted by its ability to create incentives for different federal, state, private, and nonprofit entities to come together to achieve working landscape conservation goals that would have been impossible to achieve by working alone.
- Landscape-scale conservation across public and private lands was made possible through strategic acquisitions of large, private ranches having high conservation value, high economic value, grazing permits/leases on public lands, and the potential to consolidate landholdings within watersheds.
- Scientific partnerships and research have played a key role in the company's range management planning, restoration, and grazing activities and in setting up an ecological monitoring program to assess the longer-term effects of these activities.

Conservation-oriented ranch owners are bringing changes to land stewardship practices in the West, along with new approaches to business operations and community involvement. They are demonstrating that the interspersed pattern of private lands and public lands grazing allotments, originally developed to support livestock grazing, can also be used to preserve large land-scapes and protect wildlife. For the past decade, Lava Lake Land & Livestock (Lava Lake), a private company engaged in sheep ranching and conservation work in the Pioneers-Craters region of south-central Idaho, has been at the heart of efforts to develop innovative agricultural and business practices and conservation strategies that will sustain the natural and social values of the region. Lava Lake has launched or co-convened—often in partnership with diverse coalitions of other landowners, conservation organizations, and public agencies—a number of business and conservation entities, includ-ing Lava Lake Lamb, the Lava Lake Institute for Science and Conservation, the Pioneers Alliance, the Central Idaho Rangelands Network, and the Wood River Wolf Project. This chapter reviews the major strategies utilized by Lava Lake and its partners for conserving working ranchlands.

## REGIONAL SETTING

The 2.5-million-acre Pioneer Mountains and Craters of the Moon region is a classic western landscape. Varying from lava flows and sagebrush steppe to river, foothill, and mountain ecosystems, the landscape includes a dramatic range of elevations (from 4,000 feet to 12,000 feet), long free-flowing reaches of rivers and streams and their associated riparian habitats, and extensive rugged and unspoiled country.[1]

This geographic heterogeneity and habitat diversity support herbivorous wildlife such as pronghorn (*Antilocapra americana*), sage grouse (*Centrocercus urophasianus*), mule deer (*Odocoileus hemionus*), elk (*Cervus canadensis*), and mountain goat (*Oreamnos americanus*), as well as carnivores, including wol-verine (*Gulo gulo*), wolf (*Canis lupus*), mountain lion (*Puma concolor*), and black bear (*Ursus americanus*). Large numbers of elk, deer, and pronghorn migrate seasonally between lower-elevation winter ranges and mid- to high-elevation summer ranges.[2] Their migratory corridors, however, are increasingly vul-nerable to exurban development and habitat fragmentation.

The Pioneers-Craters region has an equally rich human history and cul-ture. Originally inhabited by the Shoshone and Bannock people, the region is traversed by Goodale's Cutoff, a spur of the Oregon Trail, and its prehistoric precursors. It was settled by waves of pioneers, miners, sheep ranchers, and farmers in the late 1800s. In 1936, the resort of Sun Valley began attracting vis-itors from around the world and set the stage for the area's current prosperity and rapid growth. Today, the region supports a mix of human uses including both a thriving tourism- and lifestyle-based economy, one of the most intact long-distance migratory sheep ranching cultures in the West, and a long tradition of hunting and backcountry recreation.

The ecological significance and wilderness qualities of this region have long been recognized and have resulted in the establishment of over 40 Wilderness Study Areas, Inventoried Roadless Areas, and Areas of Critical Environmental Concern, managed by the Bureau of Land Management (BLM) and the U.S. Forest Service (USFS).[3] At the southern end of the region lies the 750,000-acre Craters of the Moon National Monument and Preserve, as well as the Department of Energy's 500,000-acre Idaho National Laboratory, which includes a 73,600-acre Sagebrush Steppe Ecosystem Reserve.[4] The middle elevations of the region are characterized by a mix of private, state, and federal lands managed by the Bureau of Land Management. The northern portion of the region lies within and forms the southern edge of a large complex of roadless and Wilderness Study Areas including the Frank Church–River of No Return Wilderness Area and the proposed Boulder White Clouds Wilderness Area.

Despite this long history of conservation on the public lands of the region, little attention has been paid to the preservation of the larger ecological landscape, as defined by watersheds or wildlife corridors between winter and summer ranges. Private lands at the middle and lower elevations, in fact, support much of the biological value of the landscape because they contain most of the major waterways. Approximately 32 landowners control over 125,000 acres of private lands in the core portions of the region. Almost 70,000 acres, or more than 50 percent of the private lands, are owned by just five landowners.[5]

Threats to the integrity of the Pioneers-Craters landscape include fragmentation of the large working ranches, conversion of properties to residential use, major infrastructure development, mining exploration, and poorly managed motorized recreational activities. Between 1999 and 2008, a number of area ranches were put up for sale by longtime family owners, and properties continue to be listed for sale and are vulnerable to being subdivided and converted to rural residential or exurban use. Because of the large amount of acreage held by a relatively small number of landowners, there is the potential for landscape change simply as the result of one or two real-estate transactions. In 2007, the Mountain States Transmission Intertie Project, a 500-kilovolt regional electrical transmission line, was proposed to run through the area. Ongoing mining exploration proposals and loosely managed motorized recreation constitute additional threats to portions of the landscape.

## THE ORIGINS OF LAVA LAKE LAND & LIVESTOCK

Lava Lake Land & Livestock was founded as a family-owned company in 1999 by Kathleen and Brian Bean with a two-fold mission: to restore and protect the native landscape of the Pioneers-Craters region and to develop business and finance strategies that support that conservation work. As the Beans began to purchase land and build the company, they quickly came to appreciate several key factors. First, they saw the importance of retaining the public lands grazing permits associated with the private lands they were purchasing. The permits greatly increased the acreage that would be under Lava Lake's management control, helped make the ranch a more economically viable grazing operation, and increased the value and manageability of the private lands. Second, the Beans became aware of the value of the region to migratory ungulates and other wildlife. A mule deer study conducted by the National Park Service showed that local deer herds moved between the sagebrush steppe and the southeast end of the Pioneer Mountains.[6] And, at the time of purchase by the Beans in 1999, Lava Lake Ranch was considered to be an important elk habitat area by the Idaho Department of Fish and Game. Third, they became aware of the extent of protected areas and roadless areas on nearby public lands. The establishment of Lava Lake was closely followed by the expansion, by presidential proclamation, of Craters of the Moon National Monument from 54,000 acres to 750,000 acres.[7] These economic, real estate, conservation, and land management dynamics served to shape how the Beans viewed the interdependence of public and private lands and the potential role Lava Lake could play in the Pioneers-Craters landscape.

### LAVA LAKE'S CONSERVATION AND BUSINESS STRATEGIES

Lava Lake's conservation and business efforts have been based on the nine key strategies described below.

#### Strategic Land Acquisitions

The goal of Lava Lake's ranch purchases was to make investments in real estate that would position the Beans to build a conservation and family legacy. Lava Lake began its purchases in 1999 with Lava Lake Ranch, a roughly 7,500-acre historic ranch crossed by Goodale's Cutoff. From 2000 to 2002, Lava Lake acquired four additional working sheep and cattle ranches that led to consolidated holdings of roughly 24,000 acres of private deeded land and almost 900,000 acres of leased land administered by the BLM, the National Park Service, the U.S. Forest Service, and the Idaho Department of Lands.

Lava Lake's purchase and leasing strategy focused on securing lands with maximum conservation potential and economic value. As shown in figure 10.1, private landholdings in this region are frequently intermingled with state and federal lands to form a complex ownership and management pattern common in much of the West. Lava Lake's consolidation of the private lands in three major watersheds provided it with a larger and more diverse land portfolio. The acquisitions also provided management control and/or leases on the public lands that neighbored and connected the private parcels. Furthermore, they allowed Lava Lake to benefit from a larger grazing base and a greater level of privacy and control on the private lands. Lava Lake's greater degree of control and influence across the upper portions of several major watersheds also gave it a potentially much bigger conservation footprint.

With its operational and conservation area in place by 2002, Lava Lake embarked on integrating the grazing operations of the five ranches and began to develop the scientific and management information it needed to define and pursue its conservation goals.

*Science-Based Grazing Management and Rangeland Restoration*

Since the company's inception, one of Lava Lake's two central goals was to achieve landscape-scale conservation in the Pioneers-Craters region. But in order to define a conservation vision, understand the ecological values of the landscape, and develop a sense of the conservation role that Lava Lake could play, the company needed to develop the appropriate supporting scientific information. At the time of Lava Lake's inception in 1999, there had been few scientific surveys of the Pioneers-Craters region. The Idaho Department of Fish and Game, the National Park Service, BLM, and the Forest Service had conducted studies on wildlife, specific habitats, and grazing use, but there had been no systematic assessment of the entire landscape.

Given the lack of an existing entity seeking to do this work across the many administrative and jurisdictional boundaries within the region, Lava Lake established a partnership with the Nature Conservancy (TNC) to begin a comprehensive effort to work with other organizations, agencies, and landowners to develop scientific information that would support conservation and grazing management planning. The overall approach consisted of the

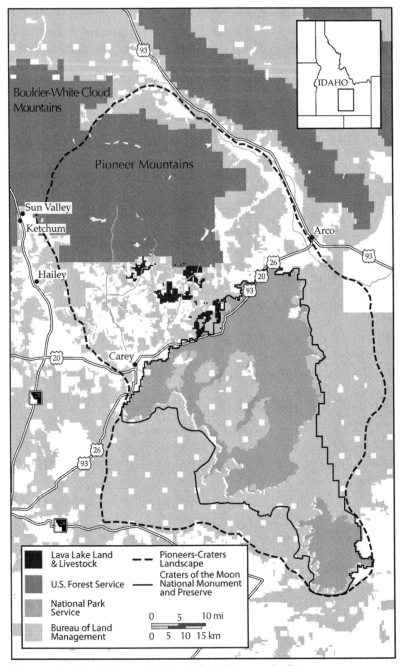

**Figure 10.1.** Lava Lake lands in central Idaho. Map by Darin Jensen and Syd Wayman, 2012.

following generalized steps: ecological studies and assessments, management and grazing planning, land and grazing management activities, habitat restoration projects, and monitoring and management adjustments.

As a means to launch the effort, Lava Lake entered into a multiyear contract with TNC that paid for a TNC staff scientist to begin approaching state and federal agencies, academic researchers, and other nonprofit organizations with proposals to conduct scientific studies and assessments on Lava Lake's private lands and public-grazing allotments. A key outcome for the Lava Lake–TNC contract was that TNC would leverage funds from partners well in excess of the funding for staff time and project costs provided by Lava Lake. The response was enthusiastic. Between 2000 and 2009, Lava Lake and its partners conducted over 35 studies and implemented a comprehensive ecological monitoring program in order to document long-term trends in ecological condition and grazing use.[8] From the outset, Lava Lake and TNC focused on developing an aggressive cost-sharing program with all interested partners. Between 2004 and 2009, $825,000 of external funding was generated for scientific projects in the region—a 10-to-1 return on contributions made by Lava Lake (principally through the Lava Lake Institute) toward those projects over that time period.

The scientific studies set the stage for the development of conservation and grazing management plans for the entire area grazed by Lava Lake. First, Lava Lake worked with TNC to develop a comprehensive conservation plan.[9] The conservation plan identified grazing management as the primary issue to be addressed. This led to a comprehensive grazing management plan, completed in 2005.[10]

In 2003, Lava Lake convened a volunteer Science and Conservation Advisory Board to provide critical review and guidance for its scientific and conservation efforts. The board reviewed the conservation and grazing management plans in detail and provided general strategic and practical advice.

These efforts paid dividends on the land and, with each successful project, helped build the capacity of Lava Lake, TNC, and other partners to secure funding for research and restoration projects. Significant improvements in riparian systems, riparian songbird populations, and aspen habitats have been documented, and Lava Lake's conservation and grazing management plans have successfully guided its work.[11] Lava Lake has been recognized with multiple awards for this work, including national rangeland management awards from the BLM and U.S. Forest Service as well as the Cecil D. Andrus Leadership Award for Sustainability and Conservation from Sustainable Northwest.

Lava Lake continues to work with a broad set of public and private partners to initiate scientific studies and to conduct habitat restoration across public and private lands. Some of the most important examples include agreements and contracts with the Natural Resources Conservation Service, the U.S. Fish and Wildlife Service, and the Idaho Department of Fish and Game to continue implementing good grazing management practices, monitor wildlife populations, and conduct sage grouse restoration and habitat enhancement projects. The funding Lava Lake has received through these programs for improving habitat, restoring waterways, and implementing grazing practices that sustain and restore vegetation and soil are the closest approximation to payments for ecosystem services that Lava Lake has been able to establish in the region.

*Conservation-Minded, Integrated Public/Private Lands Grazing*

Lava Lake's sheep bands, like those of other livestock producers in the area, begin grazing in the sagebrush steppe managed by BLM in the spring and follow the vegetation growing season into the foothills and eventually into the higher mountains managed by the U.S. Forest Service by mid-summer. In the fall, the bands work their way back to lower elevations. Of Lava Lake's close to 900,000 acres of public lands grazing permits, roughly half of those acres consist of BLM-managed allotments in the high desert, which are shared with other permittees, while the other half of the acreage consists of BLM and U.S. Forest Service allotments in the foothills and mountains grazed, with some limited exceptions, solely to Lava Lake.

The regulations governing grazing permits are production oriented. In order to secure and hold a grazing permit or lease, an owner must either graze his own livestock (required on U.S. Forest Service permits) or sublease the permits to another operator (permitted on state and BLM permits).[12] But grazing permits also represent a conservation opportunity, giving Lava Lake the potential to effect good management practices on 900,000 acres rather than the 24,000 acres owned by the company. Because sheep are herded, sheep-dominated grazing areas typically have much less fencing than a cattle-dominated landscape. And, because sheep are highly migratory and seasonal in their grazing patterns, they depend, much like migratory wildlife, on an unfragmented, undeveloped landscape. If public lands permits are to be used as part of a conservation strategy, the livestock must be grazed in a manner that supports the maintenance and, as needed, the restoration of native habitats and wildlife.

From an operational perspective, integrated use of a large public-private landscape is essential to supporting larger-scale range-based grazing operations. The elevational and seasonal gradients found in the Pioneers-Craters landscape mean that livestock must graze across a large area during the course of each year. The jurisdictional and ownership design of the landscape is based on this fact. The entire pattern of private land settlement and federal land governance in the region is based on the premise that, while the more productive valleys and bottomlands would be homesteaded and farmed, the economic viability of these homesteads required access to large areas of public lands for grazing use. Access to state and federal rangelands remains an essential economic and operational component of almost all working ranches in the region.

At its first meeting in 2003, Lava Lake's Science and Conservation Advisory Board recommended that Lava Lake "pick low hanging fruit first." Specifically, the board stressed the importance of demonstrating early successes in good land and grazing management as a means to develop support and funding for the company's work. The low-hanging fruit was, in this case, improving the condition of streams and their riparian corridors, which are the single most important habitat in the landscape. The presence of water in riparian areas speeds ecosystem responses to improved practices; put simply, plants grow faster where there is water, and improvements in grazing practices that allow plants to grow in the presence of grazing will be most evident in riparian areas next to streams. With the help of GPS collars placed on one ewe in each band and through extensive time in the field with scientists, range managers, and staff, Lava Lake developed grazing practices that would provide a period of nonuse in early summer and shift grazing in valley bottoms, riparian areas, and springs to the later part of the season, generally after most plants were dormant after the end of the growing season in early to mid-July. Lava Lake's livestock managers also reduced the time that its sheep bands spent in aspen stands. These changes brought widespread improvements to riparian areas and aspen forests across the private lands and grazing allotments without imposing significant burdens on company staff and with no impact on animal productivity and performance. In addition, Lava Lake worked with BLM, the National Fish and Wildlife Foundation, and the Idaho Department of Fish and Game to conduct active restoration projects on several severely degraded reaches of stream.

The single change in grazing management that yielded landscape-level improvements in ecological conditions was the overall reduction in sheep numbers that Lava Lake initiated starting in 2003. The reduced number of sheep and sheep bands enabled Lava Lake's staff to provide rest, use areas more lightly, and respond to drought, fire, and the presence of wolves across the operating area. But this reduction in overall numbers (from roughly 9,000 sheep to just under 6,000 sheep) also entailed a loss in gross revenues. The focus of the next section is how Lava Lake has worked to stabilize the economics of its sheep operation given this conservation-driven adjustment in sheep numbers.

*Overall Economic Strategy*

Lava Lake's overall economic strategy, integrating the company's conservation and business activities, includes the following components:

a. Run a cost-efficient livestock operation with intense focus on increasing productivity while reducing and minimizing costs and creating the opportunity to add value to products through strategies such as organic certification and grass-fed production.

b. Participate in value-added sales mechanisms for lamb (the primary ranch product) through membership in a marketing cooperative and branded sales of lamb.

c. Actively pursue investments by nonprofit organizations and state and federal agencies in land improvement and restoration projects.

d. Sell conservation easements on private lands.

e. Engage a broad collaborative of other landowners, agencies, foundations, and organizations that invest and contribute to larger-scale conservation efforts in the Pioneers-Craters region.

*Niche Marketing of Organic, All-Natural, and Grass-Fed Lamb*

In order to achieve its goal of economic sustainability, Lava Lake has, in addition to a focus on the conservation-related strategies described above, focused to date on improving the economics of its ranching operation. This effort has involved both improving the ranch's productivity (on a per head basis) and developing market mechanisms for adding value to that production.

From its inception, Lava Lake worked to stabilize and improve the economic structure of its sheep production. Lava Lake worked closely with the former owners of the ranches and existing ranch staff to ensure continuity of good operational practices and worked intensively between 2002 and 2005 to integrate the operations and grazing patterns of the five ranches and to find ways to boost the productivity of the ranch's livestock. The company also sought to add value to its production at the ranch level through organic certification of 60,000 acres of private irrigated farmland and private and public rangelands as well as a portion of the sheep herd.

The 2000–2001 collapse in lamb prices taught Lava Lake that good grazing and husbandry practices alone would not secure the economic viability of its sheep operation given ongoing volatility of the commodity lamb market. In 2002, Lava Lake became a founding member of the Mountain States Lamb Cooperative, now consisting of over 125 sheep producers throughout the West, and the controlling owner of Mountain States Rosen, one of the country's largest distributors of lamb. The cooperative has adopted a series of premiums for all-natural and quality production and its presence appears to have stabilized the lamb market. It has also provided Lava Lake with a transparent, stable, and value-added channel for its lambs, almost all of which are marketed as all-natural.

Lava Lake also established its own brand, Lava Lake Lamb, in order to sell all-natural and certified organic grass-fed lamb through its website and at farmer's markets, restaurants, and wholesale outlets.[13] The revenues associated with the sales of Lava Lake Lamb are now a central component of Lava Lake's overall business model. Lava Lake Lamb LLC now handles the processing, sales, and marketing activities associated with Lava Lake's direct sales program.

In 2010, Lava Lake Lamb partnered with the cooperative and Mountain States Rosen to launch a fresh, seasonal, all-natural, and grass-fed lamb marketing program. This program provides Lava Lake with very efficient processing and distribution networks and enables the Lava Lake–Mountain States Rosen joint venture to market Lava Lake Lamb nationally, primarily through supermarkets (which can sell large amounts of lamb efficiently) and restaurants. Given the scale of Lava Lake's operation and the inherent limits on demand (based on population) for Lava lake Lamb in Idaho, this wider reach is a crucial component of the company's long-term success. The Lava Lake–Mountain States Rosen joint venture is enabling Lava Lake to reach out to a national marketplace with its story of conservation and sustainability.

This discussion would not be complete without mention of the fact that Lava Lake Lamb receives consistent high praise for its quality—chefs, food writers, and publications such as *Sunset Magazine*, *O: The Oprah Magazine*, and the *New York Times* have given rave reviews for the taste and quality of the lamb, in addition to praising what Marion Burros of the *New York Times* called its "sustainable pedigree."[14]

While every rancher seeks to operate in as lean and cost-efficient a manner as possible, launching a value-added marketing business requires new investment and an intensive off-ranch focus on customers and supply chains. Much like its progression with conservation projects, Lava Lake made significant contributions to starting Lava Lake Lamb and building the value and recognition of the brand. Achieving long-term economic sustainability of its lamb production, much like its conservation efforts, has led the company to partnerships and compelled it to reach a broader audience.

The economic downturn of 2007–10 tested the strength of consumer demand for niche and usually more expensive meat. Lava Lake's experience has been that its customers desire (1) high quality, good-tasting meat at a reasonable price; (2) all-natural status, specifically, no antibiotics, no added growth hormones, and no animal by-products in the diet; (3) grass-fed (no feedlot) practices; (4) animals raised humanely; and (5) animals raised in an ecologically responsible manner with a special interest in predator-friendly practices. Lava Lake has found many customers unwilling to pay the necessarily higher prices required to cover the higher costs of production for certified organic meat, but strong demand remains for all-natural, grass-fed lamb. For these reasons, Lake has worked to reduce its costs of production, processing, and distribution through a partnership with Mountain States Rosen while retaining all the production characteristics valued by its customers.

*Conservation Easements*

Lava Lake's efforts to achieve economic sustainability for its conservation and lamb production efforts are focused on operational sustainability. In addition, Lava Lake has sought ways to secure the long-term conservation status of its private lands while recovering portions of the original investments made in those lands.

One of Lava Lake's specific goals is to place all of its private lands in permanent conservation status. In 2001, Lava Lake donated to the Nature Conservancy a perpetual conservation easement on 7,400 acres of private land at Lava Lake Ranch. This donation provided an opportunity for the company to generate federal income tax deductions for the owners, which is typically the driving financial incentive for the donation of conservation easements. The company also sought to develop mechanisms for generating cash revenues from the sale of conservation easements. This represents the primary mechanism for generating a direct cash return on the investment the company has made in land as a conservation buyer. Since 2009, Lava Lake has been working to sell perpetual conservation easements on the remainder of the company's private lands. The principal sources of funding for such easements are foundations, nonprofit conservation organizations, and state and federal conservation programs such as the Farm and Ranchland Protection Program and the Grassland Reserve Program administered by the Natural Resources Conservation Service. While Lava Lake's conservation easement projects are still in progress, it is their experience that sales of conservation easements can generate important value for ranch owners that may allow them to expand their operations and improve the economic stability of their ranches.

*Collaborative Conservation at a Landscape Scale*

In 2006–7, a series of speculative purchases by real-estate developers of ranches in the nearby Wood River Valley and a proposal to build a 500-kilovolt regional electric transmission line through the landscape made it clear that Lava Lake needed to be part of a larger effort to conserve the Pioneers-Craters region and to sustain its grazing operation.[15] Lava Lake knew it could neither defeat these major threats nor take advantage of the full conservation opportunity in the region alone. While Lava Lake already had been working with a wide diversity of partners, it was clear that it needed to be part of a coalition of landowners and organizations to address conservation at a regional scale.

In early 2007, Lava Lake helped form the Pioneers Alliance, a coalition of ranchers, conservationists, scientists, elected and agency officials, recreationists, and local residents working to develop a long-term vision and conservation strategy for the Pioneers-Craters landscape.[16] Lava Lake obtained start-up funding from the William and Flora Hewlett Foundation to enable five participating nonprofit organizations to devote staff time to this effort. These five organizations include TNC, the Conservation Fund, Wood River Land Trust, Idaho Conservation League, and the Lava Lake Institute for Science and Conservation (discussed below). Currently, members of the Pioneers Alliance, with significant funding from the Natural Resources Conservation Service, have protected or are in the process of protecting over 60,000 acres of private working farms and ranches as well as addressing issues related to wildlife migration routes, sage grouse conservation, implementing improved management of motorized recreation, and increasing farmer and rancher involvement in natural resource management issues. The Pioneers Alliance also appears to have succeeded in defeating the proposed regional electrical transmission line. Now it is working to protect private farms and ranches by securing the conservation and grazing status of federal lands through designations that protect both the wildlands and working lands of the landscape and garner additional funding for conservation of the region.

Lava Lake also helped form and has participated in the Wood River Wolf Project, a collaboration among several sheep ranchers, Defenders of Wildlife, Wildlife Services, the Idaho Department of Fish and Game, and the U.S. Forest Service to implement proactive methods to minimize losses of sheep and wolves in the Sawtooth National Recreation Area and other nearby grazing areas. This project successfully enabled ranchers to run thousands of sheep in established and active wolf territories with minimal losses of sheep and no lethal control of wolves from 2008 through 2010.

*Building Supportive Institutions and Organizational Structures*

In the first several years of operation, all of Lava Lake's land management, business, science, and conservation activities occurred under the umbrella of a single organization, Lava Lake Land & Livestock, a limited liability corporation—a commonly used organizational structure for private landholding entities—owned by the Bean family.

As Lava Lake's range of activities grew and the company engaged in a broad set of business and conservation partnerships, however, a more diversified organizational framework was required. It needed to distinguish between the economic and operational activities conducted by Lava Lake Land & Livestock and the growing number of conservation and scientific projects in which Lava Lake was just one partner among a number of private and public participants.

The Lava Lake Institute for Science and Conservation was incorporated as a separate nonprofit organization in 2004. The establishment of the Institute has enabled Lava Lake Land & Livestock to distinguish between the activities it conducts as part of its land ownership obligations and goals, such as weed control, grazing management, or habitat restoration, and the activities that are important contributions to scientific knowledge and conservation for the larger Pioneers-Craters landscape.

## LESSONS FROM LAVA LAKE

Lava Lake's experience shows that private landowners and public lands grazing permittees can play significant and catalytic roles in the conservation of large public-private landscapes in the West. Over the past 10 years, Lava Lake has been fortunate to be a part of a number of successful conservation efforts across the Pioneers-Craters landscape. By using an evolving set of business and conservation tools, the organization has established institutions, partnerships, funding mechanisms, and a vision that it hopes will help safeguard the future of this remarkable landscape and its people.

For the work of Lava Lake and its partners in the Pioneers-Craters region to be meaningful to other parts of the West, it must address two key questions. First, how do private landowners make conservation pay? Second, is Lava Lake's experience transferable to other landowners and landscapes?

Ranching remains a difficult business and owners of large landholdings in the Intermountain West, especially those without valuable energy resources or those not utilizing limited development as a cash-generation tool, will continue to face the challenge of how to sustain operations economically over the long term. Finding solutions to this problem is central to private landowners' ability to play an important and long-term conservation role.

Conceptually, ranchers and private landowners must address the following challenges:

a. covering the costs of holding together significant real estate and grazing permit assets;

b. generating net revenues through the sale of goods and services; and

c. generating revenue through the sale of conservation outcomes that offset the loss of revenue that may be associated with a reduction in extractive activities or the loss of the opportunity to generate revenue through real estate, infrastructure, and energy development.

Is Lava Lake's experience transferable to other landowners and landscapes? Generally, yes. In fact, the experience of Lava Lake and its partners is just one example of a wave of collaborative business and conservation efforts that has swept the West over the past two decades. From business efforts such as Country Natural Beef (see spotlight 10.1) to collaborative land management efforts such as the Malpai Borderlands Group (see chap. 4) and the Blackfoot Challenge, and the emergence of rancher-led land trusts in Colorado and California, ranchers, landowners, conservationists, and agency officials are addressing the same issues of economic viability and landscape-scale conservation of public-private landscapes that Lava Lake faces. In many cases, these efforts rely primarily on existing mechanisms, regulations, and strategies; often, the innovation is in developing incentives for groups to form and work together and thus achieve results that are impossible for single entities.

It would have been possible for Lava Lake to achieve much of its success more easily and with less initial investment of its own resources if there were a better West-wide system for sharing the lessons and tools being developed by similar efforts. A more effective means for individual landowners, especially those lacking the resources to reach a broad network of partners, to learn and gain access to funding and other resources is crucial to reducing the costs and time associated with building viable economic and conservation models that will sustain the West's iconic landscapes. Lava Lake's efforts should not be viewed as an opportunity for learning only for landowners with resources outside the ranch or for the largest landowners with sufficient management structures and economies of scale.

Lava Lake believes that the strategies of cooperative value-added marketing of ranch products and services and collaborative conservation efforts that generate funding and policy support must be married together in a way that enables ranchers and landowners to play a key role as landscape stewards and be rewarded economically for doing so. Playing this role will require ranchers to adapt to changing societal views on environmental and wildlife management and the conservation community to sustain existing and develop new mechanisms for rewarding landowners for providing desired ecological and land management services.

## NOTES

1. Valuable resources for additional information on the region can be found at www .pioneersalliance.org, www.lavalakeinstitute.org, http://www.nps.gov/crmo, http://www.fs.usda.gov/sawtooth/, http://www.nature.org/ourinitiatives/regions/ northamerica/unitedstates/idaho/placesweprotect/pioneer-mountains.xml, and http://www.idahoconservation.org/issues/land/wilderness/pioneer-mountains/ pioneer-mountains.

2. "Pronghorn Migration," Lava Lake Institute for Science and Conservation, http:// www.lavalakeinstitute.org/Wildlife_Migration_Pronghorn_Migration.php.

3. Bureau of Land Management generally manages both Wilderness Study Areas and Areas of Critical Environmental Concern, while the U.S. Forest Service manages Inventoried Roadless Areas.

4. "Sagebrush Steppe," Idaho National Laboratory, https://inlportal.inl.gov/portal/ server.pt/community/sagebrush_steppe/485. Also see "Sagebrush Steppe Conservation Project /ID National Lab," Wildlife Conservation Society: North American Program, http://www.wcsnorthamerica.org/WildPlaces/YellowstoneandNorthern Rockies/SagebrushSteppeConservation.aspx.

5. N. Welch, unpublished GIS analysis (The Nature Conservancy, Hailey, ID, 2012).

6. B. Griffith, *Ecological Characteristics of Mule Deer: Craters of the Moon National Monument, Idaho* (Moscow: University of Idaho Cooperative Park Studies Unit, 1988).

7. "Craters of the Moon," National Park Service, http://www.nps.gov/crmo.
8. See "Recent Field Studies," Lava Lake Institute for Science and Conservation, http://www.lavalakeinstitute.org/Field_Studies_Recent_Field_Studies.php, for a selected bibliography of studies.
9. M. T. O'Sullivan, A. Sands, and M. Stevens, "Lava Lake Conservation Plan" (unpublished report created for Lava Lake Land & Livestock, Hailey, ID, 2004).
10. O'Sullivan et al., "Lava Lake Conservation Plan"; J. W. Karl, A. R. Sands, and M. S. Stevens, "A Spatial Analysis Approach to Estimating Grazing Capacity and Developing Grazing Management Recommendations on Idaho Rangelands" (unpublished manuscript, Lava Lake Land & Livestock, 2009); and "Scientific Research in the Pioneer Mountains–Craters of the Moon Landscape—Supported by Lava Lake (2001–2009)," Lava Lake Lamb, https://www.lavalakelamb.com/lava-lake-conservation-studies.php.
11. M. Scheintaub and M. T. O'Sullivan, "Lava Lake Land and Livestock 2002–2008 Riparian Monitoring Summary and Analysis" (unpublished data, Lava Lake Land & Livestock, Hailey, ID, 2008); J. Carlisle, "Analysis of Riparian Bird Population Changes in the Pioneer Mountain Foothills at Lava Lake Ranch from 2001–2007" (unpublished manuscript, Lava Lake Institute for Science and Conservation, Boise, ID, 2007).
12. "Fact Sheet on the BLM's Management of Livestock Grazing," U.S. Department of the Interior, Bureau of Land Management, http://www.blm.gov/wo/st/en/prog/grazing.html; and "Laws, Regulations, and Policies," U.S. Forest Service: Rangelands, http://www.fs.fed.us/rangelands/whoweare/lawsregs.shtml.
13. In general, see www.lavalakelamb.com. For restaurants, see http://www.lavalakelamb.com/lava-lake-information/restaurants/; and for retail stores, see http://www.lavalakelamb.com/lava-lake-information/retail-stores/.
14. Marion Burros, "The Gift Is in the Mail, and on the Web," New York Times, November 28, 2007, http://www.nytimes.com/2007/11/28/dining/28gift.html?_r=0.
15. "Major Rights-of-Way Projects," U.S. Department of the Interior, Bureau of Land Management, http://www.blm.gov/mt/st/en/prog/lands_realty/projects.html.
16. "About the Pioneer Alliance," Lave Lake Institute for Science and Conservation, accessed October 15, 2013, http://www.lavalakeinstitute.org/Pioneers_Alliance_About_Pioneers_Alliance.php.

# Spotlight 10.1

# COUNTRY NATURAL BEEF

Susan Charnley and Sophia Polasky

## IN BRIEF

- Country Natural Beef's focus on ecologically and economically sustainable ranching with third-party certification of "raise well and graze well" standards takes a consumer-driven approach and ensures that meat production is environmentally sound, humane, and healthy.
- The cooperative structure of Country Natural Beef allows ranchers to sell meat directly to retailers and to establish a fixed pricing system that stabilizes beef prices, making ranchers less vulnerable to market fluctuations and providing them with more stable income, which helps keep them in business and conserve working ranches.

Doc and Connie Hatfield, founders of the Country Natural Beef cooperative, were like many independent ranchers in the United States in the mid-1980s, struggling to compete financially with industrial megaranches. They had relocated from Montana's Bitterroot Valley to eastern Oregon in 1976 in search of cheaper pastureland but continued to stretch their resources to make ends meet. Overproduction, decreasing land values, high interest rates, and dramatically fluctuating commodity beef prices took a heavy toll on farms and ranches all across Oregon—including that of the Hatfields.

Deeply in debt and on the verge of giving up, Connie sought a way to do things differently. On a trip to Bend, Oregon, she met a young man who told her that he purchased Argentine beef, despite living in the midst of Oregon's cattle country, because Argentine cattle weren't pumped full of antibiotics and hormones, nor were they crammed into feed lots. Neither were the cows on Connie and Doc's ranch, and they were only 50 miles away. But at the time, the Hatfields, along with most ranchers in the area, had no independent marketing experience and were selling their cattle to buyers on the commodity market at the going rate, which was down to 65 cents per pound.

The Hatfields began conducting consumer research in the urban centers of western Oregon and convened a meeting with their ranching neighbors. They suggested a consumer-driven beef marketing cooperative that would enable ranchers to sell their products directly to the emerging "green" market then being served by Argentine beef. The group, originally 14 families, agreed to form Oregon Country Beef (now Country Natural Beef) in 1986. Connie contacted local buyers and feedlots and arranged to have the co-op's beef slaughtered, packed, labeled, and marketed, then sold to local retailers under the Country Natural Beef label. The group's mission became to promote economically and ecologically sustainable ranching while meeting the needs of consumers who wanted "natural" beef—beef raised in a humane and environmentally sound way. The following are Country Natural Beef's goals:

· provide a sustainable means, through a group, to market profitably quality beef products desired by the consumer while maintaining every possible bit of independence;
· return proceeds realized from marketing Country Natural Beef to the rancher, rather than to the organization, for acquiring capital assets;
· remain a grassroots, producer-controlled organization, keep administrative costs to a minimum, and cover operational costs from a percentage of producers' revenue; and
· treat Country Natural Beef as an idea to be examined constantly, not an entity that can be bought or sold.

Accordingly, cattle are raised without hormones or antibiotics on a vegetarian diet and are leaner than conventional beef. They spend their first 14–18 months on open rangelands and pastures. All animals are finished on one feedlot in eastern Oregon for approximately three months, where they are fed potatoes, alfalfa, and grains. They are then transported to Washington where they are custom processed at one location, then distributed directly to retailers. Ranchers maintain ownership of the animals from the time that they are raised, through their processing, to distribution of their meat. Ranchers who are members of Country Natural Beef (CNB) must undergo third-party certification by the Food Alliance and adhere to CNB's "raise well and graze well" standards to ensure that their animals have been treated in a humane way, that grazing practices are environmentally sound, and that CNB workers are treated fairly, with respect, and with attention to the safety of their working conditions.

Connections between the ranching families who produce the beef, retailers, and consumers (actual and potential) are fostered through several approaches:

- developing personal relationships with retailers during the marketing process;
- the Country Natural Beef website;[1]
- the requirement that members volunteer their time at least once a year at a retail outlet doing demonstrations and interacting with customers; and
- the offering of periodic CNB ranch tours.

Connie's instincts about a growing demand for natural foods were synchronized perfectly with a surge in the popularity of natural food retailers. The CNB cooperative began selling to Whole Foods Market, Nature's Market, and New Seasons Market and, by 1997, had expanded to include 26 ranchers. Today, the cooperative sells to a varied portfolio of buyers, including many local restaurant chains, and counts 120 members across the western United States who together own over 100,000 mother cows and operate on more than 6.5 million acres of public and private rangelands. They earn an average of $50 million in annual sales and, in 2007, were able to cover operational costs and still pay ranchers $1.07 per pound throughout the course of the year.

Country Natural Beef has been economically successful because it takes a "value-based marketing" approach, whereby ranchers who belong to the cooperative produce a specialized beef product that is different from beef sold on the commodity market and is valued by a targeted group of consumers. These consumers are willing to pay higher prices for natural beef, and third-party verification ensures that they are getting what they pay for. Each year, CNB reviews and adjusts the model it uses for estimating the average cost of beef production, using data from a sample of ranches of different sizes. The model serves as the basis of its meat pricing system, which CNB and retailers agree on each year. Entire animal carcasses are then presold to retailers, so that the price ranchers receive for their beef doesn't change over the course of the year. Instead, adjustments are made throughout the year by changing the price of different meat cuts to balance supply and demand. Money from sales is distributed back through the production chain in a way that is proportional to each producer's contribution and risk. The result of this fixed price system is that the price of beef stabilizes for the producer instead of fluctuating with the commodity market, creating a reliable income stream that makes ranching economically viable for these families. Administrative costs are kept to a minimum, in part because profits are not converted to capital assets, but also because individual ranchers still cover the bulk of their own operational costs.

Despite the fact that members of CNB are scattered across the western United States, the cooperative remains a tight-knit community. Members attend annual meetings, during which opportunities and issues facing the cooperative are discussed and members offer input. The goal of CNB, when it comes to policy setting and problem resolution, is to reach a consensus, rather than to convince a majority. This model has created a high level of trust among members and has helped contribute to the success of the cooperative.

Country Natural Beef faces challenges, however, that members are working together to figure out how to overcome. These include:

- growing competition with grass-fed, organic, and other natural beef producers;
- a stagnant demand for red meat;
- concern among consumers about animal welfare, leading to new retail standards for humane animal treatment and animal safety that must be met in order to sell beef through some outlets, not all of which are practical for CNB;
- demand for "locally" branded meat, which calls for traceability of meat from the ranch to the plate. Currently, CNB cattle are traceable from the ranch to the feedlot, but once they are slaughtered and processed at the processing facility, their individual ranch origins are no longer distinguishable.

As Doc Hatfield writes on the CNB website, Country Natural Beef is a "community of shared values, rather than a community of place. We embrace values that are built on healthy animals, healthy food, healthy communities, and healthy land."[2] Despite some challenges, the success Country Natural Beef has enjoyed over the years suggests that the American consumer has a growing appetite for those values.

## NOTES

This spotlight was developed from the following material: D. Campbell, "The Natural: 'Brickless' Marketing Co-op Helps Ranchers Tap Growing Market for Lean, Natural Beef," *Rural Cooperatives* 73, no. 4 (July–August 2006): 4–9, http://www.rurdev .usda.gov/rbs/pub/jul06/jul06.pdf; S. Duin, "Country Natural Beef Co-op Works on Sustaining Oregon Cattle Ranchers," *Oregonian*, June 9, 2010, http://www.oregonlive .com/news/oregonian/steve_duin/index.ssf/2010/06/country_natural_beef_co-op _wor.html; M. Pullman, Z. Wu, and V. Villa-Lobos, *Country Natural Beef: A Maturing Co-op at the Crossroad, Teaching Case for Sustainable Supply Chain Management* (Corvallis, OR: School of Business, Portland State University; Portland, OR: College of Business, Oregon State University, 2009); S. Stevenson and L. Lev, "Mid-Scale Food Value Chains Case Study: Country Natural Beef," *University of Wisconsin, College of Agricultural and Life Sciences, Center for Integrated Agricultural Systems*, Research Brief 79 (2009), http://www.cias.wisc.edu/wp-content/uploads/2009/11/rb79final2.pdf; D. Wilmeth, B. Bertelsen, and D. Probert, "Evaluating Strategies for Ranching in the 21st Century: The New Market Place," *Rangelands* 30, no. 2 (2008): 15–21.

1. See the website (http://www.countrynaturalbeef.com/) for more on how connections are fostered.

2. "Our Mission: Honor the Relationship from the Land to the Customer and Back," Country National Beef, accessed December 12, 2013, http://www.countrynaturalbeef .com/mission.php.

# 11

# CONSERVATION AND DEVELOPMENT AT SUN RANCH

## The Search for Balance in the U.S. West

*Roger Lang, William H. Durham, and Josh Spitzer*

IN BRIEF

- A wide variety of conservation solutions that can continually adapt to changing social, economic, and environmental circumstances offers the greatest hope for successful ranch conservation, but even then, success is not guaranteed.
- Real-estate development can have a positive conservation value on western working landscapes, depending on the type and scale of development.
- Many ranchers may not have the capital to experiment with different enterprises to try to improve the fiscal viability of their ranchlands, but they may be able to adopt conservation models proven successful on nearby ranches, adapting them to their own circumstances to help conserve their lands.
- Private-sector conservation efforts are vulnerable to market forces and cannot meet every conservation challenge; it is important to understand how markets affect such efforts, and to seek ways of helping private-sector conservation be resilient to market forces.

This chapter focuses on the complexity and power of private-sector conservation through a case study of the Sun Ranch of Madison Valley, Montana—a 26,000-acre privately owned ranch located 20 miles west of Yellowstone National Park, about one-quarter of which is composed of leased Forest Service grazing allotments (see map, fig. 11.1). The owners of Sun Ranch and the later-formed Sun Ranch Group, a private conservation enterprise working to protect ecologically important ranchlands in the Northern Rockies,

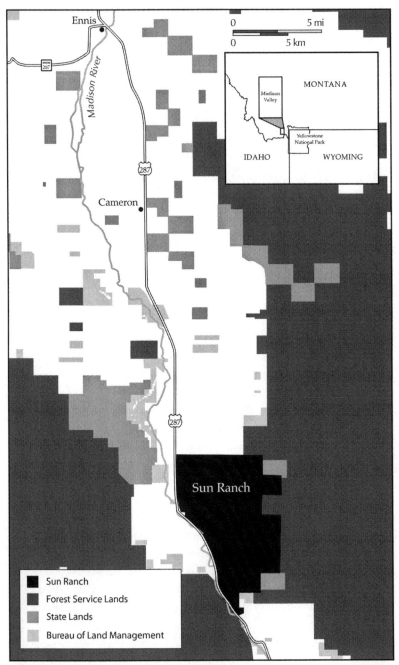

**Figure 11.1.** The location of Sun Ranch in southwestern Montana. Map by Darin Jensen and Syd Wayman, 2013.

demonstrate the great flexibility required of private-sector enterprise that places conservation at its core. With its diversified revenue sources and integrated conservation efforts—including sustainable agriculture, nature-based tourism, and carefully limited residential real-estate development—the owners were poised by the year 2000 to build a new business model with both substantial conservation outcomes and reasonable profits. From the outset, key actors in Sun Ranch and the Sun Ranch Group viewed the enterprise as an experiment, anticipating the need to recalibrate regularly. Nevertheless, the financial upheaval beginning in late 2007 challenged the principals' flexibility and taught lessons of value to future, private conservation efforts across the American West.

Prior to western settlement, the Madison Valley was a key seasonal hunting ground, used in the last few hundred years by several indigenous tribes that today still tell narratives of the valley's hunting riches. The western settled history of the region dates to the 1860s. Out of a gold rush in neighboring Alder Gulch, Montana, boomtowns Virginia City and Nevada City grew to thousands of inhabitants within just a few years. The land that became Sun Ranch started as a collection of dozens of homesteads stitched together by the Granite Cattle Company in the 1880s, when water rights and mineral rights were formally established throughout the valley. Shortly after the turn of the twentieth century, the ranch was purchased by a company called Grainger Ranches and renamed Rising Sun Ranch, which remained the name until World War II, when "Rising" was dropped because of its association with imperial Japan. Since the 1920s, the ranch has changed hands six times. Today it persists as a historic, ecologically rich, working cattle ranch—one that also has a major commitment to private conservation.

Sun Ranch is well endowed with natural capital, extending from the banks of the Madison River to the towering peaks of the Madison Range. Its four tumbling creeks—Wolf, Moose, Sun, and Papoose—descend from mountain lakes, springs, and snow pack, cutting canyons and riparian corridors over the ranch's grass-rich benchlands. Benchlands themselves are classically characteristic of Montana's geological history, in which rivers carved deeply, and at unique angles, through ancient glacial deposits.

## THE MADISON VALLEY: REGIONAL CONSERVATION ISSUES AND PARTNERS

Northwest of Yellowstone National Park, the Madison Valley is the "most ecologically intact watershed in the Greater Yellowstone Ecosystem" and consists of roughly one million acres of private lands, with approximately equal amounts of that land unprotected, under conservation easements or other forms of open-space protection, and subdivided and leaning toward development.[1] Relative to neighboring valleys like the Gallatin and Paradise, the Madison Valley has remained intact with respect to wildlife habitat. Every year large migrating herds of wildlife—primarily elk (Cervus canadensis), pronghorn (Antilocapra americana), and their predators—pass seasonally from privately owned valley bottoms serving as winter range back to the region's core, America's first national park. Yellowstone National Park itself comprises only one-tenth of the ecosystem's acreage; the rest includes state-owned sections, federal lands, and much private property.[2]

Animals living and moving throughout the ecosystem do not recognize property boundaries or political jurisdictions (except when marked by fences, roads, and other human-built obstacles). Many wildlife species transmit genetic material by migrating through habitat corridors to other ecosystems. Situated to the west of the Greater Yellowstone Ecosystem is the Bitterroot-Selway Wilderness, and to the north are the great Canadian Rockies and the Yukon Territory. Thus, if people are to manage the wildlife and habitat cohesively, private landowners must contribute to the conservation of this large landscape. Sun Ranch forms a vital link within this Yukon-to-Yellowstone ecoregion.

In the Madison Valley, as in other valleys in the Greater Yellowstone Ecosystem, nearly all of the low-lying land along the river valleys is private and, thus, was historically used for ranching and agriculture with little formal protection for other species of animals. Vast herds of elk and pronghorn, as well as their predators like wolves (Canis lupus) and grizzly bears (Ursus arctos horribills), winter in these low-lying areas each year. Human activities here, especially the development of "ranchettes," are major barriers to sustainability. Ranchettes are individual—usually fenced—family residences built at high densities on former ranchlands; they are often planned without regard to ecological function, community health, or aesthetics.[3] Habitat fragmentation from such private ranch subdivision cuts into wildlife's ability to migrate seasonally, and it crimps corridors that are vital for the dissemination of genetic material to other core regions.

One can fairly say that both biodiversity and traditional ranching culture have been pressured by rampant subdivision in the Madison Valley area—and elsewhere in Montana—since the 1970s. Today, in fact, southwest Montana is the fastest growing region in the state, and the former small town of

Bozeman one of the fastest-growing towns in the United States.[4] Complicating matters, Montana has no state-level statutes for land use or zoning. Thus, a piece of land like Sun Ranch could, in the hands of other owners, be subject to large-scale development without constraints. Indeed, under a previous owner in the 1970s, the ranch was platted for subdivision into some 200 home sites.

Formed in 1996, the Madison Valley Ranchlands Group, a 501(c)3 nonprofit, advocates for agricultural interests and ecological diversity (see spotlight 11.1). The ranchlands group has become the central, driving force behind collaborative processes among ranchers and other landowners as they address such challenging issues as the fragmentation of habitat due to ranchettes, the use of river setbacks in real-estate development, endangered species issues such as wolf reintroduction, and the spread of invasive weeds in the valley. The case of noxious weed abatement is, in fact, an instructive example of the regional effect of the group's work: when ranchers from across the region heard of the Madison Valley Ranchlands Group's success in galvanizing collaborative work to address the problem, the group was asked to educate other ranchers about how to collaborate in similar ways on significant local and regional ecological issues.[5]

When Roger Lang and his family acquired Sun Ranch in 1998, the Madison Valley Ranchlands Group helped introduce the new owners to local customs and challenges. Sun Ranch, in turn, influenced the group, helping to keep its collaborative approach alive and well. For example, Sun Ranch supported more than 20 other ranches in adopting valley-wide collaborative grazing practices by setting the model of grass sharing, or offering excess grasses to ranchers with limited supply for their cattle. Sun Ranch also took a leadership role in implementing sophisticated noxious weed-abatement practices and contributed to research on strategies and tactics for ranching with wolves and other predators. Drawing on technologies and volunteer support from environmental groups, for instance, Sun Ranch modeled the use of wolf-diverting "fladry fencing" (i.e., a line of rope mounted along the top of a fence, from which are suspended strips of fabric or colored flags that will flap in a breeze, intended to deter wolves from crossing the fence line) to help keep cattle and wolves separate, reducing the likelihood of livestock-predator conflicts. Additionally, building on the success of Lang's nascent Lodge at Sun Ranch (formerly Papoose Creek Lodge), the ranchlands group initiated a valley-wide ranch tourism business called Madison Valley Expeditions, bringing potentially important revenues to other local ranches in the valley.

SUSTAINABILITY STRATEGIES: THE SUN RANCH MODEL

North American rangeland has evolved over thousands of years in conjunction with resident bison populations, and much of the ecosystem developed
in symbiosis with their grazing and movement patterns. Cattle operations
like those at Sun Ranch can be environmentally sustainable when managed to
mimic bison ranging patterns. For instance, while cattle tend to congregate
in creek beds, polluting waterways and trampling riparian habitat in the process, they can be encouraged to use the riparian zones sparingly while moving across the short-grass prairie landscape to graze—just as did the bison
indigenous to the land. In this way, cattle may revitalize, rather than degrade,
the regional ecosystem.

The Langs quickly learned after purchasing their ranch, however, that—
ecological issues aside—cattle ranching alone would not be enough to generate a positive cash flow. While the Madison Valley offers exceptionally rich
grazing in the summer months, winters are harsh at Sun Ranch's end of the
valley, making it impossible to manage a sizable cattle herd over the winter
without high costs to ship cattle to more hospitable winter pasture. Moreover, it did not take long for the other financial demands of ranch management to become burdensome, including annual maintenance of roads and
infrastructure and costly weed-abatement efforts to preserve nutritious, native grasses. By 2001, the owners actively sought alternative business strategies for enhancing the ranch's economic viability. Long-time ranchers in the
region often find themselves in this situation—land-rich and cash-poor—
given the low rate of economic return from cattle, which graze atop highly
valued land. Many operations only manage to break even year-to-year, and
ranch owners depend on the appreciation of land values in order to build
wealth and provide for their families after retirement. Ranchers often need
an additional business component to unlock some of the value accruing to
the land. Some such innovative options include eco-lodging and ecologically
sustainable small residential development—both of which were explored by
Sun Ranch.

*Eco-Lodging*

Key advisers, including William Durham and William Bryan (founder of
Off the Beaten Path, a Montana travel company specializing in low-impact
adventure tourism), suggested exploring a sustainable lodging facility that
would attract tourists who could enjoy the ranch's rich, natural amenities.

In May 2001, the Langs launched their eco-lodge, Papoose Creek Lodge (later renamed the Lodge at Sun Ranch), modeled after sustainable destination resorts abroad. They began a foray into sustainable tourism by purchasing a property immediately adjacent to Sun Ranch that had been subdivided and slated for development. On this land, they invested over $2 million in capital improvements and staff for the new Papoose Creek Lodge, which unfortunately opened just as travel slumped in 2001. Nevertheless, the lodge was soon winning awards and garnering international media attention.[6]

Papoose Creek Lodge showed that bona fide ecotourism—a model of recreational travel that had been deployed successfully in tropical Latin America and Africa—could also work in the temperate U.S. West. The West, of course, had had a long tradition of successful dude ranches that capitalized on the mystique of cowboy life and the wide-open spaces. Papoose Creek Lodge distinguished itself from dude ranching by ensuring that all ranch practices were sensitive to local and global environmental impacts. Papoose Creek Lodge modernized the venerable western tradition of dude ranching by emphasizing conservation; in the process it showcased both the beauty of the Madison Valley region and efforts to practice sustainable grazing on the Sun Ranch.

Not only did the Madison River afford tourists with world-class catch-and-release fly fishing, Sun Ranch and adjacent lands supplied ample reach for horseback riding, naturalist-led hiking, mountain lake canoeing, wildlife photography safaris, bird watching, and other "tread-lightly" activities. The lodge aimed to hire local employees, which numbered up to 35 during the high season. Lodge guests could also venture out to the local towns of Ennis, West Yellowstone, and Virginia City, and they could take day trips to Yellowstone Park, bringing economic value to the surrounding communities. In promoting both the habitat and lifestyle of ranching, Sun Ranch was a prize-winning example of biocultural conservation.

### Sun Ranch Settlements

By 2002, Sun Ranch was dedicated to sustainable ranching and recreation, but even with these two components, financial return proved a vexing challenge. Moreover, Papoose Creek Lodge was not a model that other ranchers could easily follow, as it required considerable capital and specialized expertise in the lodging industry. After assessing a number of options, Sun Ranch's owners realized that a carefully planned, small-scale residential development could help accomplish their economic goals without compromising ecological function, while perhaps offering a usable model to other ranchers. The owner and managers recognized that people from across the world continue to move to western Montana to establish primary and secondary residences and will likely continue to do so for years to come. Indeed, the ranch's owners and a number of employees had themselves moved to Montana from other parts of the country. They realized that small-scale residential development could provide new stakeholders in the region, new minds to solve problems, and new sources of economic growth. While newcomers to Montana can be a financial boon for the state, and they bring a body of skills and civic engagement, often the residences constructed for them compound the development pressures that threaten the state's ecosystems and ways of life. As the owners of Sun Ranch began to consider how these newcomers could create potential economic benefits for the ranch, they also had to consider how these new stakeholders would live in the Madison Valley without threatening its ecological and cultural assets.

Even small-scale residential development, like that ultimately initiated by Sun Ranch, requires significant accompanying research, planning, and monitoring efforts to ensure that the venture protects wildlife habitat and supports communities in sustaining a strong, positive bond between new residents and the landscape. To carry out those important functions, an associated nonprofit—Sun Ranch Institute—was founded in 2006, funded by donations and the Langs' private capital, and staffed by three full-time employees. The institute embarked on a comprehensive research and education program to guide the planning of a new Sun Ranch "eco-enterprise" that would include Sun Ranch Settlements, a small-scale residential development on Sun Ranch. First, the institute worked with land planners and scientists to produce a multilayered geographic information system map and analysis of Sun Ranch in order to balance human and wildlife coexistence. They mapped a wide range of variables, from estimated infrastructure costs to human comfort and view preference to natural resource availability and habitat needs for roughly 30 species. A composite use map emerged, suggesting how human activity (including agriculture, tourism, and minimal residential real estate) could be balanced with the flora, fauna, creek drainages, and other natural attributes of the ranch. In the end, eight to 13 potential home sites were located in scenic areas appropriately buffered from critical wildlife habitat.

The new homes would be luxurious, ecologically unique, and sustainable, as long as the critical natural amenity—thousands of surrounding acres of key wildlife habitat—was protected from development or exploitation. By 2002, the Langs had granted the Nature Conservancy a conservation easement on 6,800 acres of the ranch for winter habitat for elk. At the inception of the new limited development strategy, the ranch placed an additional 11,500 acres into a conservation easement held by the U.S. Forest Service, after which 98 percent of the deeded acreage was protected from development of any kind. These easements ensured that no more than 20 homes could ever be built at Sun Ranch, all clustered in places with the least environmental impact.

The economic impact of conservation easements can be complicated. When donated, easements can create tax benefits for landowners who have high levels of income taxes to offset (consider the 47 Ranch experience in chap. 12). In some cases, easements can be purchased by conservation organizations from landowners. Usually, these purchases are made at prices well below the economic value that the owner forfeits as a result of land-use restrictions imposed by the easements. While the owners of Sun Ranch collected some income from the conservation easements (helping to offset the costs of maintaining the conservation property), the easements were critical to the overall eco-enterprise model. They clearly protected the conservation value of the property while preserving the value of the incredible natural amenities that would be available to guests of the lodge and residents of the settlement.

Meanwhile, a new business configuration, Sun Ranch Group, was formed to manage the entire enterprise, including creating the housing development. It included agriculture, hospitality, development, construction, and other business units. The Sun Ranch Group had the overarching goal of increasing the scale of Sun Ranch's conservation and commercial enterprise using the commercial and economic engines of its businesses to conserve the most important landscapes of the American West and beyond.

The Sun Ranch Settlement featured homesteads whose covenants controlled everything from fencing to building materials and suitable vegetation. A homestead distinguished itself from the more typical ranchettes by preventing the fencing of individual land plots while ensuring access to the entirety of Sun Ranch for homestead owners' permanent recreational use. This easement access feature would add significant value to each homestead. At the hub of the plan was a new eco-lodge, growing out of the existing Papoose Creek Lodge. The eco-lodge was designed for a new location on the ranch, following principles parallel to those for the home sites. The planned building would include dining facilities, fishing and riding opportunities, and an educational program, all available as amenities to both homeowners and visitors. Architectural plans were drawn up, and home sites were advertised for sale.

*The 2008 Economic Recession*

By September 2008, potential customers were interested in nearly all of the selected Sun Ranch Settlement sites, and closing dates appeared within reach. September 2008 also proved to be a pivotal month in the brewing financial crisis; before the month was over the federal government would take over Fannie Mae and Freddie Mac, Merrill Lynch would be sold to Bank of America, Lehman Brothers would collapse, and AIG would be bailed out by the Federal Reserve. By the end of October 2008, all prospective buyers had withdrawn from discussions, adopting a wait-and-see approach as the national economy appeared poised to collapse. Not only did the recession dry out the pipeline of possible investors in the settlement, it also reduced the healthy flow of tourists to the eco-lodge and hurt the cattle market. In response, the Sun Ranch Group began to scale back costs of all kinds, reducing staff in order to buy time. The Sun Ranch Group turned to recalibrating their financing and improving core operations on the ranch. Without revenues from the settlement, ranching and tourism operations were the two critical enterprises left to keep people employed and the enterprise viable.

Sun Ranch Group needed a whole new level of creativity in order to survive. Initially, the group had borrowed against its equity in the Sun Ranch to build the diversified enterprise. When lenders withdrew their capital, the Sun Ranch Group, like so many other businesses, found new risk capital to be especially scarce. After exhaustively exploring financing alternatives and various approaches to replacing capital, Sun Ranch Group reluctantly elected to put the ranch on the market in its entirety.

*Preservation Capital*

Even as the ranch was being listed for sale, Sun Ranch Group managers developed another new conservation model for western ranchlands that it called Preservation Capital, a fund that would purchase important ranchlands in the West, using the lessons learned from the settlement effort. Preservation Capital sought "patient capital" (for no less than eight years) in order to purchase important ranchlands and to study, restore, and conserve them, prior to developing variations of the settlement model or reselling them entirely to conservation-minded buyers. This fund was expected to offer significant conservation value over time, as larger sums of capital could be put to work while spreading costs more efficiently across many more and diverse acres.

The Preservation Capital project differentiated itself from other real-estate investment activities in two main ways: (1) it would use scientific approaches to understanding landscapes, thereby promoting the coexistence of wildlife with agriculture, and (2) it would deploy well-planned recreation services and limited residential real estate to advance conservation. In simplest terms, the scarcity of conservation-worthy ranchlands as a resource was expected to drive the appreciation of ranchland values. Unfortunately, the prevailing economic turmoil also deprived Preservation Capital of the necessary risk capital to operate at a sufficiently large scale. In the darkest days of the financial crisis, Preservation Capital generated significant interest from potential investors, ranch sellers, and ranch buyers. The lack of capital in the marketplace convinced the principals of Preservation Capital, however, to shelve the idea of a ranchland fund—though in the year after making that decision, several other similarly structured funds arose. At the time of this publication, Preservation Capital is operating as a boutique investment bank with a focus on impact investment and agricultural financing. Several principals in Sun Ranch Group manage Preservation Capital, carrying the conservation legacy of the Sun Ranch Group.

*Success of the Settlement Model Outside Sun Ranch*

In 2005, early in the development of the settlement model, Sun Ranch owner Lang carried on a discussion with a long-time rancher in the region, a prominent member of the Madison Valley Ranchlands Group, about developing his own ranch with a locally adapted version of the Sun Ranch eco-enterprise model. This rancher's 6,000-acre property was located above the valley floor, offering spectacular views that lent themselves well to plans to develop a handful of 20-acre home sites. Following a brief design period, the rancher began offering the sites at relatively high per-acre prices, with recreational rights over the broader ranch. A conservation easement was placed over the entire ranch, excluding the development envelopes for each home, thus protecting the value of each residential site. In giving up development rights, the rancher built in a long-term tax write-down against future adjusted gross income.[7] He also protected his land from wanton sprawl from a future buyer while retaining a lovely home site for his family, rights to work the land with cattle, and the opportunity for future income in the form of home site sales. In this variation on the Sun Ranch model, the rancher did not need to establish a costly lodge or club house for the home owners, as the active town of Ennis, Montana, is a mere 10 minutes away. The experience of the Ennis rancher attests to the value of the Sun Ranch formula.

## CONCLUSIONS: LESSONS LEARNED AT SUN RANCH

Despite the recession and its setbacks, a number of key Sun Ranch goals have been realized. First and foremost, more than 98 percent of Sun Ranch is protected in perpetuity by two complementary conservation easements held by the Nature Conservancy and the U.S. Forest Service. Second, ecotourism has been established in the valley and is having a positive impact on the local economy and landscape: as noted, the Madison Valley Ranchlands Group has sponsored Madison Expeditions, its own ecotourism endeavor. Third, the "conservation settlement" idea has shown that carefully designed and scaled residential real-estate development, too, has a place in the overall scheme of conservation and development of western ranchlands. In these respects, the Sun Ranch initiatives have paid off.

Just as private-sector conservation cannot meet every challenge, neither will it be immune to market forces. Conservationists need to understand how the market affects such efforts and then ensure that these ventures can respond to foreseeable changes in the business environment. Like organisms, businesses must be resilient and adaptive to change if they are to survive. Some will thrive, others will fail, and still others will adapt and provide long-term environmental value. We must not let fear of failure prevent experimental private-sector efforts at conserving the West's critical wildlife habitats, ecosystems, and working landscapes and their cultural tenants. As in nature, variation is necessary for good conservation solutions to evolve.

### NOTES

1. "A Beautiful Valley: Is It Too Beautiful for Its Own Good?" *Madison Valley Ranchlands Group Newsletter* 6, no. 1 (Spring 2007), 1, http://www.madisonvalleyranchlands.org/newsletters/Springnewsletter_2007.pdf.

2. A. Pearce, "Uncommon Properties: Ranching, Recreation, and Cooperation in a Mountain Valley" (PhD thesis, Stanford University, 2004); R. Keiter and M. Boyce, *The Greater Yellowstone Ecosystem: Redefining America's Wilderness Heritage* (New Haven, CT: Yale University Press, 1991); S. Clark, *Ensuring Greater Yellowstone's Future: Choices for Leaders and Citizens* (New Haven, CT: Yale University Press, 2008); T. Clark, M. Rutherford, and D. Casey, *Coexisting with Large Carnivores: Lessons from Greater Yellowstone* (Washington, DC: Island Press, 2005); P. Johnsgard and T. Mangelsen, *Yellowstone Wildlife: The Ecology and Natural History of the Greater Yellowstone Ecosystem* (Boulder: University Press of Colorado, 2013).

3. Pearce, "Uncommon Properties."

4. L. Sherman, "Fastest-Growing Small Towns," *Forbes*, December 7, 2010.

5. Pearce, "Uncommon Properties."

6. A. Pearce and C. Ocampo, "A Montana Lodge and the Case for a Broadly Defined Ecotourism," in *Ecotourism and Conservation in the Americas*, ed. A. Stronza and W. Durham (Wallingford, England: CABI Press, 2008), 114–24.

7. In this model, equity in the land is protected against possible future market downturns by retaining its tax write-down value at the time of assessment. In other words, at the time that the tax reduction formula is instituted, it is based on the value of the land approximated via local market comparables. The owner of the land is able to continue to claim the same tax reduction year after year, regardless of the current market value of his or her land and may eventually choose to sell when the market is right or else to pass on the land in his or her estate.

# Spotlight 11.1

# THE MADISON VALLEY
# RANCHLANDS GROUP

*Thomas E. Sheridan*

IN BRIEF

· The Madison Valley Ranchlands Group formed to address economic and
  ecological problems that threatened ranchlands in Montana's Madison Valley; the nonprofit conservation organization brings together "Old West" and
  "New West" residents to address these issues.
· The group has employed conservation easements, noxious weed control, a
  predator-protection program, land-use planning to control growth, a local
  beef marketing strategy, education programs, alternative energy production, elk hunting, and an ecotourism business as ways to promote conservation and economic vitality in the region.

Madison Valley, Montana, illustrates what Montanans are referring to when
they talk about Big Sky country. Just northwest of Yellowstone National Park,
the valley is a wide green swale flanked by snow-covered mountains. It covers
925,000 acres, most of it in private hands, making it a prime target of real-
estate developers and speculators. In the early 2000s, a rancher wryly com-
mented: "This is the most beautiful valley in the world to starve to death in."[1]
Madison Valley land prices had risen from $563 to $1,251 per acre during the
1990s—more than three times the agricultural value of the land—due to the
proliferation of hobby ranchers and ranchettes. Faced with this dilemma, in
1996 a group of local ranchers began talking about the future they wanted for
the Madison Valley. They faced numerous issues beyond rising land prices
and stagnant cattle prices: growing elk herds (*Cervus canadensis*) were con-

suming much of the forage on their private lands; wolves (*Canis lupus*) had been recently released in Yellowstone and would soon make their way into the valley; environmentalists were suing to end livestock grazing on public lands; and noxious weeds were crowding out native grasses. The ranchers therefore organized a forum to discuss community concerns and propose a means to address them, and the Madison Valley Ranchlands Group was born.

Seeking diversity in their discussions and proposals, the ranchlands group (a 501[c]3 nonprofit organization) invited government agency representatives, hunters, absentee ranch owners, and environmental groups to the table. While the notion of a bunch of ranchers forming a nonprofit conservation organization seemed unusual at the time, the group's philosophy was that, by focusing enough minds on an issue, viable solutions might be found. The Madison Valley Ranchlands Group launched multiple ventures: it formed a committee to coordinate noxious weed control efforts across the valley's nearly one million acres; it organized and managed a program called Range Riders to place cowboys on summertime grazing allotments to keep cattle safe from wolves and grizzly bears (*Ursus arctos horribilis*), with area environmental groups helping to fund the program; and it tackled the controversial topic of unregulated population growth by hosting public forums where valley growth could be discussed. The result was Madison County's adoption of a growth-policy action plan.

Each year since its founding, the group has also hosted Living with Wildlife workshops to teach new subdivision residents how to coexist with their nonhuman neighbors. The Madison Valley Ranchlands Group also has invited state wildlife-agency personnel, hunters, and landowners to discuss the valley's rapidly growing elk herd. Through a multiyear dialogue, the group began facilitating increased public elk hunting on private lands to help bring herd growth rates in line with desired objectives. The group also initiated programs to market local beef, to develop alternative energy production, and to launch an eco-tourism company—Madison Valley Expeditions. The group has learned that the collaborative process provides a meaningful way to engage people with divergent viewpoints. Community members may not always agree on every point, but they have learned to move forward together on important topics like conserving open space and controlling invasive-plant species. Madison Valley now has about 36 percent of its private lands in conservation easements, and the Madison Valley Ranchlands Group Weed Committee is one of the most active of its many endeavors, hosting a Weed Fundraiser every year and sponsoring a pilot project to investigate the effectiveness of sheep grazing on the reduction of noxious invasive plants like spotted knapweed. Without the ranchlands group, the beautiful Madison Valley would be a landscape like many others in which the New and Old West continue to collide.

## NOTES

1. A. B. Pearce, "Uncommon Properties: Ranching, Recreation and Cooperation in a Mountain Valley" (Ph.D. thesis, Stanford University, 2004), 160.

# 12

# INTEGRATING DIVERSIFIED STRATEGIES
# ON A SINGLE RANCH

From Renewable Energy and Multiple Breeds
to Conservation Easements

*Dennis Moroney*

### IN BRIEF

· The 47 Ranch demonstrates the use of a diverse set of strategies for private
  ranchland conservation, including a conservation easement, solar and wind
  power generation, raising diverse breeds of livestock and marketing them
  in innovative ways, permaculture using olive trees, engaging in numerous
  environmental restoration and wildlife habitat protection projects, and
  promoting environmental and agricultural education.
· Federal and state government programs play a critical role in providing
  financial assistance to private landowners to help them maintain working
  lands and conserve them; examples include low-interest loans to new ranchers, grants for watershed restoration and wildlife protection, and funds for
  the purchase of conservation easements.
· Conserving the skills and knowledge associated with "conservation ranching" is as important as conserving the natural resources that support it.
· Diversification of the products from and functions of a ranch to promote
  predictability and stability is an important strategy for working landscape
  conservation in light of the constantly changing social, economic, and ecological realities in which ranchers and other rural producers operate.

On the northeastern flanks of the Mule Mountains in southeastern Arizona,
not far from the border with Mexico, sits the 30,000-acre, privately owned
47 Ranch (see map, fig. 12.1). The ranch, in multiple ways, exists at a crossroads—between the Chihuahuan and the Sonoran Deserts, between northern

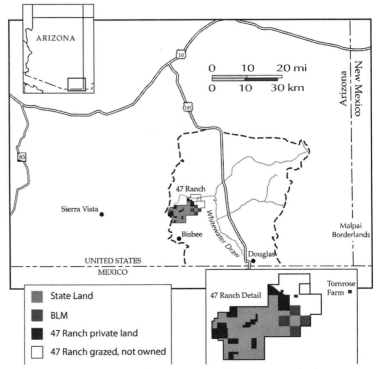

**Figure 12.1.** Location of 47 Ranch in southeast Arizona. Map by Darin Jensen and Syd Wayman, 2013.

Mexican and western American ranching traditions, and between the past and the future. As ranchers at this crossroads, my wife Deb and I have become multitasking generalists. We are simultaneously grass farmers, livestock breeders, wildlife habitat stewards, watershed restorationists, solar and wind energy generators, permaculturists, teachers, students, marketers, mentors, venture capitalists, and business partners. Ours is a unique lifestyle; a sometimes questionable business; an infinitely interesting home and family life; and a social, economic, and ecological challenge. The challenge emerges out of the fact that our work is simultaneously about the productivity of a certain piece of country; the cultural overlap of producer and consumer; urban and rural lifestyles; and traditional and contemporary practices—all set against a backdrop of changing social values, an uncertain economic future, and observable climatic change. No single strategy we've tried is enough to stitch everything we'd want in the West back together, and so we've incrementally diversified our strategies for making our ranch work in the larger natural and cultural landscapes in which we reside.

In this chapter I describe a number of innovations we are using to broaden our base of economic support and increase our long-term sustainability. These include using conservation easements (see also spotlight 12.1); using extensive solar and wind energy for water pumping and electrical generation; adding sheep and goats to our traditional beef operation; experimenting with desert-evolved and heritage livestock breeds to increase adaptation to our semiarid landscape; adding 125 olive trees of seven varieties as part of a permaculture-inspired windbreak capable of producing local cooking oil and artisanal olives; and developing alternative ways to market our products. In addition I discuss some of our work to restore degraded watershed conditions and improve wildlife habitat, as well as our role in helping to develop the next generation of responsible food producers. I also describe how we cooperate with other producers to promote local sustainable food production and protect the land and water resources that make our local agriculture possible.

## DIVERSIFICATION ON THE 47 RANCH

### Electricity Generation

The 47 Ranch is divided into 15 large pastures and numerous smaller enclosures, with relatively good water distribution in each pasture, thanks to seven solar-powered pumps and windmills located around the ranch. In addition, we graze 6,500 acres of currently undeveloped ranchette subdivision property that adjoins the ranch. Less than five miles away, we also have a small farm with grid-tied solar- and wind-powered electrical generation, which helps us pump groundwater from shallow wells to produce irrigated forage for our sheep and other livestock. Because we only irrigate during the spring and early summer before the monsoon rains arrive, surplus power generated during the rest of the year is purchased by our local electricity cooperative and helps to offset the cost associated with irrigation water pumping. Because of our relatively remote location, all electricity used to power the houses, shops, and outbuildings at our ranch headquarters is generated onsite with a combination of solar and wind generators.

### Livestock Breeds

We have been rigorously selecting our Arizona native cattle for heritable traits such as disposition, fertility, and tenderness for over 20 years. While we are able to graze our cattle year round on the desert grassland and mountain pastures found on the ranch, we know that they must be genetically adapted to thrive in these sometimes harsh conditions. We experimented with the use of seven Waguli bulls—a hybrid developed through work at the University of Arizona—with our native cows, and we liked the results we were seeing in their calves.[1] However, as we began to take their offspring to slaughter we found that although the meat from these animals was very tender, it had very little marbling, and the flavor was not particularly great.

We market a large part of what we produce directly to the consumer through farmer's markets, food cooperatives, restaurants that feature regionally produced foods, and small grocers. Because these animals are raised to slaughter weights entirely on natural rangeland forage, they have tended to be on the lean side. While many customers say they want lean meat, the fact is that the meat generally tastes better with some fat in it, and the fat produced by pastured livestock actually has some health benefits.

In the summer of 2010, we purchased 20 bred Criollo heifers, and later that year also acquired four Criollo bulls, and six two-year-old steers sourced from Chinipas, Chihuahua, by way of the U.S. Department of Agriculture's Jornada Experimental Range, located near Las Cruces, New Mexico. The Criollo cattle originated in North Africa and are direct descendants of the original cattle brought to the western hemisphere by the Spaniards in the 1500s. These cattle have a smaller frame size, exhibit high fertility, and are very well adapted to thrive under harsh desert conditions. In addition they are known to "marble" or put on slight amounts of intramuscular fat on a native rangeland diet. The meat from these Criollo steers has proven to be very popular with our farmer's market customers. In 2011, we were able to get a taste of the Criollo steers that had spent six months on our ranch before going to slaughter. We were completely amazed at the carcass quality that these animals exhibited. The meat was darker than our waguli crosses, with abundant marbling distributed throughout the muscle tissue, and the flavor was extraordinary. We felt as though we had finally found the desert-appropriate beef animal we had been searching for. Our customers rave about the flavor and tenderness, and our restaurant clients are now ordering full sides of Criollo beef at a time. We are currently breeding all of our native cows to Criollo bulls, and our herd of pure Criollo cows has grown to nearly 50 head. We are confident that the resulting genetic package is taking us in the right direction as we try to anticipate the effects of climate change on our region.

We added Navajo-Churro sheep to the mix because they produce delicious tender meat with a distinctive flavor and high-value wool favored by Navajo weavers, hand spinners, and makers of wool felt garments (see chap. 14 for economic aspects of raising these sheep). The Churro also is a descendent of the original sheep brought to the western hemisphere by the Spaniards and, like the Criollo, has survived centuries of benign neglect in the desert Southwest, adapting to its temperature extremes, periods of drought, and a diet that often has consisted of as many shrubs and forbs as range grasses. The kind of diversity the Navajo-Churro contributes to our operation helps us keep our options open in an uncertain future.

We viewed goats as a complimentary enterprise primarily because they prefer a diet consisting of shrubs and annual forbs. They are an overlay on the same high desert landscape with the cattle and produce tender and flavorful meat of fine texture and leanness. We raise Boer goats and Spanish-Boer crosses. The Boers originated in South Africa and were developed as a meat breed, with large frame size and heavy muscling. They also frequently produce twins, and sometimes triplets. We started crossing them with Spanish genetics for the same reasons we have found to use Spanish-origin sheep and cattle. Centuries of surviving and thriving in a harsh environment have winnowed out the weak and frail, resulting in a goat that is virtually immune to all diseases, is very fertile, and has excellent mothering instincts. The cross between the two breeds again seems particularly well adapted to our conditions.

*Olive Production*

Permaculture training, and a friend producing artisanal olive oils, inspired us to plant 125 olive trees of seven different varieties wrapping around the windward sides of our ranch headquarters. The trees are grown on a drip irrigation system, fertilized with goat manure, and mulched with residue from our chicken pen. Once again the Spanish connection is evident, with several of the varieties originating in Spain. Once these trees begin to produce in a few years we plan to add pork produced from pigs finished on the spent olives that remain after pressing for oil.

*Conservation Ranching and Habitat Conservation*

One of the most important factors we feel gives value to our products is our commitment to land care. Over the years we have come to understand that we are working in a landscape that has seen some pretty severe degradation over the last century. Much of what has changed is the result of a combination of factors, including changes in fire frequency, drought, inadequate grazing management, climate change, and the unintended consequences of species elimination. In general, what has resulted from these combined factors is reduced vegetative cover, accelerated erosion, the loss or reduction of key biological species, and the introduction of nonnative species. We have for 20 years been very involved in what some would term "conservation ranching," actively working to restore a functional water cycle, increase ground cover, slow or eliminate soil erosion, and restore or augment wildlife habitat. We have viewed the ranch and its surroundings in a holistic context and attempted to adapt our management to the needs of the land and use the livestock to help accomplish our goals. Specifically, we have built miles of new fencing and extended waterlines to improve livestock control and grazing management. In addition, we have constructed over 600 rock erosion-control structures in washes and gullies, replanted hundreds of acres with native range grasses, and worked with the Arizona Game and Fish Department to create year-round water for wildlife. In addition, extensive work has been done to ranch roads to enhance their role in water harvesting and reduce erosion.

We recently completed a habitat project, funded in part by the U.S. Fish and Wildlife Service, to benefit three species of migratory sparrows. The project involved selectively removing mesquite shrubs and creating small water bars, brush rows, and spreader structures to increase grass cover and reduce shrub density on about 50 acres. We also constructed 200 artificial burrows for relocation of burrowing owls (*Athene cunicularia*) affected by residential subdivision development near Phoenix and Kingman. Another project completed with Arizona Game and Fish Department and the Natural Resource Conservation Service involved drilling a well, outfitted with a solar pump, to create a pond for endangered Chiricahua and lowland leopard frogs (*Lithobates chiricahuensis* and *Rana yavapaiensis*, respectively).

Over the years, we have had interns, apprentices, graduate students, and others engaged in all manner of formal and informal educational activity on the ranch. Every summer we have one or two students here engaged in some form of study for credit. In 2011 we hosted two apprentice farmers under the Quivira Coalition's New Agrarian Program. We have had many students and researchers from the University of Arizona, Prescott College, and Cochise College, as well as international students. The ranch has served as a lab and teaching facility for countless workshops and short courses on topics ranging from permaculture, range management, alternative energy, livestock judging, horsemanship, and ranch roping, to spinning, weaving, felting, and wool processing. It has also served as the site for numerous meetings focused on land and watershed restoration, marketing locally, cross-border conservation with our neighbors in the Mexican state of Sonora, and grass-fed beef production for our neighbors on the San Carlos Apache Nation. We have also hosted numerous K–12 school groups for farm and ranch visits, field days, and so forth.

In the interest of addressing the transfer of knowledge, skills, and assets to the next generation of food producers, we sold a half interest in the cowherd to our former employee and for a couple of years he became our partner. This sale took place in 2010 and was accomplished after long and thoughtful conversations about how to keep the ranching option open to our children without burdening them with a sense of obligation regarding a career in ranching. Our partner was able to get a low-interest Beginning Farmer-Rancher Loan through the U.S. Department of Agriculture and was to have his half of the herd paid for within five years. He paid us rent for use of his share of the forage and was building equity while doing what he loves. Unfortunately, three successive years of drought and the need for herd reduction resulted in his taking his share of the cattle off the ranch, effectively ending the partnership. The arrangement did, however, give him the start he needed to become an owner of his own cow herd, and he is ranching near his parents on leased land. We still need to find practical ways to help the younger generations to transition to ownership of farms and ranches, as our current crop of producers moves into retirement age.

We could not do what we do without many cooperating partners. We have long-running working relationships with many government agencies and entities that we have worked with to accomplish goals and complete projects in which we shared a common interest. Some of these partnerships benefit wildlife conservation or habitat management, while others focus on soil erosion, improving grazing management, sharing information for research purposes, and improving our management capabilities. While we have completed many projects over the years, almost all of them were done in partnership with at least one governmental agency. Some projects involved many different agencies, nongovernmental conservation organizations, and other neighboring producers.

One of these multipartner projects serves as a good example of the kind of cooperation and coordination that have good on-the-ground results. The Hay Mountain Watershed Restoration Project began as a conversation with three neighboring ranchers and our Natural Resource Conservation Service field staff. Together, the four ranches occupy a significant part of the headwaters of the Whitewater Draw, a subwatershed of the Rio Yaqui, one of the few Arizona watersheds to drain south into Mexico. But before the Whitewater Draw crosses the border, it flows into a wildlife refuge that serves the seasonal habitat needs for upward of 20,000 sandhill cranes (*Grus canadensis*) and thousands of other migratory waterfowl, including snow geese (*Chen caerulescens*), trumpeter swans (*Cygnus buccinator*), and black-bellied whistling ducks (*Dendrocygna autumnalis*). Hundreds of other species of waterfowl, songbirds, terrestrial mammals, reptiles, and amphibians also benefit from the refuge. In addition, the cities of Douglas, Arizona, and Agua Prieta, Sonora, both draw their domestic water supply from the aquifer.

The ranchers came together in 2005 and, with assistance from the U.S. Department of Agriculture's Natural Resource Conservation Service and Agriculture Research Service, applied for a grant from the Arizona Department of Environmental Quality to implement a variety of conservation measures on about 7,000 acres of rangeland in the upper watershed. Each of the four ranches was already using rest-rotation grazing practices and had a long history of conservation ranching. The Hay Mountain Watershed Project placed well over 1,000 rock erosion-control structures and spreader dikes in drainages to slow runoff velocities, trap silt, and restore native grassland vegetation. Improvements were made to some existing stock tanks to help them better function as sedimentation traps, and efforts to manage brush encroachment allowed for native grasses to regenerate.[2] The project has reduced silt inflows to the Whitewater Draw by an estimated 22,000 tons per year. Along the way, the project has augmented other conservation work on each of the ranches and benefited both wildlife and livestock.

We have also completed several multiagency projects on the ranch benefiting threatened and endangered amphibians, migratory songbirds, and three species of quail found on the ranch. Other partners have included Bat Conservation International and Wild at Heart, both of whom worked with us on wildlife conservation measures.

Finally, the ranch is very involved in promoting sustainability, local foods, and preservation of open spaces and habitat. We work with Baja Arizona Sustainable Agriculture, Sabores sin Fronteras, and Ecoasis to promote all-natural regional food production. We've also been active in more conventional agricultural organizations. I recently served as the president of the Arizona section of the Society for Range Management, and as the president of the Cochise-Graham Cattle Growers Association. I am also on the board of the Arizona Cattle Industry Foundation and active in the Southwest Grassfed Livestock Alliance.

All of these activities feed back in some way to achieving a more resilient ranching operation, and, we hope, contributing to our community. At the same time, we have to be profitable to be sustainable. While the economics of ranching are not well understood by most people, the basic components are land, water, forage, livestock, and human knowledge or skill. Many other variables can come into play, sometimes without warning; and so while we strive to move toward clearly defined goals and toward sustainability, we also must remain very responsive to our changing economic, social, and environmental realities. Decisions about the quality or quantity of meat we offer at our local farmers' market must be made more than three years in advance; at the same time, the seasonal variation in rainfall can trump the best management and preparation; and political decisions made in Phoenix or Washington, DC, can change the price of a ton of hay overnight. The fact that there is so much variability keeps the economics of a ranching operation interesting, but elements of predictability and stability make it easier to plan for the future. In the remainder of this chapter, I discuss one of the strategies we chose—the conservation easement—to help improve our financial outlook and lend some predictability to the long-term viability of the ranching operation. With the narrow margins ranching provides, reducing or eliminating debt can be essential to long-term survival of family-scale ranches.

## CONSERVATION EASEMENTS ON THE 47 RANCH

A real-estate mortgage is a tough burden for a bunch of cows to drag around. For our ranch to be sustainable, we needed not only to diversify but also to pay off the mortgage. In 2007, almost a decade after purchasing the 47 Ranch, we completed a conservation easement on a portion of it. The agreement provided for the protection of wildlife habitat and continued active ranch management on three parcels in Abbot Canyon in the Mule Mountains. Each of the three parcels was a 320-acre homestead with a variety of terrain and vegetation types, and each had some riparian habitat and seasonal surface water. The conservation easement agreement involved the sale of development rights for each of the three parcels, which effectively eliminated the possibility of future subdivision or development for nonagricultural purposes. We retained the right to continue ranching and using the property as it has been used since settlement of the area in the late 1800s. The process took considerable time and effort. Throughout the process, we kept in mind that a conservation easement is basically a simple contract between a willing seller and a willing buyer. As with all contracts, the devil is in the details, but we made sure the details never had us giving up anything we didn't intend to offer and didn't involve our agreeing to any conditions that were inconsistent with our goals, including future enjoyment and economic use of the land. We knew we had some valuable and unique land up in the mountains that included a number of uncommon plant and animal species.

It was with these goals and values in mind that we sought like-minded partners for negotiation of the easement. In our search, we asked the rhetorical question, Who else might see value in an undeveloped ranch landscape, and who might be willing to pay to help us keep it that way? Like many people, we were concerned about the seemingly endless subdivision of Arizona's former ranch lands into ranchettes. As it turned out, the Arizona Game and Fish Department had received a grant from the U.S. Fish and Wildlife Service under their Landowner Incentive Program. The funding for this program came with certain conditions. First, there was a list of "species of concern" that would need to be present to make the land eligible for funding—we had 11 species from the list. The other condition was a 25 percent match from nonfederal funds. The Arizona Game and Fish Department did not wish to be burdened with the long-term obligation of monitoring the easement for compliance with the contract, so we asked our local land trust—the Arizona Land and Water Trust—and they agreed to hold the easement.

The next step was to have the property appraised by a certified rural appraiser with experience evaluating properties subject to the restrictions imposed by a conservation easement. Establishing the fair market value of the whole ranch and then carefully considering any diminution of value that will occur once the terms of the easement are attached to the deeds for the subject properties determine the basis of the value of a conservation easement. A value per acre is determined before the easement is in place, and then the diminution is calculated and subtracted from the ranch value on a per-acre basis. This difference is considered to be the "value of the easement." The regulations for easements completed under the Landowner Incentive Program require that the appraisal be reviewed by an independent third party approved by the federal government and that the appraisal be conducted according to the most stringent standards. Finally, the easement must be approved by the regional office of the U.S. Fish and Wildlife Service.

The 25 percent nonfederal match requirement gave us an opportunity to contribute 25 percent of the value of the easement and provided us with a significant tax deduction with a carryforward of up to 15 years. Because we were not paid the full value of the easement but, instead, accepted 25 percent less, that amount was considered a donation under the Internal Revenue Code and, thus, eligible to be used as a tax deduction. The balance of the easement value came from Fish and Wildlife funds granted to the Arizona Game and Fish Department and then paid to us just like any other real-estate transaction. Throughout the process, Arizona Land and Water Trust acted as a neutral, third party and facilitated the transaction.

When the dust settled, a Grant of Conservation Easement was recorded with Cochise County.[3] The easement allows for normal repair and replacement of agricultural improvements and specifies that ranchers may "continue the current modes and levels of ranching including the pasturing, grazing, feeding and care of livestock, including but not limited to horses and cattle, and to adapt innovative and experimental methods of ranching which may serve to further the sustainability of the agricultural enterprises and do not detract from the conservation goals of the easement." We can build agriculture-related improvements needed for the ranching operation, including roads and wells, erosion control structures, and so on. We retain all our water rights and agree to maintain a current conservation plan with the Natural Resource Conservation Service. We also agree to an annual site visit to ensure compliance with the easement terms. The easement, in turn, prevents us from building nonagriculture-related structures, subdividing the properties into two or more parcels, dumping trash, establishing a commercial feedlot, conducting commercial-scale mining, using roads for racing all-terrain vehicles, hosting billboards, or establishing athletic fields, golf courses, airstrips, or motocross tracks.

The benefits to our family ranch included a reduction in the value of our taxable estate, making it easier for our heirs to carry on ranching on the property (for more on the estate tax, see spotlight 13.1), a cash payment sufficient to pay off our mortgage, a deduction on taxable income with a significant carry-forward, no loss of usefulness of the land, and perpetual protection of wildlife habitat. The value of the development rights extinguished in this particular easement represent roughly 15 percent of the value of the ranch. However, the elimination of the mortgage placed us on more secure financial footing overall, allowing us to shift our focus to other important goals for our ranch.

In June of 2013 we completed a second conservation easement on 1450 acres of land surrounding the ranch headquarters. This easement was completed under the terms of the USDA, Natural Resources Conservation Service, Grassland Reserve Program, and resulted in nearly half of the deeded land of the ranch being protected forever from the threat of non agricultural development and subdivision. The terms and conditions of the easement were similar to the previous one, but gave protection to the lands considered most at risk for sale to real estate developers and speculators. Some 90 acres near the existing rand headquarters was excluded from to easement to allow for future expansion of ranch-related infrastructure, housing, and agriculture-related improvements.

## CONCLUSIONS

No matter what challenges arise from energy shortages, climate change, or social unrest, people will still need food and fiber to survive. There is no more important work than restoring the productive capacity of wild and open landscapes in the West and helping to pass on the skills and knowledge to manage them wisely. Along the way, collaboration is critical; reaching out to the community—the many agencies and constituencies that may share common interests—has been an important part of the 47 Ranch's holistic approach to land care. Our relationship with our community includes our role as food producers, providing a small part of the food produced and consumed within our local "foodshed" (i.e., the geographic region that produces the food for a particular population), and as active participants in promoting and practicing elements of sustainable living within our local area. Among the approaches we found most useful to our long-term stability were diversification of products and functions and the negotiation of a conservation easement on the property. The conservation easement, in particular, has ensured that our ranch will stay in ranching for generations to come and that our own heirs will be able to work the family land.

## NOTES

1. When it became possible to sequence DNA, the University of Arizona began to do some work to identify specific breeds that have both the physical attributes to thrive in a semiarid environment and also the particular gene associated with tenderness. The result of their work is a hybrid derived from a cross between the well-known Kobe beef producing Wagyu breed of Japan and a lesser-known breed from Zimbabwe called the Tuli. The Tuli brings drought tolerance and insect and disease resistance like that found in the more common Zebu or "Brahman" type cattle but with the gene for tenderness. The resulting "Waguli" seems to be a pretty good match for this environment.

2. Additional funding was obtained from Arizona Department of Agriculture, and technical assistance came from the Arizona State Land Department, USDA Natu-

ral Resources Conservation Service, and USDA Agriculture Research Service. The University of Arizona Cooperative Extension Service and graduate students from the University of Arizona's School of Renewable and Natural Resources provided vegetation monitoring.

3. The Grant of Conservation Easement reads: "The purpose of this easement is to protect the wildlife habitat, archaeological, historic, cultural, scenic and open space values to assure the property will be retained in its agricultural, scenic, and open space condition in perpetuity, and to prevent any use of the property that will significantly impair or interfere with the conservation values of the property." (Recorded under Cross U Cattle Co., an Arizona Corporation; accessible at the Cochise County Arizona Recorders Office.)

# Spotlight 12.1

# PRIVATE LAND CONSERVATION TRENDS IN THE WESTERN UNITED STATES

*Jon Christensen, Jenny Rempel, and Judee Burr*

### IN BRIEF

- Although the total acreage of developed land and unprotected private land in the West dwarfs the total acreage of private land having conservation status, more private land in the western states has been conserved than developed since the late 1990s.
- State and local land trusts, using conservation easements, are largely responsible for private lands conservation.
- Government funding to support private lands acquisition and conservation, along with tax incentives for conservation easement donations, have both declined substantially in recent years, threatening private lands conservation efforts.
- In the future, conservation policy and programs will play an important role in shaping working landscape conservation on private lands and in shaping the character of the West in general.

Private land conservation has been shaping the future of much of the American West as decisively as development in recent years. Since the late 1990s, more land has been conserved by state and local land trusts in the 11 western states than has been converted from rural land to developed properties (fig. 12.1.1).

And between 2005 and 2010—a period that included America's post-2008 recession—western land trusts conserved more land than ever, even as many environmental advocacy groups had to trim budgets and as federal funding

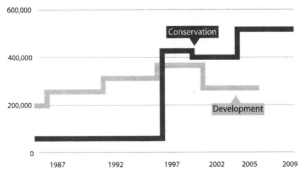

**Figure 12.1.1.** Annual average of private acres conserved or developed in the West, 1985–2009. Sources: Land Trust Alliance (conservation data), U.S. Department of Agriculture (development). Figure by Geoff McGhee, Bill Lane Center for the American West created in 2012.

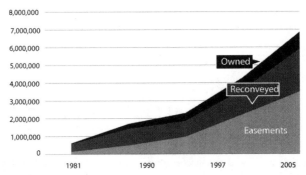

**Figure 12.1.2.** Private conservation measures, in acres, 1981–2005. In recent years, conservation easements have become the dominant tool in conservation, especially as it has shifted to working landscapes—ranches, farms, and forests. Land trusts continue to acquire and hold or own properties and, sometimes, purchase properties and then reconvey or transfer lands to public agencies. Source: Land Trust Alliance. Figure by Geoff McGhee, Bill Lane Center for the American West, created in 2012.

for conservation faltered. Most of the growth has come through conservation easements (fig. 12.1.2). Conservation easements are becoming ever more popular because they allow land trusts to protect land at a lower price compared to land acquisition or reconveyance. Trusts also avoid taking on management costs, and easements offer benefits for landowners who want to retain their land, particularly working ranches, farms, and timberlands.

The federal Land and Water Conservation Fund, which agencies once relied on to acquire private lands, suffered a 38 percent cut and protected just over 500,000 acres between 2005 and 2010.[1] During the same period, private nonprofit land trusts protected 20 times as much undeveloped land—10 million acres nationwide, according to a census of 1,700 land trusts that were

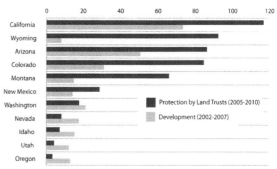

**Figure 12.1.3.** Annual average private land protected, in thousands of acres. Sources: Land Trust Alliance (conservation data), U.S. Department of Agriculture (development). Figure by Geoff McGhee, Bill Lane Center for the American West, created in 2012.

part of the national Land Trust Alliance.[2] Land trusts also grew in other ways. These included a 19 percent increase in paid employees and contractors, a 36 percent increase in operating budgets, a 70 percent increase in volunteer numbers, and a near tripling of long-term endowments. The latest gains brought the total area protected by the nation's land trusts to 47 million acres—more than twice the area covered by all of the national parks in the lower 48 states.[3]

These recent trends are particularly strong in the western states, where statewide and local land trusts conserved 2.6 million acres between 2005 and 2010, 30 percent more than they did from 2000 to 2005. The success of western land trusts put California, Colorado, and Montana among the top five states nationwide in total private land conserved. Arizona, Nevada, and Wyoming also made large gains compared to the previous period. And in Colorado, Montana, and Wyoming, so much more rural land is now being conserved than is being developed that, if these trends continue, much of their open land could likely remain undeveloped (fig. 12.1.3). In other states, such as Idaho, Nevada, Oregon, and Utah, however, rural land still appears more likely to be developed than conserved.

One explanation for these trends is that, between 2005 and 2010, specifically protecting working landscapes became a priority for the majority of land trusts nationwide, according to the Land Trust Alliance, and that trend was also particularly strong in the American West. Some of the growth in statewide and local conservation efforts in these landscapes can be attributed to the maturation of new agricultural land trusts, led by farmers and ranchers, which have sprung up all around the West. From the Malpai Borderlands Group on the border of Arizona and New Mexico, to the Blackfoot

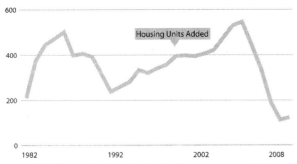

**Figure 12.1.4.** Annual building permits issued for housing in western states, in thousands, 1982–2008. Source: U.S. Census Bureau. Figure by Geoff McGhee, Bill Lane Center for the American West, created in 2012.

**Figure 12.1.5.** Distribution of land types in western states in 2010. Sources: Land Trust Alliance, U.S. Department of Agriculture. Figure by Geoff McGhee, Bill Lane Center for the American West, created in 2012.

Challenge in Montana, local collaborative efforts and agricultural land trusts have spurred community-based, watershed, and landscape-scale conservation efforts around the region. Meanwhile, construction stagnated during the recession that began in December 2007, with the number of new housing units added dropping off steeply and falling for the first time since the early 1990s (fig. 12.1.4).[4]

Still, these recent trends need to be put in the context of the overall extent of lands protected by land trusts compared to lands converted for development and the percentage of protected versus unprotected private, undeveloped rural land (fig. 12.1.5).

Lands protected by trusts totaled almost seven million acres by 2010, compared to more than 19 million acres that were developed in the western states by 2007.[5] But here, too, we find great differences among the states. In Montana, the total acreage protected by land trusts has surpassed the acreage of developed land in the state. With the current pace of conservation in

Colorado, the gap between developed lands and protected lands is also narrowing rapidly. In other states, however, the gap continues to widen. Still, the amount of unprotected, undeveloped rural land continues to dwarf the amount of land protected by land trusts in all of the western states.

Since conservation by land trusts through fee acquisition of land—and particularly easements—is usually designed to protect land for conservation in perpetuity, these trends are likely to reshape the American West permanently. However, it is important to put trends in conservation by land trusts as well as development into the overall context of a region that is still largely defined by the same three overriding spatial patterns that have shaped it for more than a century. The western United States remains predominately rural. Western states have relatively dense concentrations of population and developed land in a few metropolitan areas. And, a large portion of the land in western states is public land owned by the federal government and managed by the U.S. Forest Service, the Bureau of Land Management, or the National Park Service.

Whether the growth of private conservation in the West will continue unabated is unclear. The recession may have lagging effects. Even as the regional economy slowly began to recover, land trust leaders worried about the loss of generous tax incentives for conservation easement donations, as well as the decline in federal funding for conservation. While land trusts have more than filled the gap, large-scale efforts like the Blackfoot Challenge often depend on federal, state, and local public funds to provide crucial initial incentive to bring private donors to the table to continue to get big deals done. Land trust leaders advocate for extending tax incentives associated with donations of conservation easements to land trusts, especially for working farmers and ranchers, who often do not earn enough annual income to write off donations of large value in a single year. Some also suggest that the little federal funding available for private land conservation should be targeted specifically as seed money for the public-private partnerships that have been so successful in recent years as part of the community-based efforts led by land trusts to protect large-scale landscapes and watersheds. Private lands conservation policy and programs are likely to continue to play an important role not only in the future of western land conservation but in shaping the region as well.

## NOTES

1. "Conservation Almanac: Federal, State, Local, and Private Lands," Trust for Public Land, last updated 2011, www.conservationalmanac.org.
2. K. Chang, 2010 National Land Trust Census Report: A Look at Voluntary Land Conservation in America (Washington DC: Land Trust Alliance; Cambridge, MA: Lincoln

Institute of Land Policy, 2010), http://www.landtrustalliance.org/land-trusts/land-trust-census/census.

3. Ibid.

4. "Building Permits Survey," U.S. Census Bureau, http://www.census.gov/construction/bps/.

5. Because of the timing of surveys by the Land Trust Alliance and the Natural Resources Inventory, these figures are regularly out of step but, over time, provide useful indications of trends. After 2007, development nearly ground to a halt in much of the West, so these snapshots are likely to be as comparable as they will ever be in this particular period.

# PART FIVE

## EMERGING APPROACHES TO CONSERVING

## WORKING LANDSCAPES

Although many collaborative conservation groups arose to search for common ground amid the polarizing politics of the 1990s, most alliances today
focus on saving that common ground from development and restoring its
most degraded lands and waters. To do so, a rather small toolkit (conservation easements, mitigation banking, memoranda of understanding with
government agencies) has expanded to include an ever-widening portfolio
that includes comanagement tools, policies, and financial strategies to keep
working landscapes *working* for both the private and public good, for both
residents who live off the land and tourists who visit it, and for humans as
well as other species. This portfolio is growing even as we write this, with
novel public-private partnerships and new hybrid for-profit/nonprofit structures being proposed and provisionally used for conservation financing every
year. As a result, this book can in no way pretend to be the "last word" on the
matter. Experimentation and innovation are the names of the game.

In this final section, we explore two experiments that illustrate innovative
ways to fund conservation on a large-landscape scale. Chapter 13 addresses
how Pima County, Arizona, created the Sonoran Desert Conservation Plan to
preserve open space and biodiversity by protecting working ranches from development. The plan took shape in 1998 in response to the listing of the cactus
ferruginous pygmy owl (*Glaucidium brasilianum cactorum*) under the Endangered Species Act. Because some of the owl's critical habitat was squarely in
the path of one of the county's major growth corridors, the county realized
the listing would have an enormous impact on county projects as well as private development. Rather than dealing with the listing on an ad hoc basis,

however, Pima County decided to develop a multispecies habitat conservation plan for all species that might be vulnerable to the rapid expansion of metropolitan Tucson. It then mapped the overlapping critical habitat for 44 priority vulnerable species and came up with the Conservation Lands System that the county board of supervisors adopted as its Comprehensive Land Use Plan in 2001. Then in 2004, Pima County voters approved a $174.3 million Open Space Bond to acquire key private parcels within that comprehensive Conservation Lands System. Between the passage of the bond and 2010, when most of the funds had been spent, the county purchased every major ranch that came up for sale, preserving thousands of acres of private lands from being developed. With respect to our theme here, the county signed management agreements with the former owners of those ranches to continue running cattle on those lands, thereby honoring its commitment to one of the Sonoran Desert Conservation Plan's five major goals: ranch conservation. The estate tax is one reason ranchlands in the West are being subdivided and sold (see spotlight 13.1).

Since the Sonoran Desert Conservation Plan's inception in 1998, similar initiatives have also arisen in other stretches of the West. For example, the Laramie Foothills Mountains to Plains Project—an alliance among private landowners, public agencies, and land trusts in Larimer County, along with the Nature Conservancy and the city of Fort Collins—have protected nearly 30 square miles of working landscapes from exurban development. Additionally, Transition Colorado has been successful in forging a new financial structure to keep Grant Family Farms—the largest community-supported agriculture project in the Intermountain West—so that it continues to produce food for the local economies of the Front Range. Such experiments marshal urban financial and political support to conserve the working landscapes that surround two of the most rapidly urbanizing areas of the West and provide them with clean air, water, recreation, and wildlife.

A second innovation to emerge in the last decade is to pay rural producers directly for "ecosystem services" using market-based approaches, such as those described in chapter 14. Paying rural producers for the conservation of ecosystem services dates at least to the 1980s, when the Conservation Reserve Program was established (see spotlight 14.1). Some rural planners believe that the "restoration economy" may be the next growth frontier, generating wealth for rural producers while at the same time reversing the global decline in functioning ecosystems.[1] Payments for ecosystem services generally follow two strategies: (1) incorporating the costs of sound ecosystem management into the costs of food and fiber products or (2) offering financial incentives to farmers, foresters, and ranchers who maintain or restore ecosystems. Payments for ecosystem services is a market frontier, full of rapidly changing experiments. When this volume went to press, only three ecosystem services had been monetized to any extent: carbon sequestration, water quality and quantity (particularly where they are linked to wetlands mitigation), and habitat for species at risk. However, these markets have been and may continue to be extremely volatile; the Chicago Climate Exchange, North America's first voluntary carbon emissions cap-and-trade program, which opened in 2003, ended trading in 2010.

Mitigation banking, in contrast, has proven to be a more stable form of payments for ecosystem services. The most common form involves wetlands. If a project is going to damage or destroy wetlands on either public or private land, a developer may purchase or create equivalent habitat somewhere else to mitigate the loss, often at ratios higher than the wetlands damaged or destroyed. Pima County, among other entities, employs mitigation banking to protect species in arid or semiarid habitats. Meanwhile, ever-more innovative approaches to payments for ecosystem services continue to emerge, some of them addressing the flaws of earlier "quick fixes."

Other experiments in conservation entrepreneurship like the slow-money dialogues have appeared as well. The national Slow Money Alliance network (see www.slowmoney.org) strives to develop localized, low-interest financing for food, farming, and ranching innovations. During its first three years, Slow Money's strategies led to 86 completely local investment deals, raising more than $19 million for 139 food microenterprises. Young ranchers like Paul and Sarah Schwenessen of Double Check Ranch in Arizona are among these advocates of private/public working landscapes who are already making pitches to their peers at Slow Money–sponsored events to achieve ranching conservation and restoration goals through such "locavesting."

A more ambitious endeavor is Roots of Change (www.rootsofchange.org), a long-term planning project involving multiple counties surrounding the San Francisco Bay. Roots of Change seeks to ensure watershed quality, open-space protection, and food security through the preservation of land for future agricultural and recreational needs. Taken collectively, these experiments speak to a growing desire to bridge the nation's rural-urban divide by linking urban consumers—of food, open space, recreation, wildlife—with rural producers struggling to survive the expanding cities. They are part of what Eric Freyfogle and David Walbert call the New Agrarianism, a movement to increase personal engagement in agriculture, including ranching, among people in both the city and the countryside.[2] Whether practiced in large working landscapes in the West or by jack-hammering up asphalt and concrete to create urban homesteads, the new agrarianism, like the collaborative conservation movement, seeks to unite Americans across geographic and social divides.

## NOTES

1. S. Cunningham, *The Restoration Economy: The Greatest New Growth Frontier* (San Francisco: Berrett-Khoeler Publishers, 2002).

2. E. Freyfogle, *The New Agrarianism: Land, Culture, and the Community of Life* (Washington, DC: Island Press, 2001); D. Walbert, *Garden Spot: Lancaster County, the Old Order Amish, and the Selling of Rural America* (New York: Oxford University Press, 2002); and see, generally, David Walbert's website, www.newagrarian.com.

# 13

# THE SONORAN DESERT CONSERVATION PLAN AND RANCH CONSERVATION IN PIMA COUNTY, ARIZONA

*Thomas E. Sheridan*

## IN BRIEF

- The Sonoran Desert Conservation Plan is an innovative attempt to control urban sprawl and its negative ecological and economic consequences through land acquisition funded by open-space bonds, land-use planning, and zoning requirements.
- Planned developments that meet county planning and zoning requirements generate more tax revenue than "wildcat subdivisions" because the cost of providing county infrastructure (roads, water, utilities, sewage disposal) to these unplanned subdivisions often far exceeds the amount of property tax revenue they generate.
- Multispecies habitat conservation plans are more efficient and cost effective than species-by-species consultation and mitigation activities associated with threatened and endangered species and species of concern.
- Conserving working ranchlands adjacent to urban areas can help bridge the rural-urban divide by ensuring a locally sourced meat supply for urban residents, provided local infrastructure for processing the meat is also available.

## RANCH CONSERVATION AND URBAN SPRAWL

When Pima County, Arizona, initiated its visionary Sonoran Desert Conservation Plan (SDCP) in 1998, it designated ranch conservation as one of its five major goals—the others being critical habitats and biological corridors, riparian restoration, mountain parks, and historical and cultural preservation. In the words of the SDCP: "Historically, ranching has probably been the single greatest determinant of a definable urban boundary in eastern Pima County. While over half of our 2.4 million acre region appears to be open, unused land, virtually all of this open space is used in ranching, an extensive but low-intensity land use. Through the conservation of working ranches that surround the Tucson metropolitan area, vast landscapes of open space are preserved, natural connectivity is maintained, and the rural heritage and culture of the Southwest are preserved."[1]

Pima County goes on to identify the conversion of ranch lands to real-estate development as one of the biggest threats to landscape-level conservation and states: "Today, ranch land fragmentation is greatest within a 25-mile radius of the Tucson urban core."[2] Most of the people in Pima County live in metropolitan Tucson, and, like most western cities, Tucson has grown by sprawling outward. In 1930, when Tucson's population was 32,500, there were roughly 4,500 persons per square mile in the city. That density rose slowly through the 1960 census, and then began to fall, decade by decade as the city limits expanded, to approximately 2,300 persons per square mile by 2010. During that same period (1960–2010), however, the city of Tucson's population grew from 213,000 to 520,000—reflecting the expansion of the city outward over the last several decades. Pima County's population also swelled from 266,000 to 980,000 between 1960 and 2010. Most of this growth has occurred within the 25-mile radius cited above. Because land was generally cheaper on Tucson's outskirts, developers responded to consumer demand by providing bigger lots in new subdivisions rather than by infilling within existing municipal boundaries. The result was urban sprawl that chewed up wildlife habitat and severed wildlife corridors. During the real-estate boom of 2005–8, an average of 13 acres of desert a day—nearly 5,000 acres a year—were bulldozed for residential and commercial development.[3]

A consensus seems to be emerging among city of Tucson and Pima Association of Governments prognosticators that Pima County will have a population of about 1.7–2 million people by 2050, twice as many as live there now.[4] If sprawl continues, most will live in subdivisions built on private ranch lands or on state trust lands bought at public auction—lands that used to be leased by ranchers. All that jurisdictions like Pima County can do is attempt to channel growth through the limited tools at their disposal, especially planning and zoning requirements, or to purchase open space themselves and protect it from development.

How those tools are used reflects a community's power structure, cultural values, and political ideology. Before the late 1990s, the interest groups that dominated Pima County politics supported growth. Like much of the Sunbelt, the construction of new homes was a major industry. Whenever developers requested rezonings from the county to build new subdivisions, environmentalists and neighborhood advocates protested but to no avail. Developers packed public hearings with their employees and the employees of their subcontractors; hard hats outnumbered tree huggers. Construction Feeds My Family was a popular bumper sticker. Jobs trumped environmental protection.

Pima County officials usually supported rezoning requests in order to encourage planned developments because the alternatives were the "wildcat" subdivisions blighting Arizona's open spaces. Wildcat subdivisions are rural residential areas that have never gone through any county or municipal planning and zoning review process. According to Arizona state law, a landowner can split his or her property five times, as long as the lot-splitting does not fall below baseline zoning. In Pima County, baseline zoning is usually the rural homestead classification of one home per 4.13 acres. "Basic infrastructure such as paved roads, adequate drainage, utilities, water, and sewage disposal is essentially absent in wildcat subdivisions," a Pima County study in 1998 observed. "The residents of these areas often expect the County to provide services similar to subdivided areas. They want their roads maintained; they want adequate and safe water supplies. However, to provide such, significant investments in infrastructure must be made in these areas, often at the expense of every other taxpayer in Pima County."[5] The amount that counties spend on wildcat subdivisions far exceeds the property taxes that those subdivisions generate: that same 1998 study found that the average improved full cash value of planned developments per section was $38.5 million, compared to $8.1 million per section for wildcat subdivisions.[6] In fiscal terms, then, planned developments that meet county planning and zoning requirements generate nearly five times as much tax revenue as lot splitting.

THE BEGINNINGS OF CHANGE: CANOA RANCH

Even with better infrastructure, however, planned developments also take their toll on the Sonoran Desert. In 1999, a coalition of environmentalists, astronomers, and neighborhood associations persuaded the Pima County Board of Supervisors to reject a request for rezoning some 5,200 acres encompassing most of the southern half of the historic San Ignacio de la Canoa

Land Grant south of Tucson. At a public meeting, hundreds of opponents packed a high school auditorium, overwhelming the rows of hard hats that left promptly at 5 P.M., even though the hearing lasted for six more hours. It was the first time the board of supervisors—by a vote of four to one—had denied a major rezoning in 25 years.[7] A series of occasionally bitter negotiations eventually resulted in an approved rezoning that allowed Fairfield Homes to build homes and a golf course in one area and to develop commercial properties in another. But approximately 4,800 acres, or 80 percent of Canoa Ranch, were purchased by Pima County as a county historic park in 2001.[8]

The fight over Canoa reflected a political sea change in Pima County. The Canoa negotiations brought together astronomers who wanted to protect dark skies because of the Smithsonian's nearby Whipple Observatory and environmentalists who saw sprawl as a threat to biodiversity and wildlife habitat. But the coalition also included neighborhood groups who sought to preserve the desert around their homes. Historian Adam Rome calls this attitude the "bulldozer in the countryside" mentality: people moving into suburbs and exurbs want the development to stop at the end of their streets.[9] These suburban newcomers swelled the ranks of the environmental movement in Pima County and gave it greater clout.

## THE SONORAN DESERT CONSERVATION PLAN IS CONCEIVED

That clout expressed itself in a much more ambitious endeavor soon after the Canoa fight. In 1997, the U.S. Fish and Wildlife Service listed the cactus ferruginous pygmy owl (*Glaucidium brasilianum cactorum*) as endangered.[10] Much of the critical habitat for the owl was located in the desert ironwood (*Olneya desota*) forests northwest of Tucson, a major corridor of urban growth. The listing led to the creation of the Coalition for Sonoran Desert Protection, an umbrella organization of more than 40 environmental and neighborhood groups. It also stimulated the Pima County administrator and his staff to seek a Section 10 permit under the amended Endangered Species Act. A Section 10 permit requires the permittee to develop a Habitat Conservation Plan, which specifies the conservation measures that will be taken to minimize and mitigate impacts on the covered species and the source of funding that will allow those measures to be carried out. In return, Fish and Wildlife issues an Incidental Take Permit, which protects the permittee from being prosecuted for injuring or killing a covered species or degrading its habitat in the course of carrying out otherwise lawful activities for the life of the permit.

Rather than simply applying for a permit covering the pygmy owl, however, the County decided to develop a Multi-Species Habitat Conservation Plan (MSHCP) to avoid expensive and time-consuming species-by-species consultation and mitigation. The area covered by the permit would include the entire county, except for incorporated municipalities like the city of Tucson, federal reserves like Coronado National Forest and Saguaro National Park, and the enormous Tohono O'odham Nation, which are outside county jurisdiction. At that time, eastern Pima County was home to seven species that already had been listed as threatened or endangered and 18 more that had been proposed or petitioned under the Endangered Species Act. Pima County wanted a permit that would cover those species, but it also wanted to identify other "species of concern" that might be vulnerable as metropolitan Tucson expanded. Crafting an MSHCP would help the county channel urban growth and development into areas of lower biological diversity while protecting core areas of critical habitat for listed species and species that might be listed in the future. In addition, Pima County sought to update its comprehensive land-use plan and to conserve cultural resources, working ranches, riparian areas, and mountain preserves. It called this ambitious undertaking the Sonoran Desert Conservation Plan (SDCP).

### FIGURING OUT THE DETAILS

To determine which species needed to be protected, the county created the Science Technical Advisory Team (STAT), composed of wildlife biologists and agency personnel who manage wildlife. It also hired a biological consulting firm—Recon Environmental, Inc. of San Diego—to prepare an MSHCP that would serve as the "biologically preferred alternative"—that is, the plan best suited to ensure "the long-term survival of the full spectrum of plants and animals that are indigenous to Pima County through maintaining or improving the habitat conditions and ecosystem functions necessary for their survival."[11] The county administrator and assistant county administrator successfully erected a so-called firewall to shield the STAT and Recon from political pressure by making sure other interest groups, including developers and environmentalists, did not intrude on or try to influence its deliberations. Two independent peer reviewers of the Sonoran Desert Conservation Plan praised Pima County's "demonstrated commitment to keeping science insulated from politics. . . . The autonomy of the scientists (including the STAT, the consultants, and the expert reviewers) in this Plan allows them to exercise their best scientific judgment about what it takes to fulfill the primary goal of the conservation plan—preserving the biodiversity of the region."[12]

The county also established a Cultural Resources Technical Advisory Team to identify where vulnerable archaeological and historical sites were located. Not surprisingly, the greatest densities of such sites were along riparian corridors, which were the areas of greatest biological diversity as well. Past peoples, as well as plants and animals, settled near water in an arid land. In addition, the county set up a Ranch Conservation Technical Advisory Team.[13] Their position was that working ranches had defined metropolitan Tucson's urban boundaries for more than a century; keeping them in business, they argued, was the cheapest and most effective way to preserve open space in eastern Pima County. Some environmentalists rejected the argument that cows were better than condos. Others considered ranching a necessary evil—at least for a time.[14] Not surprisingly, some ranchers were hostile to or skeptical of the SDCP as well, viewing it as yet another regulatory land grab. Nevertheless, they took part in the process from its earliest days.

The Science Technical Advisory Team and Recon began with a list of more than 100 species that were extinct, in decline, or otherwise considered to be in jeopardy in Pima County. After two years of intense consultation with more than 200 scientists, the advisory team and Recon whittled the list down to 55 species of concern, later renamed priority vulnerable species, based on "occurrence, residency status, and opportunities for conservation in Pima County."[15] Those original 55 priority vulnerable species were reduced to 44 "covered species" in the December 7, 2012, version of the Pima County MSHCP.[16] Species experts defined the biotic and physical variables necessary to determine potentially suitable habitats for each of the priority vulnerable species. The advisory team and Recon then employed geographic information systems to model such habitats across the landscapes of Pima County. They and their scientific consultants, all of whom volunteered their time and expertise, also identified and inventoried specific conservation targets such as caves, limestone outcrops, native grasslands, and perennial and intermittent streams, which were labeled special elements and important riparian areas. The advisory team utilized accepted principles of conservation biology—including size, connectivity, and individual species' habitat requirements—to create guidelines for the establishment of a biological reserve system. Recon then applied those guidelines to Pima County landscapes to generate the biologically preferred alternative.

The resulting Conservation Lands System included biological core, important riparian, special elements, recovery, and multiple-use management areas. It encompassed more than two million acres of eastern Pima County's 2.4 million acres. Biological Core Management Areas alone contained more than 800,000 acres; they were defined as areas with "high potential habitat for five or more vulnerable species, special elements (e.g., caves, perennial streams, cottonwood-willow forests), and other unique biological features." The SDCP's Multi-Species Habitat Conservation Plan thus became the largest Habitat Conservation Plan proposed in the United States.

Even though the MSHCP had not yet been approved by Fish and Wildlife, the Pima County Board of Supervisors adopted the biologically preferred alternative and its enormous Conservation Lands System as its new Comprehensive Land Use Plan in 2001. The new plan imposed considerable restrictions on rezonings such as requests to change the zoning of a particular property to allow higher residential densities or commercial or industrial uses. If such rezonings were located within biological core areas of the Conservation Lands System, they had to set aside at least 80 percent of lands as natural open space. That figure rose to at least 95 percent on lands classified as important riparian and dropped to at least 66 percent on multiple-use lands. Golf courses were not considered natural open space. The county also established strict mitigation ratios for developers who wanted to exceed those restrictions on a particular parcel. For example, if developers wanted to build on an extra acre designated as biological core or important riparian, they had to set aside four acres of biological core somewhere else; the ratio was two to one for multiple-use parcels.[17]

## THE COMMUNITY ESTABLISHES ITS VOICE

To ensure citizen involvement, the county created a self-selected steering committee in 1999, which soon swelled to 89 members. The steering committee included environmentalists, ranchers, realtors, and even a few property-rights advocates. Major developers, who remained opposed to the SDCP, were represented by the Southern Arizona Home Builders Association, one of 15 organizations with permanent membership on the committee.

For the first year, the county force-marched the steering committee through a series of 12 monthly educational meetings colloquially called boot camp. Participants often referred to the group as the "steered committee" during this period. Some thought that the county had designed the steering committee to present the illusion of public input while being too unwieldy to make any

real decisions. After an 11-month hiatus, however, the steering committee surprised everyone by seizing control of itself and developing an organizational
structure and operating procedures that allowed it to move forward. This can
be attributed in part to the formation of a working group led by the director of
the Coalition for Sonoran Desert Protection and a local realtor. Some realtors
were beginning to realize that proximity to undeveloped areas of the Sonoran
Desert made the properties they were trying to sell more valuable. This pragmatic partnership provided the nucleus for the steering committee to examine reserve design alternatives and draft recommendations to the Pima County
Board of Supervisors.

When it issued its report on June 13, 2003, the steering committee recommended that the county submit an MSHCP and apply for a Section 10 permit lasting 20–50 years. The costs of the MSHCP were to be shared between
public and private sectors, with no more than 50 percent borne by affected
private landowners. Developers would not have to pay for the entire cost of
the MSHCP through impact fees or other assessments. The steering committee's most fundamental and controversial recommendation, however, was
"the adoption of the ecosystem approach that has resulted in a Conservation
Lands System map that protects the habitat of 55 priority vulnerable species."[18] As noted above, the number of priority vulnerable species later declined to 44 as the MSHCP was being developed.

In other words, the steering committee called for the adoption of the biologically preferred alternative reserve design, the Conservation Lands System
at its most extensive. The recommendation was not unanimous. The seven
ranchers on the steering committee joined six other members in filing a
"minority report" that dissented from the majority report on both substantive and procedural grounds. In its words, "We strongly object to egregious
procedural failures: the unequal treatment of Steering Committee members
by the facilitator who denigrated the contributions and legitimate concerns
of various members, who overtly advanced the motions made by the self-
appointed drafting group, and who permitted the drafting group to operate
as a de facto alternative steering committee."[19]

AFFORDING AND PRIORITIZING IMPLEMENTATION

Despite such conflict, the SDCP enjoyed broad community support. However, revenue was needed to implement the plan—the rough estimate was
$500 million. The county and the steering committee had already decided

that at least half the cost had to come from the public sector, so development mitigation fees, such as the $550-per-acre fee that funded the Clark
County (Las Vegas), Nevada, Habitat Conservation Plan, were not adopted.
The county looked into a number of federal, state, and local options to generate funds, including a sales tax. To explore the possibility of funding land
acquisitions through the sale of open-space bonds, which voters had passed
in 1974, 1984, 1986, and 1997, the county created a Conservation Bond Advisory
Committee in fall 2003.[20]

The biggest controversy that emerged on this committee was how to al-
locate conservation bond funds if they were approved. No one expected the
county to raise enough money to purchase the entire Conservation Lands
System, which encompassed more than two million acres. Developing a set
of priorities for acquisition therefore became the committee's most difficult
task. Fortunately, a list of the biologically most important properties had already been identified. After the Conservation Lands System had been developed, two key institutional players in the SDCP—the Arizona Chapter of the
Nature Conservancy and the Arizona Open Land Trust (now the Arizona Land
and Water Trust), the oldest land trust in the state, founded in 1978—evaluated private and state trust land parcels within the Conservation Lands System beginning in August 2002. In April 2003, they released a map of primary
and secondary Habitat Protection Priorities that contained about 525,000
acres of the Conservation Lands System, including 100,000 acres of private
land and 180,000 acres of state trust land exemplifying the "highest conservation values in eastern Pima County."[21] By July 2003, the map had been adopted
by both the steering committee and the Pima County board of supervisors.

Members of the Conservation Bond Advisory Committee who had participated in the SDCP process argued that most of the conservation bond funds
should be used to acquire habitat protection priorities, even if they were located in remote rural areas. Several other committee members disagreed,
contending that parcels with neighborhood constituencies, especially lands
in and around metropolitan Tucson, should be targeted first in order to build
public support for the passage of the bond. After considerable debate, the
committee proposed that no less than 75 percent of bond funds should be
spent on habitat protection priorities, and no more than 25 percent on what
came to be called "community open space." In May 2004, 65 percent of the
voters passed a Conservation Acquisition Bond of $174.3 million, of which $10
million was dedicated to the purchase of open space around Davis Monthan
Air Force Base and $15 million for lands requested by the cities of Tucson, Oro
Valley, and Sahuarita (jurisdictional requests). At least $112 million of the re-

maining funds had to be spent on habitat protection priorities while no more than $37.3 million could be devoted to community open space.[22]

## PIMA COUNTY ENTERS THE REAL-ESTATE MARKET

The sole rancher on the Conservation Bond Advisory Committee declined to support the committee's recommendations because she did not believe that government at any level should own more land. Nonetheless, ranchers in eastern Pima County soon benefited from the SDCP. One motivation for some ranchers to sell may have been the daunting estate tax (see spotlight 13.1). Pima County entered the real-estate market as the market was beginning to boom across Arizona and the rest of the nation. The findings of conservation biology reveal that large blocks of contiguous land offer more conservation value than smaller, disconnected parcels. This made ranches particularly attractive to the county, especially since their deeded lands tend to be more biologically diverse than surrounding public lands because they are often situated around springs or along rivers and streams.

By the time the county had funds available from the sale of conservation bonds, one major ranch in southern Arizona—the Sópori near Arivaca–had already been sold to First United Realty, which planned to subdivide it. Nonetheless, the county was able to purchase all other major ranches that came up for sale within its boundaries following the bond election. Early in 2005, the county bought the small Hayhook and King's 98 ranches in the northern Altar Valley southwest of Tucson to keep that enormous desert grassland watershed of more than 600,000 acres from being developed along Highway 86 (Ajo Highway). The county also bought the Bar V Ranch with its nearly 1,800 acres of private land and 12,700 acres of state trust grazing leases to protect portions of a major wildlife corridor linking the Santa Rita Mountains with the Rincon Mountains. Then, after months of quiet negotiations with the county and the Arizona Land and Water Trust, the Rowley family sold Rancho Seco near Arivaca to the county for $18.5 million in May of that year. The county purchased 9,600 acres of private land and a conservation easement on an additional 480 acres (table 13.1). Also included were more than 27,000 acres of state trust and Bureau of Land Management grazing leases. Because the county was committed to the conservation of working ranches, it leased all four of the ranches back to the former owner-operators for a renewable 10-year period. The King family on the Hayhook and King's 98, the Martin Cattle Company on Bar V, and the Rowley family on Rancho Seco continued

TABLE 13.1. Pima County Ranches

|  | Date Purchased | Deeded Land (acres) | Grazing Leases (acres) | Cost (millions of dollars) | Cost per Acre (dollars) |
|---|---|---|---|---|---|
| A-7* | 9/2004 | 6,800 | 33,000 | 2.0 | 300 |
| Canoa* | 2001 and 2006 | 4,800 | 0 | 8.5 | 1,800 |
| Rancho Seco | 5/2005 | 9,600 | 27,000 | 18.5 | 1,900 |
| Bar V | 2/2005 | 1,800 | 12,700 | 8.2 | 5,000 |
| King's 98 | 3/2005 | 1,000 | 3,300 | 2.0 | 2,000 |
| Six Bar | 8/2006 | 3,300 | 9,000 | 11.5 | 3,500 |
| Hayhook | 2/2005 | 800 | 0 | 1.4 | 1,700 |
| Diamond Bell | 3/2008 | 200 | 30,700 | 0.9 | 4,700 |
| Sands | 12/2008 | 5,000 | none | 21.0 | 4,200 |
| Clyne | 10/2010 | 900 | 0 | 4.9 | 5,600 |
| Sopori | 12/2009 | 4,100 | 10,500 | 18.6 | 4,500 |
| Marley, Phase I | 4/2009 | 6,300 | 0 | 20.0** | 3,200 |
| Carpenter Ranch | 8/2005 | 400 | 0 | 1.1 | 3,300 |
| Empirita Ranch | 8/2009 | 2,700 | 0 | 10.8 | 4,000 |
| Buckelew Farm and Ranch | 10/2006 | 500 | 2,200 | 5.0 | 10,100 |

Source: Data from Pima County, Arizona Land and Water Trust, Coalition for Sonoran Desert Protection. 2012.

*A-7 Ranch and majority of Canoa were purchased before the 2004 Conservation Bond election.

**Includes $2 million for two options on the remaining 23,000 acres of deeded land and 70,000 acres of grazing leases.

to run cattle and bear all the expenses of their operations on those spreads. Pima County's Natural Resources, Parks and Recreation Department worked with them to develop management agreements to conserve the biodiversity and ecosystem functions of the ranches while allowing ranching to continue.[23]

A year later, the county added Buckelew Farms in the northern Altar Valley and Six Bar Ranch along the San Pedro River, which consisted of 3,300 acres of private land and 9,000 acres of state trust grazing leases.[24] In a sense, county acquisitions with 2004 bond funds can be divided into two periods: during the real-estate boom and after the boom collapsed. During the boom, speculators were scooping up every privately owned parcel they could get their hands on. They were also bidding anywhere from four to 20 times the appraised value of state trust lands put up for auction in the metropolitan Phoenix and Tucson areas. County staff informed the Conservation Acquisition Commission—which made recommendations to the board of supervisors about which properties to purchase with conservation bond funds—that land prices in the region were rising at the rate of about 1 percent a month.[25] It was a seller's market, and if the county had not had the conservation bond funds, the ranches it purchased in 2005 and 2006 would undoubtedly have been bought by speculators and subdivided by developers. Urban, suburban, and exurban sprawl would have expanded even more explosively across the southern Arizona landscape.

In 2008, however, the boom went bust, and many speculators—especially those with liquidity problems—frantically looked for buyers as real-estate prices plummeted. By the end of 2009, the county was able to purchase about a third of the Sópori Ranch near Arivaca (4,100 acres of private land) from First United Realty, which had snatched up the entire 12,600 acres for $22 million in October 2004 before conservation bond funds became available. The rest of the ranch was in Santa Cruz County, where First United was trying to develop portions of it despite strong resistance from Santa Cruz County voters.[26] Then, in the biggest deal of all, the county entered into a three-phase agreement to buy the 30,000-acre Marley Ranch. Only Phase 1 could be completed with 2004 conservation bond funds. Phases 2 and 3 awaited a future bond election, originally scheduled for Fall 2008 but postponed because of the poor economy. If and when the rest of the Marley is acquired, Pima County will have prevented the major ranches in eastern Pima County from making the transition from ranching to real estate—from "cows to condos"—that has chopped up so many of the open spaces across the West.[27]

### BRIDGING THE RURAL-URBAN DIVIDE: WORKING RANCHES AND THE ECOLOGICAL FOOTPRINT OF PIMA COUNTY

The SDCP is very much an experiment in progress—one that is being carried out on a large and rapidly urbanizing landscape (see map, fig. 13.1). It has been very successful in accomplishing one major goal: the preservation of open space surrounding metropolitan Tucson. If Pima County's citizens continue to pass open-space bonds in the future, more of the private parcels within the Conservation Lands System will be saved from development. The county also may be able to acquire key state trust lands at public auction, even if Arizona voters fail to pass state trust land reform and set aside some of those lands for conservation. So far the county is batting 1,000 percent in open-space bond elections; voters in 1974, 1986, and 1997 as well as 2004 have approved a total of $230 million "for the purchase of critical, unique and sensitive lands."[28]

Preserving biological diversity is a more complicated task; protecting habitat does not necessarily guarantee the survival of vulnerable species. As part

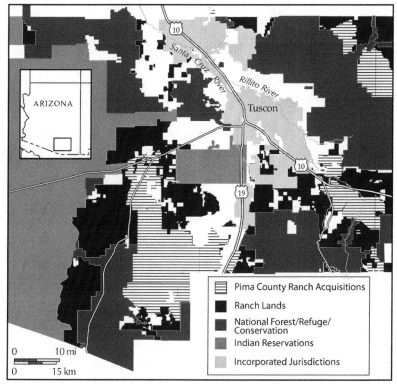

**Figure 13.1.** Sonoran Desert Conservation Plan and surrounding lands. Map by Darin Jensen and Syd Wayman, 2013.

of its MSHCP, scientists and county personnel spent several years developing a monitoring and adaptive management program for 44 of the 55 priority vulnerable species identified by STAT; 11 species were dropped from the MSHCP after it had been determined that they were unlikely to be found on any lands on which the county had jurisdiction. If and when U.S. Fish and Wildlife Service approves the plan and issues the Section 10 permit, that program will include direct monitoring of 16 species and habitat monitoring for the other 28 species that cannot be monitored directly in a cost-effective fashion.[29] The county has committed itself to conducting that monitoring for at least the 30-year life of the permit. Only time will tell if the SDCP ensures "the long-term survival of the full spectrum of plants and animals that are indigenous to Pima County through maintaining or improving the habitat conditions and ecosystem functions necessary for their survival." Global patterns of climate change threaten to make that goal ever more challenging as the Southwest grows hotter and drier in the twenty-first century.[30] Nonetheless, the conservation of large, unfragmented landscapes and the habitats they support is a fundamental step in the right direction.

Perhaps the most innovative feature of the SDCP experiment, however, is the county's commitment to conserve these landscapes as working ranches, not nature preserves. By signing management agreements with the former owners of the ranches it purchased, the county honors the SDCP's goal of ranch conservation to preserve "the heritage and culture of the Southwest." It also saves county taxpayers money: the county does not have to run those ranches itself, as it did the A-7 Ranch, which it bought from the city of Tucson.[31] Those lessees maintain the infrastructure of the ranches at their own expense—fences, windmills, stock tanks, pumps, and wells. Their water sources serve wildlife as well as cattle. Ranchers and their cowboys also provide the eyes on the ground to monitor environmental conditions and human impacts, legal and illegal, that affect the ecological health of county lands.[32]

County ranches can also partner with private ranches in southern Arizona to address an even bigger environmental challenge: reducing the ecological footprint of metro Tucson. Tucson, like most urban areas in Arizona and the rest of the United States, relies almost entirely on food produced in faraway places. Furthermore, most of the food produced in southern Arizona is exported out of state. Pima County's most important agricultural product is beef. Most of its ranches are cow-calf operations that ship their calves or steers to feedlots in the midwestern United States to be fattened on corn until slaughtered. But more and more ranchers in Pima County and elsewhere in southern Arizona are slaughtering a tiny fraction of their calves locally and selling them directly to consumers as grass-fed beef, a healthier food than corn-fed feedlot beef. They would like to sell more, but there are only two small U.S. Department of Agriculture–certified slaughterhouses in the region. If the county joined forces with these entrepreneurial pioneers, however, markets and infrastructure could be expanded. More of Pima County's beef could be eaten by Pima County consumers, reducing transportation costs and the emissions that go with them.

Linking local producers and consumers also helps bridge the rural-urban divide that isolates urban dwellers from rural people and the jobs they do. Most of us no longer have friends or family who are farmers or ranchers. We don't understand the challenges they face to produce the food, energy, or minerals our urban lives require. Such essential activities are hidden from view—out of sight and, often, out of mind—and that makes us less conscientious citizens because we don't have to take responsibility for the consequences of our consumption patterns. But if at least some of our food comes from nearby farms or ranches, then we slowly begin to stitch the urban and rural West back together. Those farmers and ranchers are no longer abstractions—they're neighbors, and neighbors look out for one another and the landscapes that they share.

## NOTES

1. Pima County, "A Vision for Ranch Conservation," Sonoran Desert Conservation Plan, Pima County, accessed October 19, 2012, http://www.pima.gov/cmo/sdcp/Ranch.html.
2. Ibid.
3. T. Sheridan, *Arizona: A History*, rev. ed. (Tucson: University of Arizona Press, 2012).
4. Ibid.
5. C. H. Huckelberry, Pima County Administrator, Memorandum re: Wildcat Subdivision Study, April 10, 1998, 1.
6. "Wildcat Subdivision Study," Pima County Administrator's Office, 1998, www.pima .gov/cmo/sdcp/reports\d11\001WIL.PDF.
7. T. Sheridan, "La Canoa: From Land Grant to Political Football," *SMRC-Newsletter* (Southwestern Mission Research Center) 34, no. 123 (2000): 2–5; D. Hadley, "San Ignacio de la Canoa Land Grant and Canoa Ranch History/Timeline," *SMRC-Newsletter* (Southwestern Mission Research Center) 34, no. 123 (2000): 11–15; T. Davis, "A Seminal Sprawl Fight Ends in Compromise," *High Country News*, June 18, 2001.
8. Pima County, "Canoa Ranch Background Report, Raúl M. Grijalva Canoa Ranch Conservation Park" (Pima County, April 2006), available at www.pima.gov/cultural/ Canoa/index_Canoa.html.
9. A. Rome, *The Bulldozer in the Countryside: Suburban Sprawl and the Rise of American Environmentalism* (Cambridge: Cambridge University Press, 2001).
10. U.S. Fish and Wildlife delisted the pygmy owl in 2006 ("Pygmy Owl Petition Filed," Coalition for Sonoran Desert Protection, March 15, 2007, http://www.sonorandesert. org/2007/03/19/pygmy-owl-petition-filed/).
11. Pima County, "Pima County Multi-species Habitat Conservation Plan (MSHCP)," administrative draft (December 8, 2010), 1. The updated public draft of the plan, December 7, 2012, can be found at: www.pima.gov/cmo/sdcp/MSCP/MSCP.html.
12. Reed Noss and Laura Hood Watchman, "Report of Independent Peer Reviewers: Sonoran Desert Conservation Plan," October 26, 2001, p. 1, http://www.pima.gov/cmo/ sdcp/reports/d10/020RES.PDF.
13. I served as chair of the Ranch Conservation Technical Advisory Team.
14. In 2011, e.g., the Coalition for Sonoran Desert Protection still included on its website a report critical of livestock grazing.
15. Pima County, "Multi-species Habitat Conservation Plan," administrative draft, 8.
16. Pima County, "Multi-species Habitat Conservation Plan" (2012).
17. Information on the Conservation Lands System can be found in numerous reports accessed on Pima County's Sonoran Desert Conservation Plan home page, but a succinct summary of it is located on the Coalition for Sonoran Desert Protection's website: www.sonorandesert.org/learning-more/conservation-land-system/.

18. Sonoran Desert Conservation Plan Steering Committee, "Report to the Pima County Board of Supervisors," June 13, 2003, http://www.pima.gov/cmo/sdcp/reports/d50/ SC61303.PDF.

19. Sonoran Desert Conservation Plan Steering Committee, "Appendix B, Minority Views," submitted by Pat King on behalf of S. Chilton, L. Vitale, J. DuHamel, C. Coping, H. Fox, R. Harris, L. Harris, M. Darling, A. McGibbon, M. McGibbon, M. Miller, and A. Lurie, in Sonoran Desert Conservation Plan Steering Committee, "Report to the Pima County Board of Supervisors," 5.

20. I served as a member of the Bond Advisory Committee.

21. V. Bechtol, "In Arizona, GIS Mapping Prioritizes Habitat Conservation Targets," Arizona Land and Water Trust (n.d.), http://www.aolt.org/images/pdf/ARCNews%20 article.pdf.

22. Summaries of the 2004 Pima County Open Space Bond can be found on Pima County Bond Programs Approved by Voters in 1997, 2004 and 2006 at www.bonds.pima.gov/ approved.htm; and the Coalition for Sonoran Desert Protection's Open Space Preservation at www.sonorandesert.org/learning-more/conservation-land-system/.

23. Pima County ranch acquisitions are summarized in Chuck Huckelberry, Bill Singleton, Mark List, and Mark Probstfeld, *Protecting Our Land, Water, and Heritage: Pima County's Voter-Supported Conservation Efforts* (Tucson, AZ: Pima County, 2011), http:// www.pima.gov/cmo/admin/Reports/ConservationReport/. Individual reports on all of these ranch purchases can be found on Pima County's Sonoran Desert Conservation Plan home page at http://www.pima.gov/cmo/sdcp/reports.html.

24. Huckelberry et al., *Protecting Our Land, Water, and Heritage*.

25. I have served as a member of the Conservation Acquisitions Commission since its inception.

26. In December 2007, the Santa Cruz County Board of Supervisors voted 2–1 to approve an amendment to its 2004 comprehensive plan that would have permitted the development of Sópori Ranch and another project at higher densities than the plan allowed. Enraged county residents successfully petitioned to place referendums on the ballot to reverse the amendment. In November 2008, voters rejected both rezonings by more than two-to-one margins. The Sópori Ranch plan had called for the construction of 6,800 homes.

27. Pima County's acquisition of portions of the Sópori and Marley ranches are summarized in Huckelberry et al., *Protecting Our Land, Water, and Heritage*.

28. C. Huckelberry, "Foreword," in ibid., 3.

29. Pima County, "Multi-species Habitat Conservation Plan" (2012).

30. W. deBuys, *A Great Aridness: Climate Change and the Future of the American Southwest* (Oxford: Oxford University Press, 2011).

31. In 2013, the county signed an agreement with a local rancher along the San Pedro River to manage the A-7 Ranch.

32. Information on the management of Pima County ranches can be found at "Natural Resources, Parks and Recreation," Pima County, http://www.pima.gov/nrpr/parks/ nrparks.htm.

# Spotlight 13.1

# RANCHING AND THE "DEATH TAX"

A Matter of Conservation as Well as Equity

*Thomas E. Sheridan, Andrew Reeves, and Susan Charnley*

## IN BRIEF

- The estate tax can cause working forests and ranches to be subdivided and converted to real-estate development if those who inherit the land cannot afford the estate taxes.
- Estate taxes are especially high in the West, where private forest and ranch holdings are large and have been subjected to rapidly escalating land values in amenity-rich areas; landowners who are "land rich and cash poor" are particularly vulnerable.
- Some people believe the estate tax law should be repealed or rewritten for conservation reasons. Those who support this law believe estate taxes should be high so that owners of hobby forests and ranches don't use them as an excuse to protect their assets from taxation; so that there is an economic incentive to donate conservation easements; and, to help control land values.
- The estate tax law enacted in 2001 to provide some tax relief will expire in 2013, creating an opportunity to consider whether working lands should be exempt from estate taxes unless they are sold for nonagricultural purposes.

As the chapters in this volume point out, one of the biggest threats to both working landscapes and the conservation of biodiversity in the West is the conversion of ranch and forestlands into real-estate developments. Between 1964 and 1997, the total acreage of farm and ranch lands in eight of the western states dropped from 268 million to 228 million acres, a loss of 15 percent.[1]

TABLE 13.1.1. Acreage in Farmland in the West, 1997, 2002, and 2007 (in thousands of acres)

| State | Year | | | Acreage Change | Percent Change |
| --- | --- | --- | --- | --- | --- |
| | 1997 | 2002 | 2007 | | |
| Arizona | 27,170 | 26,587 | 26,118 | −1,052 | −3.8 |
| California | 28,796 | 27,589 | 25,365 | −3,431 | −11.9 |
| Colorado | 32,350 | 31,093 | 31,605 | −745 | −2.3 |
| Montana | 58,445 | 59,612 | 61,389 | +2,944 | +5.0 |
| Nevada | 6,398 | 6,331 | 5,865 | −533 | −8.3 |
| Oregon | 17,658 | 17,080 | 16,400 | −1,258 | −7.1 |
| Washington | 15,789 | 15,318 | 14,973 | −816 | −5.2 |
| Wyoming | 34,302 | 34,403 | 30,170 | −4,132 | −12.0 |
| Total | 220,908 | 218,013 | 211,885 | −9,023 | −4.1 |

Source: USDA Agricultural Censuses, available at http://www.agcensus.usda.gov/Publications/index .php. 2013.

During the following decade, that trend continued: more than nine million acres of agricultural farmland went out of production, representing a decline of 4.1 percent (see table 13.1.1)

There are many reasons why a farming or ranching household decides to sell part or all of its private lands. Prices for agricultural products, including beef and lamb, have risen much more slowly than the costs of fuel, feed, veterinary care, and other inputs. The average rate of return for a viable working cattle ranch—300 or more mother cows for a cow-calf outfit—is about 2 percent. And since the initial investment in a ranch with a 300-cow breeding herd is at least $1 million and often considerably more, few investors are attracted to the business and many young couples who would like to get into ranching simply cannot afford it.[2] That makes the intergenerational transfer of ranches a critical variable—perhaps *the* critical variable—in the survival of working ranches in the West. But the intergenerational transfer of these working lands has a major constraint: the estate tax.

Few topics in the West generate as much hyperbole as the estate tax, better known as the Death Tax among most ranchers, farmers, and family forest owners. In 2012, Senator John Thune (R-SD) and Representative Kevin Brady (R-TX) introduced the Death Tax Repeal Permanency Act—identical legislation in both the U.S. Senate and House of Representatives—to abolish the estate tax. In the words of J. D. Alexander, president of the National Cattlemen's Beef Association, which supported the legislation, "Our priority is to keep families in agriculture and this tax works against that goal. The appraised value of rural land is extremely inflated when compared to its agricultural value. Many cattle producers are forced to spend an enormous amount of money on attorneys or sell off land or parts of the operation to pay off tax liabilities. This takes more open space out of agriculture and usually puts it into the hands of urban developers."[3]

Organizations as diverse as the National Cattlemen's Beef Association, the Public Lands Council, the Environmental Defense Fund, and the Rural Voices for Conservation Coalition support legislative efforts either to repeal the tax altogether or to exempt working forests, farms, and ranches from it as long as they remain in production. "It is impossible to separate private working lands and healthy ecosystems in the western United States," Dan Grossman of the Environmental Defense Fund has said. "Good private land stewardship preserves critical wildlife habitat and the nation's natural resources. When the estate tax forces ranchers to sell private land, the environment suffers too."[4]

Opponents dismiss such arguments as "a carefully crafted myth," in the words of Howard Gleckman. Writing in *Forbes*, Gleckman continues, "Start with the cowboy. Some Wyoming ranchers have been on their land for generations—property stolen fair and square from Native Americans in the 19th century. Others are grazing their herds on federal land while paying cut-rate fees for the privilege. And still others are L.A. doctors, lawyers, and actors who spend a few weekends a year on their ranchettes, where they enjoy both spectacular scenery and generous income tax subsidies for grazing a few head of someone else's cattle."[5]

Gleckman glibly trots out the usual antirancher stereotypes—the "welfare rancher" and the "hobby rancher" among them. In his opinion, ranching is just another way the wealthy protect their assets from taxation. But he cites more serious research as well, especially that conducted by the Tax Policy Center of the Urban Institute and the Brookings Institution. Based on a complex statistical model, the Tax Policy Center argues that only a miniscule number of family farms and ranches pay any estate tax at all. "If we define those estates as ones in which farm and business assets total less than $5 million and make up at least half of the gross estate, only 2,000 will have to file estate tax returns in 2008 and nearly three-quarters (73 percent) of those will have no estate tax liability," their 2008 report states. "About 550 farm and business estates will owe any estate tax liability and more than three-quarters of those will owe less than $500,000."[6]

Like many such statistical analyses, the Tax Policy Center report fails to see the forest for the trees—or, in this case, the ranches for the ranchettes. Treating ranches as any other small business, the analysis fails to recognize the critical conservation value of working ranches across the West. Ranches may be a tiny percentage of the total number of estates affected by the estate tax, but those ranches are critical pieces of the conservation puzzle. Once their private lands become residential subdivisions, wildlife habitat is lost, wildlife corridors are fragmented, and the reintroduction of fire as a natural disturbance becomes difficult, if not impossible, to achieve. Any discussion of the impact of the estate tax on ranching, then, has to focus on conservation as well as economic equity.

But economic equity is an issue as well. The Tax Policy Center's report does not address the differences between farms and ranches and other small businesses. In 2003, the U.S. Department of Agriculture estimated that approximately 76 percent of the total value of farm assets consisted of land and buildings.[7] For many farmers and ranchers, the value of their lands is a rollercoaster ride that soars or plummets according to real-estate markets rather than agricultural markets. Diane Holly's Overlook Ranch, six miles west of the ski resort town of Steamboat Springs, Colorado, is a case in point. When her mother died in 1983, the 2,000-acre ranch was worth $800,000. Nine years later, when Holly became sole heir after her brother died, the value had climbed to $1.2 million. That figure had nearly doubled to $2.3 million by 1997, when her father died in a car crash. Facing an estate tax bill of $400,000, Holly "essentially had a fire sale," selling off her "most beautiful, emotional parcel," 200 acres of river bottom land. Her annual income from the ranch was about $28,000 a year.[8]

In 2008, when she was interviewed by a reporter from *High Country News*, Holly's 1,800 acres were probably worth about $22 million based on sales of adjacent agricultural land. "It's so grandiose. It's so unbelievable. I can't wrap my brain around it," the rancher said. "My kids would owe $4 million if I died tomorrow. Now I understand why Aspen and Vail ranchers had to sell out."[9]

In 1997, only estates worth $600,000 or less were exempt, while the top estate tax rate was 55 percent. That burden eased somewhat after Congress passed the Economic Growth and Tax Relief Reconciliation Act in 2001, which gradually increased the exemption to $5 million and decreased the top tax rate to 35 percent by 2010, when it expired. When President Obama signed the Tax Relief, Unemployment Insurance Reauthorization and Job Creation Act in 2010, those rates were extended for two more years. The top tax rate remained at 35 percent while the exemption increased to $5,250,000 in 2013 and was scheduled to rise to $5,340,000 in 2014.[10]

Even the current rates present a formidable challenge to many ranching families, especially in the arid and semiarid West where viable ranches have to be much larger than elsewhere in the country. In May 2005, Pima County purchased Rancho Seco for $18.5 million with open-space bond funds as part of its Sonoran Desert Conservation Plan (see chap. 13). Triggered by the death of the ranch's owner, the sale occurred as real-estate prices were skyrocketing in southern Arizona. The county was competing with land speculators and developers scrambling to buy every ranch that came on the market. We don't know what kind of estate planning the owners of Rancho Seco had in place before their decision to sell the ranch. But if the heirs faced a full estate tax bill at the going rate in 2005, they would have only been able to deduct $1.5 million from the total price. Their check to the Internal Revenue Service would have been a staggering $7.99 million—47 percent (the top rate in 2005) of that $17 million. Without Pima County's open-space bond funds, Rancho Seco's private land—nearly 10,000 acres—would have been subdivided.

Not many heirs can afford huge estate tax bills and remain in ranching. They may be millionaires on paper, at least when the real-estate market is bullish, but those millions are a perverse abstraction as long as they raise cattle, not condos. With most outfits barely turning a profit, the only way ranchers can become part of the wealthy elite—the elite the estate tax is supposed to target—is to leave the land, and livelihood, they love.

Nonindustrial private forest owners face similar pressures. The first study to quantify the effects of the estate tax on nonindustrial private forest holdings was conducted in 1999 and focused on estate transfers that occurred between 1987 and 1997, when $600,000 of an estate's value was shielded from the federal estate tax.[11] This study found that some 77,200 forest estate transfers occur each year throughout the United States on the death of their owners, encompassing the transfer of an estimated 79.1 million acres of family forestland nationwide. Although 62 percent of owners reported that no federal estate tax was due in the transfers with which they were involved, 38 percent of forest owners had to pay estate taxes (compared with 2 percent for estates in general during this period). In most cases where estate tax was owed, insurance or other assets were used to pay the tax. However, in 22 percent of the cases, timber was sold to pay part or all of the tax, and in 19 percent of cases land was sold to pay part or all of the tax. The majority of owners who made these sales did so because their other assets were insufficient to cover the cost of the tax. In 29 percent of the cases in which land was sold, it was converted to developed uses.

Based on data from the 1987–97 period, the study estimated that about 2.4 million acres per year of family forestland must be harvested to pay the federal estate tax (constituting unplanned harvests), and 1.3 million acres per year must be sold, with roughly 400,000 of these acres converted to other, more developed uses annually. As described above, since the time of this study, the Economic Growth and Tax Relief Reconciliation Act of 2001 has caused the amount of the estate tax exclusion to increase; however, land values have also increased. Moreover, the impacts of the estate tax on family forestland may be more pronounced in the western United States than in the eastern United States.

Although the West has a much smaller amount of family forestland than the eastern United States, the average size of a family forest holding is larger in the West than in the East.[12] The likelihood of land transferring from one owner to another increases with the size of the forest holding, and the average value of an estate is substantially higher in the West than in the East.[13] Given that roughly one-third of the primary decision makers for family forestlands are 65 and older, and two-thirds are 55 and older, the problem of what will happen to the West's family forestlands looms large in the near future.[14] In sum, federal estate taxes not only lead to the proliferation of ranchettes and condos; they can also cause forest fragmentation, conversion of forestland to other uses, and unplanned timber harvests that may have undesirable environmental effects.

Estate taxes have existed in the United States since 1916, and calls for their repeal are just as old. Before the Tax Relief Bill Act kicked the can down the road, no fewer than 10 bills were introduced in 2010 (seven in the House and three in the Senate) to address the issue. Most notable of these was Congressman Mike Thompson (D-CA) and John Salazar's (D-CO) Family Farm Preservation and Conservation Estate Tax Act. That act would have exempted working lands from the estate tax unless they were sold for nonagricultural purposes.

Supporters of the estate tax like the Tax Policy Center argue that passage of such legislation "would give wealthy individuals who expect to pay the estate tax a huge incentive to convert assets into qualifying farms or businesses before they died. Ironically, this could endanger many existing small farms and businesses, as wealthy people would bid up the price of such properties to claim their tax benefits. (How much of Iowa could Bill Gates buy with his fortune?)."[15] They also contend that exemption would reduce the economic incentive for ranchers to sell or donate conservation easements to conservation-interested third parties.

Land trusts, which hold many of those easements, aren't so sure. As land prices escalate during booms, nongovernmental organizations like the Nature Conservancy and land trusts can never cobble together enough money to buy the development rights on all of the private hotspots of biodiversity that come up for sale. That's one of the reasons the Land Trust Alliance, which represents 1,700 land trusts across the United States, endorsed Thompson and Salazar's family farm–preservation bill. "Estate taxes can undermine important conservation values by forcing the breakup, sale and development of family-owned farm, ranch and forest lands, even when those lands provide important resources for the public, including high-quality watersheds, wildlife habitat, and food and fiber production," Rand Wentworth, president of the alliance, wrote in a letter to Representative Thompson. "Your bill's provision for deferred estate taxes to be paid upon any subsequent sale of such lands is fair to all taxpayers, and ensures that this deferral is not used to simply avoid taxation."[16]

Looking forward, the reintroduction and passage of similar legislation would provide tremendous tax benefits for keeping working landscapes intact, whether they are family farms, forests, and ranches transferred from one generation to another or operations owned by wealthy individuals who want to conserve those landscapes while raising food or fiber. Without the passage of such legislation, the conversion of farm, forest, and rangeland into other types of real estate will continue. If a ranch were to come up for sale in Pima County today, the county would have to sit on the sidelines. Its open-space bond funds have been spent. There's no more money in the pot, and voters aren't likely to approve another open-space bond as long as times are hard. Even when the economy improves, land speculators and developers will be back, driving land prices up once again and exacerbating the death tax problem for owners of working farms, forests, and ranches.

## NOTES

1. M. J. Sullins et al., "Lay of the Land: Rich Land and Ranching," in *Ranching West of the 100th Meridian*, ed. Richard Knight et al. (Washington DC: Island Press, 2002), 25–32.

2. L. Butler, "Economic Survival of Western Ranching: Searching for Answers," in *Ranching West of the 100th Meridian*, ed. Knight et al., 195–202.

3. "Senate Bill Would Repeal Estate Tax: Identical to House Version," *Southeast Farm Press*, March 29, 2012, posted on the website of Congressman Kevin Brady, http://kevinbrady.house.gov/kevin-brady-in-the-news/senate-bill-would-repeal-estate-tax-identical-to-house-version/.

4. "Unlikely Allies Support Bill to Exempt Working Farms and Ranches from Estate Tax," South Llano Watershed Alliance, August 20, 2009, http://southllano.org/2009/08/unlikely-allies-work-to-exempt-working-ranches-from-estate-tax/.

5. Howard Gleckman, "Death Tax Revival Murders Yeoman Ranchers," *Forbes*, November 9, 2010, http://www.forbes.com/sites/beltway/2010/11/09/death-tax-revival-murders-yeoman-ranchers/.

6.  L. E. Burman, K. Lim, and J. Rohaly, "Back from the Grave: Revenue and Distributional Effects of Reforming the Federal Estate Tax" (paper written for the Tax Policy Center, Urban Institute and Brookings Institution, October 20, 2008), 15–16, http://taxpolicycenter.org/UploadedPDF/411777_back_grave.pdf.

7.  R. R. Korves, "Land Values and the Federal Tax System" (FairTax White Paper, written for Americans for Fair Taxation, 2007), http://www.fairtax.org/PDF/Land_values_and_the_federal_tax_system.pdf.

8.  J. Lay, "Death, and Taxes," *High Country News*, August 11, 2008.

9.  Ibid.

10. "What's New—Estate and Gift Tax," IRS website, last reviewed or updated September 3, 2013, http://www.irs.gov/Businesses/Small-Businesses-&-Self-Employed/What's-New—-Estate-and-Gift-Tax.

11. J. L. Greene, S. H. Bullard, T. L. Cushing, and T. Beauvais, "Effect of the Federal Estate Tax on Nonindustrial Private Forest Holdings," *Journal of Forestry* 104, no. 1 (January–February 2006): 15–20.

12. B. J. Butler, *Family Forest Owners of the United States, 2006*, General Technical Report NRS-27 (Newtown Square, PA: USDA Forest Service, Northern Research Station, 2008).

13. Ibid.; Greene et al. "Effect of the Federal Estate Tax."

14. Butler, *Family Forest Owners.*

15. L. E. Burman, W. G. Gale, and J. Rohaly, "Options to Reform the Estate Tax," *Tax Policy Issues and Options*, no. 10 (March 2005).

16. C. Souza, "Estate Tax Proposal Would Help Farm Families," *AgAlert: The Weekly Newspaper of California Agriculture*, September 2, 2009.

# 14

## PAYMENTS FOR ECOSYSTEM SERVICES

### Keeping Working Landscapes Productive and Functioning

*Gary P. Nabhan, Laura López-Hoffman, Hannah Gosnell,*
*Josh Goldstein, Richard Knight, Carrie Presnall, Lauren Gwin,*
*Dawn Thilmany, and Susan Charnley*

IN BRIEF

· Payments for ecosystem services are a mechanism for simultaneously
protecting and restoring ecosystems and the services they provide while
generating economic benefits for those involved in ecosystem restoration,
investors, and ultimately society at large.
· For payments for ecosystem services to be successful there must be tangible
needs that create a demand for an ecosystem service, a supply of projects that can
meet the demand with a reasonable return on investment, and financial mechanisms for linking supply and demand, which are most often market based.
· The two main mechanisms for compensating land stewards—such as forest
and ranch owners—who provide ecosystem services are to embed the costs
of providing these services in the price of products produced sustainably
and to give financial incentives and rewards to those who protect, enhance,
and restore ecosystem services.
· A significant challenge for payments for ecosystem services schemes is
creating a means for scaling them up to the landscape level with strategic
and targeted investments in conservation and restoration.

INTRODUCTION

The "restoration economy" has been called the next growth frontier on the
assumption that it may simultaneously generate wealth while reversing the

global decline in functioning ecosystems and the degradation of the essential life-giving services they provide to humanity.[1] Because there are millions of dollars of tangible economic benefits to be gained by restoring our degraded watersheds, forests, and rangelands, we have recently witnessed the development of novel, creative, and cost-effective ways of restoring natural, agricultural, and ranchland systems through the use of social impact investing strategies that are often associated with the "slow money" movement.[2] These strategies often draw on "patient capital" (financial investments that have long-term returns and provide social or environmental benefits rather than maximizing profits and short-term returns), in a manner that has begun to provide one of several nonregulatory means for payment for ecosystem services (PES).[3]

Historically, landowners who have carefully stewarded their land have provided society with diverse benefits, such as crop pollination (supporting food and forage production), regulation of water quality and quantity (for drinking water and other uses), and climate stabilization (through sequestering and storing carbon in plants and soils). Landowners have typically provided these benefits, collectively termed "ecosystem services," without compensation for management costs incurred, largely because these efforts reaffirm their social values and land stewardship ethics and, pragmatically speaking, make economic sense because they maintain a productive resource base having market value.

In the United States, one of the first PES programs was the Conservation Reserve Program, which was nonmarket based (see spotlight 14.1). More recently, a growing number of strategies have emerged that "monetize" the maintenance or restoration of ecosystem services for ranchers and private forest owners so that these landowners may begin to receive better compensation for management practices that provide environmental benefits to the public. There are also many individual investors and corporations now voluntarily investing in the restoration of ecosystem services because of their social and environmental values and their interest in being seen by the public as a "good neighbor." Through such means, formerly degraded or fragmented working landscapes can first be restored and then sustainably managed to higher levels of ecological and economic functioning, in concert with rural social values. For these types of investment strategies to be successful in restoring ecosystem services and generating profits, however, there must be tangible needs that create demand, a "supply" portfolio of projects that can meet this demand with a good return on investment, and sound market mechanisms for linking supply and demand.

There are two general approaches to compensating land stewards for ecosystem services: embedding the costs of providing these services in the price of products like food, fiber, and wood products that are produced sustainably and offering financial incentives and rewards to ranchers, foresters, or farmers working to protect, enhance, or restore ecosystem services. These two means of PES are being implemented through a broad range of financial mechanisms and are applied to ecosystem services with benefits at the local, regional, national, and international scales.[4]

In the past, it has been difficult for society to imagine why the private and public sectors might pay ranch and forest owners for something that they had theretofore been provided for "free." Recently, however, the *New York Times* devoted nearly an entire page to the topic of "ranch as investment" for venture capitalists and social impact investors with their eyes on the future of carbon credit exchanges and mitigation banking.[5] We suggest that the sectors of society that had not previously supported ranching and forestry are now seeing the convergence of their own interests with investment in ecosystem services in the American West. Governmental agencies, for-profit businesses, nonprofit land trusts and foundations, and consumers are now demonstrating clear interest in paying for the maintenance and restoration of ecosystem services provided by working landscapes.

Moreover, certain financial mechanisms for restoring degraded ecosystem services are already legislatively sanctioned, if not mandated (for instance, through the Clean Water Act). Other nonregulatory payment mechanisms may be voluntary or optional means of environmental mitigation but remain market driven (such as Forest Stewardship Council certification). Although voluntary markets began to falter as economic and political uncertainty hit the United States and Europe in 2008 and 2009, some began to recover by 2010 and to achieve a greater number of transactions than they had accomplished in any of the previous five years.[6]

We wish to underscore that the terrain of environmental markets and PES is rapidly shifting. We would not be surprised if the mechanisms being used today are superseded by other, more refined means within as little as five years. At this early stage in the development of such markets, we clarify here which of the many ecosystem services are uniquely or characteristically provided by working landscapes in the American West. This chapter identifies some potential beneficiaries (venture capitalists, social impact investors, philanthropists, public agencies, and corporations) that may have a particularly strong interest in paying for these services. We also outline a few of the legally mandated, social, public relations–oriented, or financial rationales for them. We describe a few current and emerging means of implementing PES, while cautioning ranchers and private forest owners against considering PES to be a "silver bullet" solution to any of their problems. Like conservation easements, PES may not work for all scales or management goals of particular ranch and forestry operations. Finally, we propose some strategies and principles for adapting PES mechanisms to best work for saving entire landscapes, rather than single ranch or forest landscapes.

It is important to point out that the commodification of natural systems and processes that contribute to the public good is not without risks, both philosophical and practical, as both activists and scholars have pointed out. For example, Sally Fairfax and her colleagues have noted that paying landowners for conservation sets up an expectation among them and society at large that payment mechanisms are required for conservation actions.[7] At the same time, indigenous peoples and their legal counsels have objected to payment for ecosystem services as a further privatization of the common good that may unintentionally generate perverse consequences. Journalist Jeff Conant documented such potential consequences in an interview with Larissa Packer, a Brazilian lawyer with *Terra de Direitos*, which works to secure land rights for landless communities: "Payment for environmental services," Packer said in the interview, "posits that the actions of nature—the water cycle, the carbon cycle, the pollination of flowers by bees—are commodities, subject to the law of the market. In the current market, prices are based on supply and demand, that is, on scarcity. As petroleum becomes scarce, its value goes up. The green economy will follow the same logic. . . . If we put a price on forests, on biodiversity, on other common goods, those prices will be driven up by scarcity, so, for investors in these things, the greater the scarcity of ecosystem services, the greater their value. Where do we think this will lead?"[8] These philosophical issues will continue to be debated for some time.

## ECOSYSTEM SERVICES OF WORKING LANDSCAPES IN THE AMERICAN WEST

The Millennium Ecosystem Assessment surveyed the state of ecosystems and their services around the world. According to the assessment, 15 of 24 different kinds of ecosystem services surveyed have been degraded over the past 50 years.[9] This finding suggests that remaining functional ecosystem services provided by large, intact working landscapes have become all the more important to human health and ecological integrity. In the American West, many large landscapes dominated by ranching and forestry provide:

1. A diversity of products that provide ranchers and forest owners with financial compensation. These products include beef, lamb and goat meat, fiber, timber, hay, fence posts, firewood, hides, fertilizer, compostable manure, fish, game, and nontimber forest products such as harvestable berries and mushrooms. Scientists now include the generation of these products under the rubric of *ecosystem goods* and/or *provisioning services*.[10]

2. *Supporting services*, such as primary production of plant biomass, maintenance of soil moisture-holding capacity and fertility, and supply of habitat for wildlife.

3. *Regulating services*, such as the regulation of water quality and quantity, carbon sequestration, erosion control, crop pollination, and others. Society has seldom put a price tag on most of the regulating services generated within working landscapes from which people benefit, although pollination services have had an explicit market value in some food production systems such as California almond orchards for some time.

4. *Cultural services*. Many visitors to the West pay handsomely for access to the intangible aesthetic and heritage values that they "discover" in working landscapes: beauty, solitude, reconnection to familial or historical ties, spiritual peace, or communal celebration in landscapes. Ranchers, farmers and forest owners may receive financial benefits through agritourism and ecotourism for maintenance of cultural services, including local ecological knowledge, skills, management strategies, and folk crafts used in livestock and land stewardship; western history; and cowboy lore.

Of these ecosystem services, only three are currently being monetized to any extent with PES programs: carbon sequestration, water quality and quantity, and habitat for imperiled species. Next, we briefly highlight some ongoing PES efforts to support these services in working landscapes.

*Carbon Sequestration*

Plants and soils in terrestrial ecosystems contain over three times as much carbon as is currently found in the earth's atmosphere. About half of this terrestrial carbon is stored in forest ecosystems.[11] Rangelands, including savannas, pinyon-juniper woodlands, and natural "park" meadows in forests, store an estimated one-quarter to one-fifth of all terrestrial carbon.[12] The capacity of forests and rangelands to sequester carbon can be augmented in several ways. The main strategies for carbon sequestration in forests include planting trees, avoiding deforestation and forest degradation, and altering forest management practices to extend harvest rotation intervals, use partial harvest techniques, and reduce the removal of dead biomass (e.g., dead trees and forest floor debris).[13] For rangelands, where most of the carbon is found in soils, strategies focus on increasing carbon inputs to soils or decreasing carbon losses from soils. Reducing soil erosion and converting the production of annual forage on a piece of land to perennial grasses are two approaches.[14]

The Chicago Climate Exchange (CCX), which opened in 2003, was North America's first voluntary, legally binding carbon emissions cap-and-trade program. The CCX provided a mechanism through which forest owners and ranchers could generate and sell carbon credits from forest and range management offset projects and was significant because it was the only emissions trading platform with a protocol for carbon offsets from rangelands. In 2007, Montana's Sun Ranch (see chap. 11) became the first working ranch in the United States to generate carbon credits through the CCX's Sustainably Managed Rangeland Soil Carbon Sequestration Offset program. Sun Ranch sold those credits for $30,000. Subsequently, approximately 1,000 ranchers enrolled in the CCX program, and many received some level of compensation.[15] On December 31, 2010, the CCX ended its second legally binding commitment period for emissions reductions by participating entities due to inactivity in U.S. carbon markets. No new round of commitments was made, meaning that trading for compliance reasons had ended, though voluntary transactions can still be processed.

A few years prior to the economic downturn of 2007, carbon-offset transactions had begun to interest many landowners and some investors but that was short-lived. After the CCX rose and then crashed with the downturn and the failure of the federal government to develop a greenhouse gas emissions trading program, it became painfully clear that there were virtually no federal or state laws that explicitly described how carbon credits and their transaction should be considered from a property rights perspective. Carbon credits appeared to be the property right of the surface owners of forest and ranchland properties but not of federal lands lessees or people who harvest timber or remove biomass from federal lands; nor could they be considered mineral rights of below-ground property holders. In *Roseland Plantation LLC v. United States Fish and Wildlife Service et al.*, a federal court judge in Louisiana reaffirmed the "right to report, transfer, or sell carbon credits as part of the bundle of rights associated with [surface] property ownership."[16] While carbon credit exchanges have been essentially moribund in the West since the CCX closed (apart from over-the-counter transactions), some ranchers and forest owners have nonetheless been assessing the potential value of carbon sequestration on their lands to be well positioned when and if the demand for carbon credits rises once more.[17] The opening of California's greenhouse gas emissions trading program (a compliance market) in 2013 should jump-start carbon-offset projects from forestry.

*Water Quality and Wetlands Habitat*

The management of soils, woody vegetation, and livestock has direct effects on water quality, as well as on the quantity of water available to downstream users and associated habitats.[18] In a number of cases, ranchers have been compensated for mitigating impacts to aquatic resources, and by managing lands in ways that sustain, restore, or enhance water quality and wetlands. One of the earliest PES schemes in the U.S. was wetlands mitigation banking. Since the mid-1980s, Section 404 of the Clean Water Act (passed in 1970) and Section 10 of the Rivers and Harbors Appropriation Act of 1899 have been interpreted to sanction environmental restitution through what is known as wetlands mitigation banking. However, the Environmental Protection Agency, Army Corps of Engineers, and U.S. Fish and Wildlife Service did not establish their final policies regarding wetlands mitigation banks until 1995. If a development project threatens to damage or destroy wetlands habitat on either private or public lands, arrangements must be made to purchase and create or protect equivalent habitat or to restore an equal or larger area of habitat to comparable quality. In their policy directives, wetlands mitigation banks are characterized as innovative "offset" programs. Landowners can develop formal habitat conservation plans for units of restored, created, or preserved wetlands or other aquatic habitats, which are then marketed as "compensatory mitigation credits" that can be used to offset losses allowed at other sites. This has been accomplished for vernal pools in California, for example.

Perhaps the most widely heralded example of mitigation banking to support the restoration of stream and wetlands habitat for the preservation of wildlife species is at Nevada Spring Creek in the Blackfoot Valley of Montana.[19] Listed in 1990 by the U.S Fish and Wildlife Service's Montana Partners for Fish and Wildlife Program as a priority for habitat recovery for three trout species (native westslope cutthroat [*Oncorhynchus clarkii lewisi*], native bull [*Salvelinus confluentus*], and introduced brown [*Salmo trutta*]), the Nevada Spring Creek watershed had historically been plagued by unsustainable logging, poor grazing practices, siltation, pollution, erosion, and wetlands fragmentation. An assessment of its riparian health determined that summer stream temperatures had become too high and that it lacked suitable spawning areas and adequate feeding pools to support the two native and one introduced species of trout. In 2001, the Nature Conservancy purchased 1,900 acres of the watershed and sought a conservation-oriented buyer for the property who would work with them to restore it along with the U.S. Fish and Wildlife Service, Montana Fish, Wildlife and Parks, and a coalition of non-governmental organizations, including Blackfoot Challenge, the Montana Land Reliance, and Trout Unlimited.

The eventual buyer—Fred Danforth, a financier and avid fly fisherman—had some novel ideas about how to make conservation pay for itself. In his own words, Danforth became "aware of the possibility of using market based mechanisms to provide some return for the large amount of capital that is required for this kind of restoration. . . . Eventually we came across the whole notion of wetlands and stream mitigation banking," on which he and his colleagues anchored a new integrative approach to private conservation financing. They established the first ever entrepreneurial mitigation bank designed to coordinate with legal guidelines developed by the Army Corps of Engineers.[20] It eventually restored habitat that attracted the trout back to spawn, and this restored fishing resource was then successfully marketed to conservation-oriented buyers of the offset credits.

*Habitat for Fish, Wildlife and Other Species at Risk*

Similar to mitigation banking under the Clean Water Act, impacts to habitat of species listed under the Endangered Species Act (ESA) can be offset through the use of "conservation banking," which generates credits that can be sold. Conservation banking, overseen by the U.S. Fish and Wildlife Service and the National Oceanic and Atmospheric Administration, has been implemented in several western states since the 1973 passage of the ESA. The conservation banking concept has recently been extended to critical habitat for state-listed species considered to be at risk, and is particularly popular in California.

There have been numerous efforts to involve private investors in both conservation and mitigation banking for ecosystem service restoration. Beartooth Capital, for example, matches investors with ranch properties in Idaho, Montana, Colorado, and California and finances their restoration using such programs when possible to increase the economic, ecological, and agricultural value of these properties.

Banking and other PES programs are being launched globally, and recent reports provide a snapshot of their financial scale. The estimated value of forest carbon market offset transactions in compliance and voluntary markets combined during 2012 was $216 million.[21] Substantial growth is expected to continue. For conservation banking, there are at least 100 species banks operating in the United States, many of which are in California, and an estimated 600 banks or more operating globally. Financial information was only available for about one-fifth of these banks, but a conservative estimate of the financial value of these habitat markets was set at $1.8–2.9 billion per year.[22]

## IMPLEMENTATION: COLOCATION AND "BUNDLING" OF ECOSYSTEM SERVICES

Ecosystem services do not necessarily occur in isolation from one another. It is critical that land-use planners, ranchers, forest owners, and policymakers recognize the colocation and synergisms among various services.[23] Strategically planning to protect, enhance, and/or restore multiple services by bundling them together enables land managers and planners to leverage the conservation of services, particularly regulating services, that are not usually the focus of conservation efforts. Some counties in California, Colorado, and Florida have mapped where these ecosystem services are colocated and have provided zoning recommendations and financial support for keeping them in undeveloped open space.[24]

Potential investors, both public and private, should not exclusively focus on single targets such as carbon credits but should broaden their attention to opportunities to bundle together multiple ecosystem services derived from the same working landscape. The Willamette Partnership, for instance, is advancing efforts in this area through its Counting on the Environment program. This program is focused on developing the Ecosystem Credit Accounting System for bundling credits from four currencies: wetlands, salmonid habitat, prairie habitat, and riparian habitat.[25] Another example of efforts to market bundled credits involves the 3,283-acre Sooes Forest on Washington's Olympic Peninsula, one of four forests owned by Ecotrust Forests LLC (a forestland investment fund) and managed by Ecotrust Forest Management (together referred to as "Ecotrust" below).

Ecotrust hopes to achieve the "triple bottom line" of economic, ecological, and social sustainability by managing the forest for timber and biomass production, ecosystem services, forest health, and community benefit and to provide a model for other forest owners in doing so.[26] Since 2007, Ecotrust has been developing an improved forest management project in the Sooes Forest for carbon offsetting under the Climate Action Reserve standard. This project extends harvest rotations, expands riparian reserves, establishes steep slope reserves above salmon streams, and protects unique forest features.[27] The first carbon credits were sold in March 2010 to the Eco Products Fund, a private equity group that will in turn sell the credits to businesses and government agencies that want to offset their carbon emissions.[28]

Although the greatest opportunity for forest landowners to be paid for ecosystem services was once perceived to be in carbon markets, carbon sequestration is only one ecosystem value that forests provide. Managing forests and rangelands to enhance carbon sequestration can produce environmental cobenefits such as biodiversity protection, increased water storage and filtration, and improved soil quality. Ecotrust plans to bundle still other ecosystem services from the Sooes Forest stands with carbon encumbrances and market them through water-quality trading credits, conservation banking, and other means.[29]

POTENTIAL BENEFICIARIES AND ALLIES FOR
PAYMENTS FOR ECOSYSTEM SERVICES

*Public Beneficiaries*

Ecosystem services produced by working landscapes benefit not only the rural communities in their immediate vicinity but millions of people living downstream as well. Downstream beneficiaries can pay upstream providers to manage their upstream land in ways that continue to provide or increase specific ecosystem services. In parts of the United States, urban water users now pay for the conservation (nondevelopment) of working landscapes in the headwaters of their watersheds. The Adirondack Forest Preserve of New York, established in 1885, was one of the earliest examples of landscape-level conservation intended, in part, to benefit downstream water users—in this case, those in the New York metropolitan area. More recently, in 1997, New York City further committed itself to watershed stewardship by purchasing conservation easements in the hinterlands of its watersheds to improve and protect water quality for the city's water users.[30]

Similar logic was used for the 2000 establishment of the Las Cienegas National Conservation Area, spanning Arizona's Pima and Santa Cruz Counties and the Cienega Creek Natural Preserve in Pima County, Arizona. The majority of the land in the conservation area continues to be grazed on contract with local ranchers in Pima and Santa Cruz Counties. These areas were set aside not only to protect Cienega Creek, a critically important riparian habitat, but also to preserve the headwaters that recharge the Tucson Basin aquifer. Tucson, the major metropolitan center within Pima County, relies on the Tucson Basin aquifer for much of its drinking water. Protection of the headwaters, accomplished in this case by using perennial cover on grazed lands to mitigate the impacts of flash floods, provides flood storage capability, reducing erosion and property damage during periods of intense rainfall. These measures can also increase the natural recharge of an already overtapped aquifer. Such conservation measures for working landscapes are far more cost-effective than constructing and maintaining numerous water treatment facilities, flood retention basins, and erosion control structures. Payments for ecosystem services have therefore generated cost-effective means for maintaining and restoring ecosystem functions.

In Colorado, recent assessments have determined that the 1.4 million acres already placed under conservation easements will return $3.5 billion dollars of ecosystem services as public benefits. This estimate is based on calculating the replacement costs of ecosystem services that would be lost if these lands were not maintained for productive range and forest uses.[31] In other words, for every dollar invested in keeping private range and forestlands out of development, the citizens of Colorado receive a six-dollar return in ecosystem services on this public investment. This return on investment will undoubtedly increase through time, particularly where neighboring lands are developed and their capacity to produce comparable ecosystem services are lost (especially where those easements were strategically targeted in "at risk," high development pressure areas).

*Corporate, Nonprofit, and Foundation Beneficiaries*

Many kinds of businesses, nonprofits, and foundations recognize that their images and constituencies—and, in some cases, their primary operations—may be increasingly dependent on the healthy functioning of ecosystem services. For example, in 1989, outdoor apparel retailer Patagonia, Inc., co-founded the Conservation Alliance, composed of outdoor adventure and travel companies that rely on scenic views and access to recreational areas and waterways. Similarly, insurance companies rely on coral reefs and mangroves for protection of coastal communities.[32] Global beverage companies such as New Belgium Brewing and Coca-Cola rely on supplies of clean water; to mitigate impacts related to point source pollution from their factories, some have begun to fund watershed protection programs. In France, Nestlé pays farmers to manage animal waste and reforest sensitive areas to protect the mineral water used in its Vittel line of bottled water.[33]

Nonprofits and foundations are also now involved in such transactions. The World Resources Institute has developed the Corporate Ecosystem Services Review, a toolkit to help businesses proactively manage risks and opportunities related to ecosystem services.[34] Several foundations, including the Heron Fund and the CS Fund /Warsh-Mott Legacy Fund, provide low- or no-interest loans for land preservation and restoration. The Big Spring Creek case study cited above is one in which nonprofits such as Trout Unlimited and the Nature Conservancy have played important roles in restoring ecosystem services.

## PES SCHEMES THAT EMBED THE COST OF ECOSYSTEM SERVICES

In July of 2009, Butch Mayfield of the Mayfield Ranch in Hidalgo, New Mexico, became the first American rancher successfully to sell carbon credits attached to the livestock he auctioned. These credits were directly related to the land where Mayfield's carbon sequestration efforts had been independently verified. The sale of "carbon credits embedded in the cost of meat" took place at the Superior Livestock auction in Winnemucca, Nevada, and was facilitated by Integrated Systems Information, Inc.'s VerifiedGreen program of third-party verification. Today, nonprofits such as Animal Welfare Approved and for-profits such as Integrated Systems Information, alike, are directly embedding environmental, health and welfare values to establish market-derived prices for meat bundled with publicly oriented claims.

The earliest examples of embedding the costs of ecosystem services conservation in pricing date to the early 1990s. Forestry certification systems were developed as market-based mechanisms for curbing environmentally destructive forestry practices by setting standards of operation for forest management to assure consumers that certified, third-party verified wood products had been produced in a manner that was ecologically sound and socially responsible. Forestry certification schemes targeted consumers who were willing to pay more for these wood products.

Without actually calculating and purposefully embedding the cost of maintaining ecosystem services in their products, some ranchers have begun to market, along with their meat, the "back story" of how it was produced sustainably while maintaining wild species, cultural traditions, and environmental amenities. Survey after survey confirms that consumers are willing to pay more for eco-labeled and landscape-labeled wood products, meats, eggs, and dairy products for several reasons: to be economically supportive of their neighbors and the early adopters of best practices; to have transparency about food sources should food safety issues arise; to reduce carbon footprints; to optimize freshness, seasonality, and heritage values; and to support socially and environmentally sustainable production practices.[35] The value-based priorities of consumers surveyed in these studies can help ranchers decide how, where, and for what price they should market their meat.

An example that highlights the success of marketing beef expressly associated with land stewardship practices for the enhancement of ecosystem services emerges in the oldest rancher-environmentalist collaboration in the West: the Diablo Trust. Out of this collaboration has emerged the award-winning Diablo Burger restaurant, in the heart of downtown Flagstaff, Arizona, featuring natural range-raised beef and other local food products grown within the Diablo Canyon Rural Planning District. Since opening in March of 2009, thousands of local residents and tourists have come to Diablo Burger to partake of the edible "goods" from this working landscape. While at the restaurant, at countless public events, and on websites, these residents and tourists have learned about land stewardship practices in the foodshed. A similar venue, Local Burger, in Lawrence, Kansas, markets beef, lamb, and bison from surrounding ranches and provides consumers with stories of local land stewardship associated with its meat production. Another successful initiative is Country Natural Beef (see spotlight 10.1), a co-operative of ranchers who are third-party certified by the Food Alliance for several values, including their efforts to conserve soil and water and protect biodiversity and wildlife habitat.[36] Although these stewardship practices are not directly framed as enhancing ecosystem services, they are directly used in marketing.

Sustainable Northwest Wood, Inc., a wholesale distribution center in Portland, Oregon for wood products produced in the Pacific Northwest, sells wood products grown and milled for the purpose of creating a sustainable local forest economy. The wood products are produced from forests managed to meet Forest Stewardship Council certification standards, or are derived from restoration projects designed to restore native ecosystems (e.g., juniper removal to restore grasslands in places where juniper encroachment has occurred as a result of wildfire suppression, and is lowering the water table).

The case of meat sales from Navajo-Churro sheep provides a compelling example of the potential for incorporating both the real and symbolic values associated with cultural heritage and genetic conservation in product pricing. After marketing materials began to highlight the cultural and conservation values of the Navajo-Churro sheep, lamb prices soared from less than $0.50 per pound (in 2001) to $6.50 a pound (in 2007). These price increases were the result of collaborative marketing and promotional initiatives of the Dine Bei'ina (a Navajo Sheep and Lifeways nonprofit), American Livestock Breeds Conservancy, Navajo-Churro Sheep Association, Renewing America's Food Traditions Alliance, and Slow Food Flagstaff. Established as a market recovery project under Slow Food International, the Navajo-Churro sheep presidia returns economic and cultural benefits to Navajo sheepherders who market their lamb meat and wool for its cultural heritage value to their own people, as well as its nutritional, genetic resource, and historic preservation values to the larger community.

Value-added niche markets for wood products (certified, made from certain species or from certain locations, unique products, made with special technologies, etc.) and meats (grass-fed, organic, free-range, chemical-free, predator-friendly, etc.) have been rapidly growing for more than a decade, though the economic downturn starting in 2008 somewhat slowed their growth. As an example of rapid growth, current estimates are that direct-market sales of value-added, niche meats comprise 3–5 percent of the total meat market, and this portion has grown significantly over the last quarter century.[37] According to several sources, the current estimated growth rate of local and sustainable food markets in the United States is more like 22–25 percent, suggesting that there is more room for meats with natural or cultural services associated with them to grow— obviously, it takes longer to bring grass-fed beef to market than it does an heirloom carrot.[38] Both these current growth rates and the current unprecedented high price of beef suggest additional opportunities for marketing "embedded ecosystem services" along with beef.

## STRATEGIES AND PRINCIPLES: LANDSCAPE LABELING AND PES ON CONSERVED LANDSCAPES

In some cases, niche products with embedded ecosystem values have brought better returns to ranchers and improved their bottom line; but in other cases, the additional time and energy costs of direct marketing have been sobering. While most of the consumers who purchase specially labeled meats claim that they are willing to pay more for such products that reflect their environmental and ethical values and satisfy their health criteria, it remains unclear how many risk-averse ranchers will target the sales of their products for these rather volatile, dynamic markets.[39]

One potential flaw in current PES plans concerns the fact that individual forest owners and ranchers may be granted payments from the public or private sectors for their stewardship practices on private lands, but these payments can seldom be justified on federal, state, or county lands leased by ranchers or harvested by forest workers. In addition, the quality of ecosystem services is seldom affected by only one forest owner's, or rancher's, stewardship practices—however diligent the individual may be; unless neighboring forests and ranches adopt the same practices, ecosystem services are seldom maintained at a landscape level.

A landscape-scale approach to PES would enable strategic and targeted investments in conservation and restoration, rather than opportunistic reliance on the participation of individual landowners to initiate restoration projects in isolation from one another. Such an approach would rely on consensus building and collective action among forest owners and ranchers. Group certification allowed under the Forest Stewardship Council certification system is one approach to addressing this challenge. Under group certification, one organization or person holds a Forest Stewardship Council certificate, but other forest owners or managers can participate as members of the group and become certified under the single certificate, as long as their management practices conform to council rules. Group certification reduces the cost of certification for individual forest owners, making it easier for landowners to participate.

Another approach that has been used in the context of marketing carbon offsets is for forest or ranch owners whose landholdings are not large enough to make carbon market participation financially feasible to work together through an aggregator. The CCX developed the first aggregation schemes, whereby an aggregator combines the carbon management activities of multiple landowners or carbon project developers into a single project so that they can access the market as one big project, streamlining project development and reducing transaction costs.[40]

Alternatively, landowners on conserved landscapes could collectively select the most vulnerable or degraded areas within their landscapes to restore. Rather than simply parceling out payments to individual owners or lessees, these collectives would invest in strengthening the weakest links in service maintenance across the entire landscape. Ranchers or forest owners would share the costs of restoration and maintenance work with consumers who purchase their products at premium prices and/or with other stakeholders seeking ecosystem services. In this way, the costs—as well as the benefits— would be shared across the foodshed or watershed; no longer would ranchers or forest owners have to bear alone the financial burden of maintaining good stewardship practices. Government agencies as well—from local to federal—could participate in restoring degraded lands through cost-sharing programs. Government and nongovernmental organizations could also assist bridge-building activities between rural and urban communities through development of ecosystem service marketplaces in local communities that would complement farmers' markets and processing facilities.

Some forest owners in Europe have recently suggested such a strategy— one that would function at the level of the entire working landscape: "Specifically, we propose that managed rural landscapes delivering valuable ecosystem services should be awarded a 'Landscape Label' that would be used to identify products produced from the landscape. A Landscape Label could also represent and publicize ecosystem service delivery as well as cultural and symbolic attributes of the landscape as defined by local communities. . . . Thus a Landscape Label has the potential to improve market recognition, secure premium payments and gain access to niche markets."[41]

Using the Diablo Trust of northern Arizona in a hypothetical example of this strategy, all products derived from the Diablo Canyon Rural Planning District would be labeled with the Diablo Trust brand, indicating their producers' membership in the Diablo Trust collaborative land management alliance. All products from this working landscape—meat, timber, firewood, and so forth—would be promoted, marketed, and sold for a premium price to consumers and industries wishing to support the protection of ecosystem services at a landscape level. A percentage of that premium would go back to the nonprofit, Diablo Trust, in order to fund habitat restoration and species recovery projects aimed at improving the maintenance of ecosystem services. This has already occurred to some extent, as the Diablo Burger restaurants in Flagstaff and Tucson, Arizona use the cachet of the "locally produced" and "grass-finished" designations to market their burgers.

## SUMMARY AND CAUTIONS

The ecosystem services provided by private, public, and tribal forests and rangelands in the working landscapes of the American West are increasingly being economically valued by a wide array of beneficiaries. While funding from the private, public, and nonprofit sectors might partially help to conserve ranching and forestry as ecologically and economically sustainable livelihoods, there is no single silver bullet solution to keeping working landscapes intact. Market-driven strategies are not panaceas for maintaining the integrity of working wildlands; and all strategies have economic as well as ecological and ethical risks.[42] One such risk involves reducing working lands to the sum of their marketable products, in the process undervaluing other dimensions of these places that have little or no monetary value. A second risk is encountered in considering current market prices for ecosystem services to be reflective of society's valuation of them; in this way, we may be undermining efforts of future generations of ranchers and forest owners to obtain the true value of multigenerational investments, which are currently defined and assessed by the most rudimentary means.[43] Finally, many forest owners and ranchers may not have access to more lucrative markets because their holdings are not large enough to make participation financially feasible, they don't have the technical assistance needed to gain market entry, and so on.

Nevertheless, the rapidly evolving mechanisms for paying ranchers and foresters for their efforts in sustaining the public good may soon be able to help them in further diversifying their income-generation portfolios. These diversified revenue streams may someday offer them more economic resilience, just as the maintenance of the ecosystem services themselves may instill greater resilience to the very landscapes from which we obtain a considerable portion of our food, fiber, lumber, and fuelwood.

## NOTES

1. S. Cunningham, *The Restoration Economy: The Greatest New Growth Frontier* (San Francisco: Berrett-Koehler Publishing, 2002).
2. For a description of this movement, http://slowmoney.org/.
3. Millennium Ecosystem Assessment (Program), *Ecosystems and Human Well-Being* (Washington DC: Island Press, 2005); S. Engel, S. Pagiola, and S. Wunder, "Designing Payments for Environmental Services in Theory and Practice: An Overview of the Issues," *Ecological Economics* 65, no. 4 (2008): 663–74; W. Tasch, *Inquiries into the Nature of Slow Money* (White River Junction, VT: Chelsea Green Publishing, 2008).
4. S. Wunder, "Payments for Environmental Services: Some Nuts and Bolts," *CIFOR Occasional Paper* no. 42. (2005); Engel et al., "Designing Payments"; but for an alternative view, see K. H. Redford and W. M. Adams, "Payment for Ecosystem Services and the Challenge of Saving Nature," *Conservation Biology* 23 (2009): 785–87.
5. P. Sullivan, "Ranch as Investment: Be Prepared to Work," *New York Times*, August 4, 2011.

6. M. Peters-Stanley, K. Hamilton, T. Marcello, and M. Sjardin, *Back to the Future: State of Voluntary Carbon Markets, 2011* (Washington, DC: Ecosystem Marketplace; New York: Bloomberg New Energy Finance, 2011) http://www.forest-trends.org/documents/files/doc_2828.pdf.

7. See: R. Leigh, and S K. Fairfax, "The 'Shift to Privatization' in Land Conservation: A Cautionary Essay," *Natural Resources Journal* 599 (2002): 42.

8. J. Conant, "The Dark Side of the 'Green Economy,'" *Yes! Magazine*, August 23, 2012, www.yesmagazine.org/issues/its-your-body/going-against-the-green.

9. Millennium Ecosystem Assessment (Program), *Ecosystems and Human Well-Being*.

10. R. Skaggs, "Ecosystem Services and Western U.S. Rangelands," *Choices* 23, no. 2 (2008): 37–41; G. C. Daily, S. Polasky, J. Goldstein, P. M. Kareiva, H. A. Mooney, L. Pejchar, T. H. Ricketts, J. Salzman, and R. Shallenberger, "Ecosystem Services in Decision Making: Time to Deliver," *Frontiers in Ecology and the Environment* 7, no. 1 (2009): 21–28.

11. Intergovernmental Panel on Climate Change, *Climate Change 2007: The Physical Science Basis Contribution of Working Group 1 to the Fourth Assessment Report of the Intergovernmental Panel on Climate Change* (New York: Cambridge University Press, 2007).

12. K. M. Havstad, D. P. C. Peters, R. Skaggs, J. Brown, B. Bestelmeyes, E. Frederickson, J. Herrick, and J. Wright, "Ecological Services to and from Rangelands of the United States," *Ecological Economics: The Journal of the International Society for Ecological Economics* 64, no. 2 (2008): 261.

13. D. D. Diaz, S. Charnley, and H. Gosnell, *Engaging Western Landowners in Climate Change Mitigation: A Guide to Carbon-Oriented Forest and Range Management and Carbon Market Opportunities*, General Technical Report PNW-GTR-801 (Portland, OR: U.S. Department of Agriculture, Forest Service, Pacific Northwest Research Station, 2009).

14. G. P. Nabhan, D. Blair, and D. Moroney, "Ranching to Produce Tacos Sin Carbon: The Low Carbon Foodprint of Grass-Fed Beef and Sheep Production in the Semi-Arid West," *Quivira Coalition Journal*, no. 35 (February 2010), 28–34, http://quiviracoalition.org/images/pdfs/2055-Journal35.pdf.

15. H. Gosnell, N. Maness, and S. Charnley, "Engaging Ranchers in Market-Based Approaches to Climate Change Mitigation: Opportunities, Challenges, and Policy Implications," *Rangelands* 64, no. 6 (2011): 20–24, and "Profiting From the Sale of Carbon Offsets: A Case Study of the Trigg Ranch," *Rangelands* 64, no. 6 (2011): 25–29.

16. Roseland Plantation LLC v. United States Fish and Wildlife Service et al., 2006 U.S. Dist. LEXIS 29334.

17. Sullivan, "Ranch as Investment."

18. K. T. Weber and B. S. Gokhale, "Effect of Grazing on Soil-Water Content in Semiarid Rangelands of Southeast Idaho," *Journal of Arid Environments* 75 (2011): 464–70.

19. "Project Summary: Nevada Spring Creek Ranch," Ecosystem Investment Partners, http://www.ecosystempartners.com/nevadaspringcreekranch.

20. J. N. Levitt, "The Effective Use of Conservation Incentives" (working paper, Lincoln Institute of Land Policy, Cambridge, MA, 2005).

21. M. Peters-Stanley, G. Gonzalez, and D. Yin, *Covering New Ground: State of the Forest Carbon Markets 2013* ([Washington, DC:] Ecosystem Marketplace, 2013), http://www.forest-trends.org/documents/files/SOFCM-full-report.pdf.

22. B. Madsen, N. Carroll, and K. Moore Brands, *State of Biodiversity Markets Report: Off-*

*set and Compensation Programs Worldwide* (Washington, DC: Ecosystem Marketplace, 2010), http://www.ecosystemmarketplace.com/documents/acrobat/sbdmr.pdf.

23. K. M. Chan, M. R. Shaw, D. R. Cameron, E. C. Underwood, and G. C. Daily, "Conservation Planning for Ecosystem Services," *PLoS Biology* 4, no. 11 (2006): 2138–52; C. Raudsepp-Hearne, G. D. Peterson, M. Tengo, E. M. Bennett, T. Holland, K. Benessaiah, G. K. MacDonald, and L. Pfeifer, "Untangling the Environmentalist's Paradox: Why Is Human Well-Being Increasing as Ecosystem Services Degrade?" *BioScience* 60 (2010): 576–89.

24. T. Kroeger, "The Economic Value of Ecosystem Services in Four Counties in Northeast Florida" (Conservation Economics Working Group Paper no. 2, Defenders of Wildlife, Washington, DC, 2005); Chan et al., "Conservation Planning."

25. "Counting on the Environment," Wilamette Partnership, http://willamettepartnership .org/ongoing-projects-and-activities/nrcs-conservation-innovations-grant-1/ counting-on-the-environment.

26. Ecotrust Forest Management, Inc., "Forest Management Plan," February 2009 (updated and amended August 2011), http://www.ecotrustforests.com/docs/2011_EFM _Forest_Mgmt_Plan_082311_final.pdf.

27. B. von Hagen "Developing Forest Carbon Projects: Observations from the Bleeding Edge" (paper presented at the conference Ecosystem Markets—Making Them Work, Portland, OR, June 18–19, 2009).

28. "Ecotrust Sells First Carbon Credits from Olympic Peninsula Forest," *Seattle Times*, March 19, 2010.

29. Von Hagen, "Developing Forest Carbon Projects."

30. D. Warne, "New York City," Information Center for the Environment (ICE), accessed March 16, 2010, http://ice.ucdavis.edu/node/133.

31. Trust for Public Lands, *A Return on Investment: The Economic Value of Colorado's Conservation Easements* (Boulder: Trust for Public Lands, 2010).

32. C. Hanson, J. Ranganathan, C. Iceland, and J. Finisdore, *The Corporate Ecosystem Services Review: Guidelines for Identifying Business Risks and Opportunities Arising from Ecosystem Change*, version 1.0 (Washington, DC: World Resources Institute, 2008).

33. Ibid.

34. Ibid.

35. For example, D. J. Thilmany, J. Grannis, and E. Sparling, "Regional Demand for Natural Beef Products: Urban vs. Rural Willingness to Pay and Target Customers," *Journal of Agribusiness* 21(December 2003): 149–66; D. Thilmany, W. Umberger, and A. Ziehl, "Strategic Market Planning for Value-Added Natural Beef Products: A Cluster Analysis of Colorado Consumers," *Renewable Agriculture and Food Systems* 21(September 2006): 192–203; G. T. Tonsor and R. Shupp, "Valuations of 'Sustainably Produced' Labels on Beef, Tomato, and Apple Products," *Agricultural and Resource Economics Review* 38, no. 3 (2009): 371–83; F. Solup, K.K. Hagen, and K. Lyon, *Center for Sustainable Environments Community Foods Focus Group Study* (Flagstaff: Social Research Laboratory, Northern Arizona University, 2003).

36. The latest certification standards are published by the Food Alliance ("FA Sustainability Standard for Livestock Operations," 2012, http://foodalliance.org/livestock/ FA-SS-02-%20livestock1113.pdf).

37. G. P. Nabhan, "Making Food Systems Work in Ranching Country," in *State of Southwestern Foodsheds*, ed. G. P. Nabhan and R. Fitzsimmons (Tuscon, AZ: Sabores Sin

Fronteras, 2011), 14–15, http://saboressinfronteras.files.wordpress.com/2011/02/state-of-sw-foodsheds.pdf.

38. For example, K. Meter, "Growth in Local Food Economies" (paper presented at the conference Home Grown Economy, Crookston, MN, March 31, 2008), available at http://www.crcworks.org/crcppts/petersonKM08.pdf.

39. Thilmany et al., "Strategic Market Planning."

40. S. Charnley, D. Diaz, and H. Gosnell, "Mitigating Climate Change through Small-Scale Forestry in the USA: Opportunities and Challenges," in "Climate Change," special issue of *Small-Scale Forestry* 9, no. 4 (2010): 445–62; Diaz et al., *State of Forest Carbon.*

41. J. Ghazoul, C. Garcia, and C. G. Kushalappa, "Landscape Labeling: a Concept for Next-Generation Payment for Ecosystem Services," *Forest Ecology and Management* 258, no. 9 (2009): 1889–95.

42. D. Ehrenfeld, *The Arrogance of Humanism* (Oxford: Oxford University Press, 1985).

43. For more cautions see Redford and Adams, "Payment for Ecosystem Services" (n. 3 above, this chapter).

# Spotlight 14.1

# THE CONSERVATION RESERVE PROGRAM

*Steven E. Kraft*

IN BRIEF

- The Conservation Reserve Program (CRP) is an early example of a payments for ecosystems services program: the federal government pays farmers to remove marginal agricultural land from production to reduce soil erosion and promote conservation practices on the land instead.
- The CRP focus has shifted from its original focus on providing farm income, conserving highly erodible land, and reducing surplus commodities to providing environmental benefits; it is a successful example of a flexible government program that has responded to changing needs over time.
- Participants in the CRP —both buyers and sellers—need good information about the ecosystem services being traded, how those services are assessed, and how prices for these services are determined so that they can be fully informed program participants and ensure that they obtain or pay a fair value for the ecosystem services being provided.

During the world food crisis of the early 1970s, then Secretary of Agriculture Earl Butz exhorted American farmers to plant "fencerow to fencerow." Farmers responded by bringing marginal land into cultivation to increase the amount of cropland in production. While there was an increase in the amount of crops produced, there was also an increase in the level of off-field impacts, or negative externalities, such as reduced surface water quality through soil erosion and sedimentation and loss of wildlife habitat. With the strong export markets of the 1970s, farmers expanded the size of their farms through the purchase of additional farmland. Such purchases were frequently financed by higher valuation of land they already owned and the expectation that increasing farmland values and high commodity prices would support their land acquisitions. This world abruptly changed on December 27, 1979, when the Soviet Union invaded Afghanistan and President Carter embargoed exports of American grain to the Soviet Union, a major importer of American grain. Commodity prices started to fall, and worldwide recession in the early 1980s further reduced export demand.

The low commodity prices coupled with increasing interest rates on cropland loans created an economic crisis with environmental overtones: farmers going out of business unable to service their loans, rural banks with an increasing number of nonperforming loans, and the 1977 and 1982 National Resources Inventories for the first time permitting the tying of soil erosion to specific land areas.[1] Thus it became possible to identify agricultural areas that were responsible for significant levels of soil loss and to develop policy tools targeting those areas. Consequently, various policy initiatives were undertaken to deal with these challenges. One set of policies of particular interest is Title XII of the 1985 Food Security Act, known as the Farm Bill. One provision of Title XII is the Conservation Reserve Program (CRP)—a long-term land retirement program having a goal of removing 45 million acres of highly erodible land from production over time and shifting its use to one or more conservation practices to reduce soil erosion significantly while providing additional conservation or environmental benefits. The CRP is an early example of paying landowners for the conservation of ecosystem services.

The legislative objectives of the CRP were to reduce erosion from wind and water, protect the long-term productive capacity of the nation's soils, reduce sedimentation, improve water quality, create improved habitat for fish and wildlife, curb production of surplus commodities, and provide income for farmers. Landowners/farmers participating in the program generally receive an annual rental payment over the 10–15-year life of the typical CRP contract and a cost-share payment to help in the establishment of an approved conservation vegetative cover or use. The annual rental payment for a particular parcel of land and a specific cover or use is generally determined through a competitive bidding process among landowners offering their land to the program during sign-up-specified bidding periods. The actual specifics of the CRP have changed through the different Farm Bills since it was established in 1985, as have the specifics of sign-up. For example, during the fourth sign-up in February 1987, successful landowners whose bids were accepted by program managers and whose land was enrolled in the CRP received a one-time payment of $2.00 per bushel of established corn yield on each acre of corn base retired. To date, there have been 41 sign-up periods.

Starting in 1986/87 with the first four sign-ups, the CRP program became an effective tool to retire highly erodible land from crop production while providing landowners/farmers with a dependable stream of income and the public with a flow of environmental benefits, or ecosystem services, from conservation of vegetative cover and reduced erosion. Beginning in 1990, the U.S. Department of Agriculture (USDA) began using an Environmental Benefit Index (EBI) to account for multiple environmental concerns used as criteria to evaluate the eligibility of parcels of land offered to the program. Over time, the components of the EBI and their relative weights have varied. For example, in sign-up 39 ending on August 27, 2010, the EBI incorporated six major factors: (1) improvement of wildlife habitat, (2) water quality, and (3) air quality (including carbon sequestration), as well as (4) reduced on-farm soil erosion, (5) benefits that will endure beyond the contract, and (6) cost effectiveness. Frequently, these factors are composed of a number of subfactors. Unlike in 1986/87 when implementation was focused chiefly on highly erodible land, farm income, and surplus commodities, the focus of the CRP in 2010/11 was almost entirely on environmental benefits. Indeed, because of their high environmental value, there is a continuous sign-up for former cropland entering conservation uses such as riparian buffers, filter strips, grass waterways, and shelterbelts. As of September 2012, roughly 29.5 million acres were enrolled in the program.[2] In 2007, the program reached its maximum enrollment of 36.8 million acres.[3]

In just over 25 years, a program that developed in response to the farm crisis of the early 1980s has been transformed into a far-reaching environmental program for working private lands in rural America. Concerns over the expansion of production onto marginal croplands and the need to get cash into the hands of income-strapped farmers have broadened to ensure the protection of vital ecosystem services for the general public. The CRP will undoubtedly remain an integral part of farm policy and its emerging focus on broadening and enhancing the multifunctional role of America's working lands in our economy and culture.

It also provides one example of a successful, nonmarket based, payment for ecosystem services program from which new initiatives being developed to reward landowners for the conservation of ecosystem services can potentially learn. Between 1986 and 2011, U.S. taxpayers spent approximately $37.9 billion through the CRP to retire highly erodible land from agricultural production and to secure the provisioning of ecosystem services from those rural lands able to score high enough on the EBI at an acceptable cost to the USDA.[4] Unlike a traditional market in which there are many buyers and sellers, the CRP had one "buyer," the USDA. Initially, farmers submitted a bid to the USDA reflecting the minimum annual rental rate they would be willing to accept for their land in exchange for not farming it. The country had been divided by the USDA into bid pools reflecting areas of relatively homogeneous productive conditions. In theory, at least, the farmers were bidding against each other for a limited amount of dollars from the USDA to pay CRP rents. In practice, a bid cap emerged that reflected the expected cash rent equivalent of land eligible for the CRP in an area. However, information about (1) the bid cap—the primary financial incentive for a farmer to participate in the CRP— and (2) the eligibility of his or her land for the CRP did not move very quickly through the agricultural community thus limiting participation in the program.[5] Currently, a farmer offers his or her land for inclusion in the CRP, an EBI is calculated, the farmer indicates a rental rate she or he is willing to accept, and the agency administering the program calculates a cost relative to the EBI. This cost in conjunction with available budget and the relative costs of other offered land is used to determine if an offered parcel is accepted.

In any payments for ecosystem services program, participants (both buyers and sellers) need information about the ecosystem services being traded and how they are assessed, and they need an understanding of how prices are determined for the service(s) or offers being evaluated. Otherwise they have no foundation for being a "fully" informed participant in the exchange; they enter it with asymmetrical information.[6] Initially, this seems to have been the case with the CRP. Eligible landowners did not understand the program well enough to submit an informed bid that was not disadvantageous to them but was fair relative to what existing cash rents were in their area. The situation was made more difficult in many areas by the lack of a cash rental market for agricultural land. The introduction of the EBI with its multiple factors and their corresponding subfactors and weights created similar challenges.[7]

From the perspective of the taxpayer, one of the issues associated with the CRP is: Has the government (as buyer) paid too much to rent the annual cropping rights from the landowners? A landowner with a greater knowledge of her land might have an advantage vis-à-vis the local government program administrator in formulating a bid. The literature on the CRP is extensive, and the program has gone through a transformation from the time that the first rules were written and the first sign-up took place in 1986 to its current rules and form in 2012. During this time, the interest in payments for ecosystem services has emerged as an active area of policy development and implementation. The implementation history of the CRP can be plumbed for relevant lessons that can be used in the design of future payments for ecosystem services schemes involving working rural landscapes as well as other landscapes rich in ecosystem services.

## NOTES

1. American Farmland Trust, *Soil Conservation in America: What Do We Have to Lose?* (Washington, DC: American Farmland Trust, 1984).
2. U.S. Department of Agriculture, Farm Services Agency, "Conservation Reserve Program—Status: End of September 2012," http://www.fsa.usda.gov/Internet/FSA_File/septonepager0912.pdf.
3. U.S. Department of Agriculture, Farm Services Agency, *Conservation Reserve Program—Summary and Enrollment Statistics FY 2007* (Washington, DC: USDA Farm Services Agency, 2007), http://www.fsa.usda.gov/Internet/FSA_File/annual_consv_2007.pdf; U.S. Department of Agriculture, Farm Services Agency, *Conservation Reserve Program—Cumulative Enrollment by Year: 1986-2011* (Washington DC: USDA Farm Services Agency, 2012).
4. U.S. Department of Agriculture, Farm Services Agency, *Conservation Reserve Program—Rental Payments by state 1986-2011* (Washington DC: USDA Farm Services Agency, 2012).
5. J. D. Esseks and S. E. Kraft, "Why Eligible Landowners Did Not Participate in the First Four Sign-ups of the Conservation Reserve Program," *Journal of Soil and Water Conservation* 43 (1988): 251–56.
6. See J. Stiglitz, and M. Rothschild, "Equilibrium in Competitive Insurance Markets: An Essay on the Economics of Imperfect Information," *Quarterly Journal of Economics* 90 (1976): 629–49; and P. J. Ferraro, "Asymmetric Information and Contract Design for Payments for Environmental Services," *Ecological Economics* 65 (2008): 810–21.
7. See U.S. Department of Agriculture, Farm Services Agency, *Conservation Reserve Program: Summary and Enrollment Statistics—FY2008* (Washington DC: USDA Farm Services Agency, 2008), http://www.fsa.usda.gov/Internet/FSA_File/annualsummary2008.pdf 7; and M. O. Ribaudo, D. L. Hoag, M. E. Smith, and R. Heimlich, "Environmental Indices and the Politics of the Conservation Reserve Program," *Ecological Indicators* 1 (2001): 11–20.

# CONCLUSIONS AND POLICY IMPLICATIONS

*Thomas E. Sheridan, Gary P. Nabhan, and Susan Charnley*

As we move forward, it is valuable to remember that the first revolution in western land management created the federal public lands system a century ago. That revolution protected millions of acres of forested lands, rangelands, and critical wildlife habitat in the West from forces like the development of housing subdivisions, irrigated agriculture, and strip-mining. But it also created a mosaic of federal, state, and private lands that has made managing western landscapes at the watershed level, as wisely advocated by John Wesley Powell at the onset of that revolution, extremely difficult.

The case studies chronicled in this volume reinforce and expand the notion, already in circulation, that a second revolution has already begun—watershed by watershed—that aspires to stitch the West back together, both ecologically and socially. Emerging from these chapters is a set of insights critical to the future of working landscapes, rural livelihoods, and biodiversity in the West, with implications for federal agencies and their local partners. We see the following principles as some of the guideposts needed to move this discourse forward over the next quarter century.

## 1. True, long-term but dynamic partnerships are key to achieving meaningful and lasting change.

Most land managers and others invested in the survival of western working landscapes and biodiversity concede that the second revolution cannot be a top-down, command-and-control campaign run by the federal government. The federal agencies that manage and regulate so much of the West—the Forest Service, Bureau of Land Management, National Park Service, Bureau of Reclamation, and U.S. Fish and Wildlife Service, among others—should be encouraged to continue to ensure that federal laws are honored in spirit and pragmatically enforced. But it has at times been difficult for agency personnel to come to the table as partners with ranchers, farmers, and foresters when they are seen not just as neighboring land managers but also as regulators or enforcers. Nevertheless, substantive participation and investment in collaborative efforts on the part of federal agencies is an important way for them to support and engage in collaborative conservation.

For the second revolution to be successful, both private and public partners will need more than token or temporary roles in reshaping how the West is to work. And because innovations often happen at the grassroots, by those having their boots on the ground in an agency or community, it is also critical that the voices of people who have felt rather powerless in the forest and rangeland debates be heard as loudly as those who have been in power—in agencies, forestry and cattlemen's associations, or elected offices. Adaptive management plans for local landscapes and watersheds are more likely to be effective when they are developed in partnership with local stewardship groups that include people who represent the full diversity of rural producers who make their living from those landscapes and watersheds than management plans that do not include these groups.

We emphasize the word "partners" rather than "stakeholders" because we believe that the traditional process of public input into federal decision making can be too passive. All too often, "public input" has pitted interest groups against one another and perpetuated a zero-sum-game attitude toward public lands management. More to the point, it does not encourage the search for common ground and principled compromise that is the foundation of a healthy democracy. We believe that further fostering of a "radical center" (see chap. 5) of ranchers, foresters, environmentalists, scientists, sportsmen, agency personnel, and consumers is critical to a West that works in the twenty-first century.

## 2. Stitching the West back together involves changing the culture of public lands management to one of collaboration rather than confrontation.

Changing the culture of land management to one of collaboration calls for changes in the cultures of institutions from the Forest Service, Bureau of Land Management, National Park Service, and state land departments to nongovernmental organizations, farm bureaus, and cattlemen's associations. Those institutions that merely defend the status quo may see their memberships diminish, their current ranks age without much recruitment, and their ideas and values become ossified. Those that reach out to collaborate with a wider range of partners will gain rather than lose traction with their values. Cultures of collaboration need to replace cultures of confrontation.

It will take repeated real-life engagement in the processes and protocols of community-based collaboration for individuals and institutions to be inoculated with that culture, which cannot be learned merely through books or online short courses. Emphasizing the identification and accomplishment of shared conservation goals across local watersheds will help us find common ground before there is no ground left for any of us to walk on.

A culture of collaboration does not mean a culture of co-optation. Nor does a culture of collaboration mean the disappearance of differences. Conflicts between rural producers, environmentalists, sportsmen, or agency personnel will not simply evaporate. Although it is unlikely we will return to the time when single-interest groups or industries dominated the management of public lands, considerable effort and time are needed to develop and nourish the institutional frameworks and cultural processes that bring people together to discuss those differences. It is far more difficult to demonize someone you know—someone you have collaborated with in the past—than to demonize an abstraction.

### 3. Long-term commitments of agency personnel, land ownership, land-use activities, and conservation efforts foster sounder management decisions and policies.

Partnerships depend on trust, and relationships of trust develop over time. It also takes time to get to know a landscape. If agencies provide incentives for their professionals to "stay in place" long enough (e.g., by making it easier to be promoted in place rather than having to transfer to other jobs and places to be promoted), they will become more functional partners in community-based collaborations. One of the most persistent complaints of rural producers is that federal employees rarely remain long enough to get to know them or the lands on which they depend. When personnel change roles, locations, and job descriptions too frequently, locally engaged citizens become disappointed or even cynical about government employees' personal commitment to place-based collaborative conservation. Nongovernmental conservation organizations might also adopt similar incentives and employment practices, since, after all, the promotion of place-based education and ethical land stewardship are among their missions; one cannot accomplish these goals merely by moving up the career ladder through changing locations every three to four years.

Ranchers and foresters often represent multigenerational commitments to the land and hold multigenerational funds of knowledge about that land.

- but, ideally, want a mix.
- need new ideas + perspectives, balanced with longer term, deep understanding of a place+organization

It is difficult for them to trust, educate, and collaborate with federal and nongovernmental-organization partners who rotate in and out every few years. Collaborative conservation requires patience, mutual understanding and respect, and a commitment to the long haul.

By the same token, rural producers need to make, and be allowed to make, long-term commitments to the federal forests and ranges they utilize. Speculators who run rent-a-cow operations until they can develop private lands are not likely to treat their public allotments with as much care as a rancher who intends to stay in the ranching business and pass that business on to his or her heirs. Serious consideration should be given to the length of time that rural producers are granted rights of resource access and use to federal lands. For example, extending the allotment periods of ranchers having a demonstrated commitment to sustainable practices from 10 years to 25 years or establishing a minimum of 25 or 40 years for contracts to farm, produce shellfish, or ranch on public lands where those activities are permitted could have positive outcomes for both people and natural resources. Greater security of tenure often means greater investment in sound land management practices and infrastructure.

**4. Public and private lands in the West are economically and ecologically interdependent, calling for an "all lands" approach to forest and range management that crosses ownership boundaries and encourages investment in private as well as public lands management.**
Much of the West contains a patchwork of interspersed public and private lands—and private lands that surround public lands serve as a buffer against exurban development that occurs nearby. Private and public working forests and rangelands are economically and ecologically interdependent. Ecologically, tribal and private working lands contain a significant component of the biodiversity in the West because formally designated protected areas are not large enough, diverse enough, or located strategically enough to support viable populations of many species. Some species also move across ownership boundaries, as do ecological processes that affect biodiversity—such as fire, disease, and the spread of invasive species. Working landscapes on private and tribal lands are therefore important conservation areas that deserve protection and investment, yet their contributions to maintaining biodiversity are often overlooked.

Economically, many of the dynamics that affect production from public lands also affect production from private lands. Moreover, a number of foresters and many ranchers are dependent on both for their livelihoods. For example, if local mills close because of insufficient timber harvests from federal lands, private forest owners also lose a local market for their timber. Likewise, ranchers who lose access to federal grazing allotments may have to fold if their private lands are not big enough to accommodate the large herd sizes needed to run a viable ranching operation and alternative forage sources are unavailable.

Although collaborative conservation groups provide an institutional structure for developing and implementing landscape-scale land management practices and policies that cross ownership boundaries, implementing large-scale collaborative conservation projects using an all-lands approach remains challenging. Policies, programs, funding sources, and outreach efforts that encourage such approaches have started to emerge, but there is a long way to go before all-lands management becomes a common way of doing business.

**5. Enabling institutions and policies that support innovation, adaptive management, flexibility, and resilience are needed.**
Land health, human health, and community economic health are mutually reinforcing. Such health depends on land management that is truly adaptive. Flexibility, innovation, and experimentation at all levels of western land management—from federal agencies headquartered in Washington, DC, to local collaborative conservation organizations—will encourage adaptive management. Rewarding administrative independence in solving problems at the lower levels of federal, state, and county agencies (as long as it is ethical and conforms to law) and cultivating agency personnel who are creative problem solvers rather than gatekeepers can help foster flexibility, innovation, and experimentation. In contrast, excessive legal and regulatory requirements and administrative complexity will stifle them.

Although this book features innovative individuals, families, organizations, and communities that have found solutions to many working landscape and biodiversity conservation dilemmas in specific places, we will ultimately need broad-scale institutions and policies to be less static and more flexible if such solutions are to be mainstreamed. Many of the sustainable practices and processes highlighted in this book cannot make a lasting difference if only implemented on a single ranch rather than over an entire watershed. Unless concepts like stewardship contracting and the bundling of ecosystem services are applied beyond their points of origin, their innovators will have won a battle but lost the war. But if agencies reward field personnel and administrators who find ways to sow the seeds of proven innovations over a wider area, their adoption is more likely to be fast-tracked and democratized in ways that foment additional refinements and fresh innovations.

**6. Seeking out local knowledge, expertise, and workforces in public land management activities can improve project outcomes and long-term project success.**
Despite the increasing efficiency of remote sensing tools, radio collars for tracing animal movements, and other technological advances, first-hand observations and local ecological knowledge remain essential to good land and resource management. We believe that large landscape conservation and restoration efforts benefit by drawing on local knowledge and including local economic incentives that in turn help promote healthy rural communities. For instance, when management plans on federal lands call for restoration projects such as erosion control, prescribed burning, or forest thinning, local contractors with the expertise to carry out such projects might be given preference in the bidding process, especially if those local contractors are affiliated with local grassroots collaborative conservation organizations.

The stewardship contracts issued by the Forest Service and Bureau of Land Management, described in chapter 9 and spotlight 9.1, could serve as models and be extended to other federal agencies; at the very least, they warrant permanent authorization by Congress. And since cost-effective monitoring is the foundation of adaptive management, federal agencies would do well to develop results-driven monitoring protocols in partnership with scientists, rural producers, and community-based organizations who know intimately the species or the landscapes to be monitored. Again, when local scientists or technicians possess the expertise to conduct such monitoring, they could be given preference or at least priority points in the awarding of contracts to do so.

It may be unnecessary for federal agency personnel to carry out some land restoration tasks when local workforces are willing and able to do so at less expense to U.S. taxpayers. Long-term conservation and restoration efforts across the West and the world at large are likely to be most successful when and if local people recognize that it is in their best long-term interests to promote rather than oppose them. After all, the ecological conditions and processes needed to support healthy ecosystems and economies are the same— whether for purposes of sustaining wildlife or livestock, timber production or habitat.

An effective way to create such a culture is to give local communities vested economic interest in the sustainable long-term management of public lands. Because the very process of restoring habitats to ecological integrity and economic viability usually takes longer than the duration of a single contract, communities are one of the few institutions that have the longevity to see such work through to fruition. As such, they could make excellent long-term stakeholders and partners. Providing them with opportunities to do so has the added benefit of maintaining the local capacity and resources needed to engage in restoration work on both public and private lands in the future. This lesson is as applicable to the American West as it is to Africa, Asia, and Latin America.

**7. Diversification of commodities and services from working landscapes is becoming increasingly critical to the success of rural operations.**
Working landscape conservation depends on having viable financial mechanisms to support it. Rural producers are diversifying their income streams in order to stay in business because the prices of beef, lamb, and other commodities they produce have seldom been sufficient to reward them adequately for their management investments. Dude ranching, pioneered in the late nineteenth century, was an early form of diversification on working ranches that incorporated cultural and ecological tourism into ranchers' portfolios.

Twenty-first-century ranchers and foresters now seek to diversify economically the portfolio of goods and services they produce from their lands in ways that are consistent with their values and are valued by society at large. Some portfolios today include alternative energy production and sales, niche marketing of value-added products such as grass-fed meat and certified wood products that incorporate the value of sustainable land management practices, payments for ecosystem services produced such as carbon sequestration, development of recreation and ecotourism-based businesses, offering environmental and agricultural education classes, and developing "conservation settlements" on their lands.[1]

It is likely that ranchers, farmers, and foresters who collectively brand their products, which are derived from the same landscape, will be increasingly able to ask consumers to pay a premium price to help subsidize their ongoing restoration efforts. This emerging strategy is known as landscape-level bundling and branding of (ecosystem) goods and services and can allow private landowners to choose collectively where to invest funds and efforts to restore the weak links in watershed function or in wildlife corridor connectivity. Diversification of products and functions from ranch and forestlands promotes the predictability and stability needed for working landscape conservation in the face of the ever-changing social, economic, and ecological realities that rural producers operate within.

**8. Diverse strategies for working landscape conservation are often needed in one place—landowners and managers need a broad toolbox from which to choose.**

Landowners and managers can diversify the portfolio of commodities and services they produce as one way of making working landscape conservation more financially viable using market-based approaches (as described above). But as several chapters in this book illustrate, they also often draw on a diverse set of conservation tools, trying them out consecutively to see which works best for their situation, adopting a number of them simultaneously to meet their needs, or adapting existing tools and pioneering new ones to best address their circumstances. These tools include conservation easements, land acquisition programs, land-use planning and zoning, multispecies habitat conservation plans, low-interest loans and grants, government funding and tax incentives, private capital funds, conservation-oriented bonds, grass banks, and stewardship contracts and agreements.

Having a number of tools available to support working landscape conservation is important because not every tool is appropriate to every circumstance, and what works in one place may not work in another. Yet the tools and funding mechanisms currently available to support working landscape conservation in the West are limited given the scale of the need. Diverse tools are needed to address diverse circumstances and to give landowners options for finding what works best given their ecological, economic, and social realities. Laws, policies, and programs that continue to support existing tools will make it possible for them to be sustained, adapted, and improved on; and incentivizing and supporting the development of innovative new tools will help facilitate working landscape conservation into the future. At the same time, laws, policies, and programs that may work against working landscape conservation (e.g., the estate tax) should be identified, scrutinized, and revised or eliminated if appropriate.

**9. Rural communities that find and foster allies in urban and exurban communities, and from the local to the national level, will have more clout to influence the public policy of western working lands for the better and for the long haul.**

With less than 1.5 percent of Americans identifying themselves as farmers or ranchers, and only 16 percent living in rural areas, it is likely that major land and water policy initiatives that rural communities care about will be decided by interests other than their own, unless they establish cross-sectoral linkages to support their political interests and initiatives.

For example, over the last half century, the political power of farmers and ranchers in state legislatures and the U.S. Congress has shrunk relative to that of land developers and manufacturers. Without forging new alliances, their political power is likely to shrink further. And unfortunately, in recent years, the aging population engaged as leaders in the American Farm Bureau Federation has drawn a line in the sand against consumer activists and other urban dwellers rather than positively engaging them as supporters who can work together with them for healthy lands and the delivery of healthy food choices. With two demographic strikes against them—a diminishing population size and an aging population—ranchers, foresters, and farmers will need to strive to recruit more youth to their ranks and more allies to their fold.

The chapters in this book illustrate two approaches to doing this: (1) having community-based organizations form political alliances with local, regional, state, and national interest groups to advance their agendas around issues that resonate with people in other places and (2) bridging the rural-urban divide through initiatives like the local food movement. Regarding the former, we have seen that when local and regional organizations partner with established state and national entities that share their interests, they can be more successful at achieving conservation goals and policy change, as the Swan Valley case illustrates (chap. 7). Local groups that join broader coalitions and form a shared organizational structure also have a better chance of gaining access to national leaders and advocating for laws, policies, and programs that support their interests.

Regarding the latter approach, the local food movement has already bridged the urban/rural divide in some western landscapes and elsewhere in the United States by linking local and regional producers with local and regional consumers through community-supported agriculture and farmers' markets. But these grassroots efforts need infrastructural support to spread. For instance, ranchers interested in selling more of their beef or lamb to local markets face enormous challenges. Few areas of the West have enough slaughterhouses to process more than a tiny percentage of the livestock raised nearby. And even when there are sufficient slaughterhouses, ranchers often have to sell their beef or lamb at farmers' markets or online. Although local and regional food markets are among those with the most rapid economic growth, there are few supermarkets even in rural food-producing areas where consumers can buy local produce. In fact, the U.S. Department of Agriculture has declared many rural counties in the West to be "food deserts" because the food producers in those counties export their produce elsewhere.

Producing, efficiently distributing, and consuming more food, fiber, forest products, and other goods and services within the regions in which they are produced may be key for the second revolution in the West to proceed.

This process helps to bridge the rural-urban gap and builds constituencies that support both conservation and sustainable production. It creates multiplier effects in rural communities that help existing infrastructure to survive and expand. It encourages entrepreneurs or local governments to invest in new and necessary infrastructure like local slaughterhouses, mills that process smaller-diameter trees, and biomass utilization facilities.

The local food movement is just one of many possible means to form new alliances that democratize public-policy decisions facing the West. Positively engaging and investing in groups such as the National Young Farmers Coalition, the Greenhorns, the Young Agrarians, and FarmFolk/CityFolk is another means—one that would be far more advantageous to established organizations like the American Farm Bureau Federation than ignoring or opposing such groups.

There are now many voices—from the freshly resurrected Grange and the Quivira Coalition to the Rural Voices for Conservation Coalition—that provide examples of these growing alliances. We encourage people to both celebrate and coalesce around these diverse voices, rather than seeing them as distractions or detriments to the future of rural western communities. A tiny and ever-shrinking fraction of westerners make their living as ranchers, foresters, and farmers; if the 1.5 percent of Americans who self-identify as rural food and fiber producers are the only constituency in America to vote for policies that nurture the integrity, health, and wealth of working landscapes, they will surely lose at the polling booth in every election. Instead, these rural producers need to reach out to other constituencies for political support. If rural dwellers and those interested in its future want the rural West to be more than a playground or a dumping ground for city dwellers, they need to foster relationships between rural producers and urbanites, suburbanites, and exurbanites near and far. If and when the second revolution in the American West takes place, rural economies will benefit, and the ecological footprints of western cities will decrease. Then ecological and economic health will begin their reintegration, stitching back together a West where we are defined more by the ground we share than by the interests that have divided us.

## NOTES

1. A conservation settlement is a carefully designed, small-scale, ecologically sensitive residential real estate development that may be established by landowners who subdivide and sell a small portion of their holdings to obtain revenue without selling off their entire property.

# ACKNOWLEDGMENTS

This book began a decade ago when two of the coeditors—Tom Sheridan and Susan Charnley—met at an American Anthropological Association annual meeting and shared their stories about the struggles facing ranchers and foresters in the twenty-first-century West. Sheridan was a founding member of the Arizona Common Ground Roundtable, a state-wide forum that organized ranchers, government agency personnel, and environmentalists to identify issues that brought them together rather than tore them apart. Charnley worked with timber communities reeling from the sharp cuts in commercial logging on federal lands. They decided to invite ranchers, foresters, academics, and conservationists to meet and talk about the challenges facing working landscapes in the West, particularly those composed of public as well as private lands.

The result was the Saving the Wide Open Spaces workshop held at the White Stallion guest ranch just north of Tucson, Arizona, in May 2005. In addition to the three coeditors, more than 25 participants came together for three days of structured conversation. That workshop included ranchers, foresters, scholars and students, and staff from nongovernmental organizations across the West: Ernie Atencio, Emily Brott, Julie Brugger, John Crumley, Mac Donaldson, Bill Durham, Vernita Ediger, Connie Falk, Penny Frazier, Martin Goebel, John Heaston, Kim Hendrick, Ross Humphreys, Eric Jones, Dean Lueck, Bill MacDonald, Rob Marshall, Dennis and Deb Moroney, Barron Orr, Ray Powell, Nathan Sayre, Bill Shaw, Diane Snyder, Lois Stanford, Nita Vail, Peter Warren, Colin West, and Courtney White.

At the end of the meeting we pondered potential products from the workshop that we could create, but after the initial meeting, other responsibilities diverted our attention from this project. Then, at the 2008 American Anthropological Association annual meeting, Susan resurrected the idea of pushing the project forward. By then, Tom had met Marc Miller, a professor at the University of Arizona's College of Law, who had started The Edge: Environmental Science, Law, and Policy book series, then at the University of Arizona Press. The Edge series offered a focused way of creating an edited volume by bringing together contributors at the beginning and end of book projects to encourage more cohesion and direction than most edited volumes evinced. It also provided funds and staff time to facilitate volume development and peer review and broader distribution of the final product. The goal of this series was to address substantial environmental issues in a nondisciplinary fashion, to be attentive to science, law, and policy solutions, to write and edit books for policy makers and other general readers, and then to get the books into those readers' hands. Marc was enthusiastic about the idea of adding a volume on working landscape conservation in the West to the series, and thus the book project took off. The series moved from the University of Arizona to the University of Chicago Press and changed names from The Edge to Summits in 2012.

The initial gathering of book participants occurred at Stanford University in 2009 with support from the Roger and Cynthia Lang Fund for Teaching and Research in Anthropological Sciences at Stanford. The gathering was hosted by Bill Durham, a professor in the Department of Anthropology, whose proposal to the fund made possible the summer workshop at Stanford. We are extremely grateful to Bill for providing logistical, financial, and intellectual support for the 2009 gathering of many of this book's authors and for helping us further develop a vision for the book.

Funding for different stages of the project came from a number of different sources. We express deep thanks to the following organizations for their financial support: the Southwest Center of the University of Arizona, the Desert Southwest Cooperative Ecosystem Studies Unit of the National Park Service, the Anthropology and Environment Section of the American Anthropological Association, the Center for Sustainable Environments at Northern Arizona University, New Mexico State University, the Department of Geography at the University of California, Berkeley, the Quivira Coalition, the U.S. Department of Agriculture Forest Service's Pacific Northwest Research Station, the Cooperative Forestry Staff of the Forest Service's State and Private Forestry branch, the Bill Lane Center for the American West at Stanford University, and Stanford University's Department of Anthropology. In particular, we thank Joe Wilder, director of the Southwest Center, Pat O'Brien and Larry Norris of the Desert Southwest Cooperative Ecosystem Studies Unit, Susan Stein of the Forest Service's Cooperative Forestry unit, the Bill Lane Center at Stanford for providing a visiting fellowship to support Susan Charnley while working on the book, and Jamie Barbour of the Pacific Northwest Research Station for his support and review.

Additional thanks go to Laurie Bower, Tanya Denckla Cobb, Dan Daggett, William deBuys, Diablo Trust, Francesco di Bernardone, Michael Dimock, Peter Forbes, Lauren Gwin, Darin Jensen, Kevin and Nancy Lunny, Mandy Metger, Luther Probst, Judy Prosser, Tom Sisk, Derrick Widmark, and Charles Wilkinson, whose ideas, activities, and/or direct contributions added greatly to the depth and breadth of this book.

A number of individuals also provided critical assistance in developing and coordinating the book project, facilitating communication among authors, overseeing the peer review and revision processes, editing chapters, and pulling the book together into a cohesive whole. Emily McGovern and Betsy Woodhouse stand out in this regard, with additional assistance from Leah Stauber; we thank them all profusely. Marc Miller and Barbara Morehouse, series editors, also provided significant intellectual contributions throughout the book's development. And Camille Cope, Sophia Polasky, Scott Turnoy, and Kendra Wendel provided research assistance. Finally, we thank three anonymous reviewers for their extremely helpful comments on the draft manuscript.

The challenge of working landscapes—their role in Western society, including as part of efforts at conservation of lands and natural resources—is the kind of problem sometimes identified by social scientists as a "wicked problem." Such problems are hard to define, resist simple solutions, often include difficult trade-offs, and tend to change over space and time. The solutions to such problems—knowledge and effective change—are most likely to come from a long, informed exchange between those who spend most of their time studying and those who spend most of their time living these issues. It is through this "life up, theory down" interaction that successful solutions may emerge and be replicated over time. This volume, and the insights of the many people who work on and live these issues, reflects this perspective. Such an approach requires patience and openness by all, and for that spirit, demonstrated so many times in the gestation of this text, we are grateful.

# CONTRIBUTOR BIOGRAPHIES

**Mark Andre** is environmental services director for the city of Arcata, California (http://www.cityofarcata.org/), responsible for the natural resources, water/wastewater and recreation divisions that include activities such as forest management and open space protection. He has served as the city forester for Arcata's community forest for the past 26 years and also provides forestry-consulting services for nonindustrial and tribal timber landowners. Prior to working for the city of Arcata he worked as a hydrologist for the U.S. Forest Service.

**Carrie Balkcom** is executive director of the American Grassfed Association (AGA, http://www.americangrassfed.org), which is the national, multispecies entity organized to protect and promote grass-fed and pasture-based farmers and ranchers. The association has a third party on-farm certification program. Balkcom grew up on a Florida cattle ranch and has stayed connected to the agriculture and livestock industry. She has spoken, presented, or coordinated numerous regional and national conferences and is well known in agricultural, culinary, and sustainable agricultural circles.

**Judee Burr** is a policy associate with the Frontier Group (http://frontiergroup .org/) in Santa Barbara, California, where she does research and writes policy reports on public interest issues with a focus on the environment. She attended Stanford University, where she worked as a research assistant for the Carnegie Institute for Global Change and as a student researcher for the Bill Lane Center for the American West (http://west.stanford.edu/).

**Susan Charnley** is a research social scientist and team leader with the U.S. Department of Agriculture Forest Service's Pacific Northwest Research Station in Portland, Oregon (http://www.fs.fed.us/pnw/). As an environmental anthropologist, her research focuses on natural resource use and management among rural producers and on the institutions needed to help support sustainable livelihoods and healthy ecosystems on public, private, and communal lands. She has published in a range of anthropology, forestry, and natural resource journals; this is her first book. Charnley works mainly in the western United States and Africa.

**Jon Christensen** is an adjunct assistant professor in the Institute of the Environment and Sustainability and History Department at the University of California, Los Angeles (http://environment.ucla.edu, http://christensenlab .net). He spent 20 years as an environmental journalist and science writer in the American West, writing for *High Country News*, the *New York Times*, and others before coming to Stanford University as a Knight Journalism Fellow in 2002. He stayed on to work on a PhD in environmental history, the history of science, and western history. He was executive director of the Bill Lane Center for the American West at Stanford before moving to the University of California, Los Angeles.

**William H. Durham** is senior fellow in the Woods Institute for the Environment and Bing Professor in Human Biology and Anthropology at Stanford University. He is also Stanford director of a tourism research organization, the Center for Responsible Travel (http://www.responsibletravel.org). His career has focused on causes and consequences of environmental degradation in Latin America and on ways to reverse them and build a sustainable future. As one example, Durham is coleader of Stanford's Osa-Golfito Initiative (http://inogo.stanford.edu), helping Costa Ricans to develop a sustainability plan for that region.

**Maia Enzer** served as policy program director at Sustainable Northwest (http://www.sustainablenorthwest.org/) from 2000 to 2012. She is now with the U.S. Forest Service as a partnership and collaboration specialist. Over the last 20 years, Enzer has worked on issues related to forest restoration and community economic development with a focus on federal lands policy. She has extensive experience bringing diverse stakeholders together around federal lands management issues. Enzer also founded the Rural Voices for Conservation Coalition, a national coalition of community-based organizations and other conservation leaders dedicated to promoting solutions to the eco-logic, economic, and social problems facing the rural West.

**Pat Frost** was district manager at the Trinity County (California) Resource Conservation District (http://www.tcrcd.net) for over 13 years and is an adjunct faculty member at Shasta Community College, teaching biology and natural history. He has extensive experience in natural resource conservation and has worked in natural resources management for 30 years. He has served on a number of advisory committees, including the Weaverville Community Forest Steering Committee, Trinity County Resource Advisory Committee (U.S. Forest Service), and California Fire Safe Council. Frost and his wife also are the stewards of their own forestlands in northern Trinity County.

**Martin Goebel** has a BSc in forestry from Oregon State University and MAg in natural resources from Texas A&M University. He has worked throughout Latin America and Germany in natural resources management and community development since 1979. In 1994, he founded Sustainable Northwest (http://www.sustainablenorthwest.org/), based in Portland, Oregon, and has served as its president. Sustainable Northwest partners with a variety of rural communities and national, regional, and local leaders to advance the community-based collaborative conservation and sustainable development movement through community capacity building and enterprise development, network learning and policy, and linking rural producers to urban consumers and markets.

**Joshua H. Goldstein** is an ecosystem services scientist with the Nature Conservancy. He received his PhD in environment and resources from Stanford University. His expertise is in ecosystem services, ecological economics, and conservation finance. He has conducted research on working landscapes in the United States and Latin America, working collaboratively with private landowners, academics, government agencies, nonprofits, and corporate partners.

**Hannah Gosnell** is associate professor of geography in the College of Earth, Ocean, and Atmospheric Sciences at Oregon State University. Her research interests have to do with agricultural landscape change, water resource management, climate change and environmental governance in the rural American West, and the ways in which laws and institutions might evolve to better reflect changing geographies. Gosnell earned MA and PhD degrees in geogra-

phy from the University of Colorado in 2000 and a BA in American civiliza-
tion from Brown University in 1988.

**Lauren Gwin** is assistant professor and extension food systems specialist
at Oregon State University and associate director of the university's Center
for Small Farms and Community Food Systems (http://smallfarms.oregon-
state.edu). Her extension and research focus is on policy and regulations,
small-scale processing, and distribution and marketing within local and regional food systems. She coordinates the national Niche Meat Processor Assistance Network (http://www.nichemeatprocessing.org), an eXtension Community of Practice (http://www.extension.org/), and teaches on-campus food
systems courses. Her PhD is in environmental science, policy, and management from the University of California, Berkeley.

**Michael E. Jani** is president and chief forester of Humboldt Redwood Company in Humboldt County, and Mendocino Redwood Company, in Mendocino and Sonoma County, California. The two companies collectively consist of 440,000 acres of Forest Stewardship Council–certified forestlands.
Previously, he worked for 24 years for Big Creek Lumber Company in Santa
Cruz County, the first company in California's redwood region to be certified
under the Forest Stewardship Council program. Jani participated in the development of the council's Pacific Coast Standards, is a member of its U.S.
Standards Committee, and currently cochairs the FSC-US board. Jani is also a
board member and past chair of the California Forestry Association. He has a
BS in Forestry from the University of California, Berkeley.

**Chris Kelly** is California program director for the Conservation Fund (http://
www.conservationfund.org), where he has led efforts that have protected
over 125,000 acres of forest lands in northern California through projects pioneering nonprofit ownership of working forests, use of low-interest public
financing, and accessing carbon markets to generate capital for conservation.
Prior to joining the fund, he served as director of conservation programs for
the Nature Conservancy of California. He previously practiced land use, real
estate, and corporate law and holds a JD from the University of California,
Davis, and a BA in philosophy from the University of California, Berkeley.

**Richard Knight** is interested in the nexus of land use and land health in the
American West. A professor of wildlife conservation at Colorado State University, he received graduate degrees from the University of Washington and

University of Wisconsin. He sits on boards including the Colorado Cattle-men's Agricultural Land Trust, the Malpai Borderlands Group Science Board, the Diablo Trust, the Rancher's Stewardship Alliance, and Resources First Foundation. He has served on editorial boards for the journals *Conservation Biology* and, currently, *Ecological Applications*. In 2007, Colorado State University honored him with the Board of Governors Excellence in Teaching award. With Courtney White, he edited *Conservation for a New Generation* (Island Press, 2008).

**Steven E. Kraft** is retired from the Department of Agribusiness Economics at Southern Illinois University Carbondale. He was in the department for 30 years, the last 15 as department chair. He is a founding codirector of the PhD program in environmental resources and policy at Southern Illinois University Carbondale. His research areas include soil and water conservation policy, watershed planning, and ecosystem services. Recent publications include *The Law and Policy of Ecosystem Services* with J. B. Ruhl and Chris Lant (Island Press, 2007). He is a fellow of the Soil and Water Conservation Society.

**Roger Lang** owns and operates Lang Studios, LLC, a private fund that seeds and helps high tech start-ups grow, ranging from Internet communications to emerging "green trends." Previously, Lang was owner and founder of the Sun Ranch Group, using a market-driven model for conserving important tracts of the American West. Passionate about the need to conserve important tracts of land for wildlife habitat and migration corridors, Lang developed a model for a market-driven conservation approach that included sustainable lodging and recreation, cattle ranching, and biology-driven residential real estate development appropriate for open space. Earlier in his career, Lang was named one of the 50 most influential people in the history of financial risk management by *Risk* magazine. He is currently chairman of the Pollinator Partnership, the goal of which is to influence policy relating to food security and the vital role of pollinators to human populations. Lang holds a BA and MA from Stanford University.

**Laura López-Hoffman** is assistant professor in the School of Natural Resources and the Environment and assistant research professor at the Udall Center for Studies in Public Policy at the University of Arizona. López-Hoffman obtained a PhD from Stanford University in biological sciences and a BA from Princeton University. She uses interdisciplinary and comparative approaches to conduct research on the development of environmental policies and institutions that protect ecosystems and sustain human well-being. She coleads a U.S. Geological Survey John Wesley Powell Center working group studying how migratory species facilitate the sharing of ecosystem services between Canada, the United States, and Mexico and developing new approaches to protect migratory species. López-Hoffman is lead editor of *Conservation of Shared Environments: Learning from the United States and Mexico* (University of Arizona Press, 2010).

**Curt Meine** is a conservation biologist, environmental historian, and writer who serves as senior fellow with the Aldo Leopold Foundation and the Center for Humans and Nature and as associate adjunct professor at the University of Wisconsin—Madison. He has written and edited several books, including the biography *Aldo Leopold: His Life and Work* (University of Wisconsin Press, 1988; reprint, 2010) and *Correction Lines: Essays on Land, Leopold, and Conservation* (Island Press, 2004). He is also the on-screen guide for the documentary film *Green Fire: Aldo Leopold and a Land Ethic for Our Time* (http://www.aldoleopold .org/greenfire/).

**Dennis Moroney** grew up on the rural fringe of Phoenix, Arizona, in the 1950s and 1960s. He has BS degrees in animal science and agriculture education from the University of Arizona and a Masters in education from Central Washington University and has taught courses on agriculture, permaculture, and renewable natural resources at Prescott College, Yavapai College, and Cochise College. In 1992, he and his wife Deb purchased a ranch near Prescott, Arizona. Since 2002, the Moroneys have been raising quarter horses, cattle, goats, and Navajo Churro sheep on the 47 Ranch in Cochise County, Arizona. Moroney is past president of the Cochise Graham Cattle Growers Association and the Arizona Section of the Society for Range Management. His home is 18 miles from the border with Sonora, Mexico.

**Gary Paul Nabhan** is the Kellogg Endowed Chair for Sustainable Food Systems based at the Southwest Center of the University of Arizona (http://swc .arizona.edu/). He is author or editor of six books and coauthor of over 100 articles about agriculture, land stewardship, and the environment. His work in collaborative conservation has been honored by the Quivira Coalition, MacArthur Foundation, and Society for Conservation Biology. He has served on the boards of the Navajo-Churro Sheep Association, Wild Farm Alliance, Slow Money, National Park Service, and Seed Savers Exchange and Native Seeds.

**Melanie Parker** is a resident of Montana's Swan Valley, where her family operates a traditional outfitting and guide business. She is cofounder and executive director of Northwest Connections (www.northwestconnetions.org), a regional education and conservation organization offering field courses in working lands conservation for undergraduate and graduate students from across the country. Parker initiated and facilitated much of the collaborative process that led to the Montana Legacy Project.

**Sophia Polasky** is a forestry research technician with the U.S. Department of Agriculture Forest Service Pacific Northwest Research Station, supporting the Communities and Forest Management Team (http://www.fs.fed.us/pnw/about/programs/gsv/cfm.shtml). She is currently a PhD student in the College of Forestry at Oregon State University in the Department of Forest Ecosystems and Society (http://fes.forestry.oregonstate.edu/). Her research interests include social vulnerability to environmental change and the political ecology of natural resource use and management.

**Carrie Presnall** is a graduate research assistant at the University of Arizona focusing on the potential of ecosystem services to improve communication and decision making. Her Master's thesis assesses federal agency staff's familiarity with and opinions about incorporating ecosystem services into National Environmental Policy Act documents. She has published in *Environmental Law Reporter* and *Rangelands*.

**Andrew Reeves** is a recent University of Arizona Law School graduate, currently clerking for the Honorable Christine Quinn-Brintnall of the Washington State Court of Appeals. He was the 2009–10 Sol Resnick Water Resources Fellow at the University of Arizona and a fellow with the Rocky Mountain Mineral Law Foundation in 2010–11. He is interested in energy and environmental law and has published on the subject, including an article he coauthored titled "Solar Energy's Cloudy Future" (*Arizona Journal of Environmental Law and Policy*, vol. 1 [2010]).

**Jenny Rempel** is a Tom Ford Fellow in Philanthropy at the Jessie Smith Noyes Foundation (http://www.noyes.org/), where she helps identify, cultivate, and support organizations working to create a more just and sustainable food system. She has a degree in interdisciplinary environmental science and policy from Stanford University, and she has conducted research for the Bill Lane Center for the American West (http://west.stanford.edu), the U.S.

Department of Agriculture Forest Service, the National Oceanic and Atmospheric Administration, and the San Francisco Estuary Institute (http://www .sfei.org/). She was a founding member and leader of Generation Anthropocene, a student-led podcast, and the Stanford Farm Project, an advocacy and outreach organization.

**Nathan F. Sayre** is associate professor of geography at the University of California, Berkeley. He has worked on rangeland conservation with the Quivira Coalition, the Altar Valley Conservation Alliance, the Malpai Borderlands Group, the Redington (Arizona) Natural Resources Conservation District, and the Sonoran Desert Conservation Plan. He is author of three books about the history and conservation of Southwestern rangelands, including *Working Wilderness* (Rio Nuevo, 2006) and *Ranching, Endangered Species, and Urbanization in the Southwest: Species of Capital* (University of Arizona Press, 2006). He serves on the board of directors of the Malpai Borderlands Group and is affiliated social scientist with the U.S. Department of Agriculture–Agricultural Research Service–Jornada Experimental Range and the Jornada Basin Long-Term Ecological Research site in Las Cruces, New Mexico.

**David Seibert** has been working in collaborative conservation and ecological restoration in the U.S. Southwest for the past 20 years. He currently runs a small restoration consulting business and is the conservation director for Borderlands Habitat Restoration Initiative and for Borderlands Restoration L3C, the first limited-profit business registered in Arizona, which fosters collaborative conservation work. Seibert is a PhD candidate in anthropology at the University of Arizona and resides in the borderlands region of Patagonia, Arizona.

**Thomas Sheridan** is research anthropologist and professor of anthropology at the Southwest Center and School of Anthropology at the University of Arizona. He has written and edited 12 other books and monographs, including *Arizona: A History* (University of Arizona Pres, 2012; rev. ed.) and *Where the Dove Calls: A Political Ecology of a Peasant Corporate Community in Northwestern Mexico* (University of Arizona Press, 1996). A past president of the Anthropology and Environment Society, he serves on Pima County's Conservation Acquisition Commission and is community representative and member of the board of the Altar Valley Conservation Alliance, a grassroots organization of ranchers southwest of Tucson. He received the Sonoran Institute's Faces of Conservation: Sustainable Communities Award in 2007.

**Josh Spitzer** is chief executive officer of Schedulicity, Inc., an online business appointment and scheduling platform. Previously, he was cofounder and partner of Preservation Capital, a boutique financial advisory and investment firm that specializes in high-growth companies accessing new capital markets. Spitzer holds an MBA from Stanford University's Graduate School of Business and BA from Cornell University and has numerous publications on innovative finance.

**Michael S. Stevens** is Washington State director for the Nature Conservancy, based in Seattle (www.nature.org/washington). From 2002 to 2011, he was president and chief operating officer of Lava Lake Land & Livestock in Hailey, Idaho. Under his leadership, Lava Lake received national and regional awards for its conservation leadership and developed Lava Lake Lamb into a nationally distributed brand. In 2006, Stevens cofounded Pioneer Mountain Group, a conservation and natural resources firm. He received an MS from the Field Naturalist Program at the University of Vermont and a BA in biology from Middlebury College.

**Johnny Sundstrom** has served local, state, and national Conservation District Associations for 20 years. He is founder and coordinator of the Siuslaw Institute and manager and part-owner of a family ranch and forestlands in Oregon's Coast Range. He also coaches the local high school's track and field teams.

**Shiloh Sundstrom** is a native of Deadwood, Oregon, and currently serves as program manager with Resource Innovations at the University of Oregon. Sundstrom has an MS in forest resources from Oregon State University and has conducted research in southern Kenya and worked for the Siuslaw Institute and the Siuslaw Watershed Council. He also trains horses and comanages a small cow-calf and sheep operation in Deadwood.

**Dawn Thilmany** is professor of agribusiness and agribusiness extension economist with Colorado State University and specializes in analyzing markets and consumer behavior surrounding local, organic, and other value-added food market segments (http://dare.colostate.edu/dthilmany/index.aspx, http://buildingfarmers.colostate.edu/). She holds or has held leadership positions with the Agricultural and Applied Economics Association, the Western Agricultural Economics Association, several U.S. Department of Agriculture regional research committees and has served as a Farm Foundation Fellow.

**Amanda D. Webb** is a graduate student in the School of Natural Resources and the Environment at the University of Arizona (http://www.snr.arizona .edu/). Prior to coming to the university, she worked for government agencies and nongovernmental organizations in natural resources conservation and education and in vegetation management. Her research interests include ecology and the relationships between peoples and landscapes.

**Courtney White** is a former archaeologist and Sierra Club activist who, in 1997, cofounded the Quivira Coalition (www.quiviracoalition.org), a nonprofit dedicated to building bridges among ranchers, conservationists, public land managers, scientists, and others around the idea of land health. Today, his work with Quivira concentrates on building economic and ecological resilience on working landscapes, with a special emphasis on carbon ranching and the new agrarian movement. In 2010, White was given the Michael Currier Award for Environmental Service by the New Mexico Community Foundation. In 2012, he was the first Aldo Leopold Writer-in-Resident at Mi Casita, in Tres Piedras, New Mexico.

**Charles F. Wilkinson** is Moses Lasky Professor of Law at the University of Colorado Law School in Boulder, Colorado. His primary specialties are federal public land law and Indian law. In addition to his many articles in law reviews, popular journals, and newspapers, his 14 books include *Crossing the Next Meridian: Land, Water, and the Future of the West* (Island Press, 1993) and *Fire on the Plateau: Conquest and Endurance in the American Southwest* (Island Press, 2004).

# SUMMITS

## SUMMITS BOARD OF ADVISERS

**Gregory H. Aplet**
Senior Science Director
The Wilderness Society

**Rainer Bussmann**
Director, William L. Brown Center
and Curator for Economic Botany
Missouri Botanical Garden

**Andrew C. Comrie**
Senior Vice President for Academic
Affairs and Provost
University of Arizona

**Henry F. Diaz**
Research Climatologist
National Oceanic and Atmospheric
Administration (NOAA)
Climate Diagnostics Center, Boulder,
Colorado

**Karl W. Flessa**
Director, School of Earth and
Environmental Sciences
University of Arizona

**Lisa J. Graumlich**
Dean, College of the Environment
University of Washington

**Lance H. Gunderson**
Professor, Department of
Environmental Studies
Emory University

**Kevin Harrang**
Metajure, Inc.

**Charles F. Hutchinson**
Office of Arid Lands and School
of Natural Resources and the
Environment (emeritus)
University of Arizona

**Travis E. Huxman**
Center for Environmental Biology,
School of Biological Sciences
University of California, Irvine

**Douglas Kysar**
Joseph M. Field Professor of Law
Yale Law School

**Laura López-Hoffman**
Assistant Professor, School of Natural
    Resources and the Environment
Udall Center for Studies in Public Policy
University of Arizona

**George Matsumoto**
Senior Education and Research
    Specialist
Monterey Bay Aquarium Research
    Institute

**Sharon B. Megdal**
Director, Water Resources Research
    Center
C. W. and Modene Neely Endowed
    Professor Professor, Department
    of Soil, Water, and Environmental
    Science
University of Arizona

**Pierre Meystre**
Director, Biosphere 2 Institute
Regents Professor of Physics & Optical
    Sciences
Editor-in-Chief, Physical Review Letters
University of Arizona

**Mary Kay O'Rourke**
Associate Professor, Community,
    Environment and Policy Division
College of Public Health
University of Arizona

**Carol M. Rose**
Lohse Chair, Water and Natural
    Resources, College of Law
University of Arizona

**Joaquin Ruiz**
Dean, College of Science
Professor of Geosciences
University of Arizona

**Glenn L. Schrader Jr.**
Dean of Research and Graduate
    Education
Professor of Chemical Engineering
College of Engineering
University of Arizona

**B. L. Turner II**
Gilbert F. White Professor of
    Environment and Society
Arizona State University

**Scott Whiteford**
Professor, Center for Latin American
    Studies (retired)
University of Arizona

**Robert A. Williams**
E. Thomas Sullivan Professor of Law
Professor of American Indian Studies
Faculty Co-Chair, Indigenous Peoples
    Law and Policy Program, College of
    Law
University of Arizona

**Xubin Zeng**
Director, Climate Dynamics and
    Hydrometeorology Center
University of Arizona

# INDEX

ACF. *See* Arcata Community Forest (ACF)

Adirondack Forest Preserve of New York, 284

Afghanistan, 296

agriculture, Conservation Reserve Program and, 295–99

Agriculture Research Service, 234, 238–39n2

Alexander, J. D., 268

Alsea Stewardship Group, 168

Altar Valley Conservation Alliance (AVCA): authors as participants in, 55; Buenos Aires National Wildlife Refuge and, 60; erosion control and, 62–63, 64; fire management and, 74n18; focus of, 70; funding for, 74n29; membership of, 69; staffing of, 71; as successful CBCC, 53; as watershed group, 58; western range wars and, 58–59

American Grassfed Association, 96

American Livestock Breeds Conservancy, 288

America's Great Outdoors initiative, 116

Animal Welfare Approved, 97, 286

Animas Foundation, 62

Animas River Stakeholder Group, 54

*Antilocapra americana* (pronghorn). *See* pronghorn (*Antilocapra americana*)

Arcata Community Forest (ACF), 121, 137–40

Areas of Critical Environmental Concern, 187, 200n3

Arizona Cattle Industry Foundation, 234

Arizona Common Ground Roundtable, 50, 59

Arizona Department of Agriculture, 238–39n2

Arizona Department of Environmental Quality, 234

Arizona Game and Fish Department, 231–35, 236–37

Arizona Land and Water Trust, 183, 236–37, 259, 260

Arizona Open Land Trust, 259

Arizona State Land Department, 238–39n2

Arizona Water Protection Fund, 74n29

Army Corps of Engineers. *See* U.S. Army Corps of Engineers

A-7 Ranch, 261, 264, 266n31

*Athene cunicularia* (burrowing owl). *See* burrowing owl (*Athene cunicularia*)

AVCA. *See* Altar Valley Conservation Alliance (AVCA)

Babbitt, Bruce, 181

Baja Arizona Sustainable Agriculture, 234

Bank of America, 151

Bar V Ranch, 260, 261

Bat Conservation International, 234

Baucus, Max, 132

Bean, Brian, 182, 188

Bean, Kathleen, 182, 188

Bean family, 198

Beartooth Capital, 282

Berry, Wendell, xiii

Bingaman, Jeff, 110

biodiversity: collaborative stewardship and, xiv; contiguous ranch land and, 260; development and, 33, 44, 272–73; in ecological buffer zones, 33; exurban development and, 39–40; fire management and, 62;

biodiversity (*continued*)
  fragmentation of forestlands and, 25–26,
  134–35; generalist versus specialist species
  and, 40; geography conducive to, 37; habitat projects and, 262–63; limitations of protected lands and, 33, 34; most biodiverse
  ecoregions in US and, 36; multispecies
  habitat conservation and, 254–58, 262–63;
  in private and tribal lands, 34–35, 37–38,
  39; protected versus nonprotected lands
  and, 2, 34; riparian corridors and, 256;
  variation by land ownership and, 36–38,
  41–43; working landscapes and, xiv–xv, 39
biotic integrity, 89
Bitterroot-Selway Wilderness, 212
black bear (*Ursus americanus*), 186
black-bellied whistling duck (*Dendrocygna
  autumnalis*), 234
Blackfoot Challenge, 199, 243–44, 245, 282
BLM (Bureau of Land Management). *See* Bu-
  reau of Land Management (BLM)
bobcat (*Lynx rufus*), 40
Bob Marshall Wilderness, 124
Boise Cascade, 145–46
Bonneville Power Administration, 130
Boulder White Clouds Wilderness Area, 187
*Brachyramphus marmoratus* (marbled murre-
  let). *See* marbled murrelet (*Brachyramphus
  marmoratus*)
Bradley, Nina Leopold, xiii
Brady, Kevin, 268
Brookings Institution, 67, 269
Brown, D. E., 36
brown trout (*Salmo trutta*), 281
Bryan, William, 214
Buckelew Farm and Ranch, 261
Buenos Aires National Wildlife Refuge, 59, 60
bull trout (*Salvelinus confluentus*), 123–24, 281
Bureau of Land Management (BLM): Areas
  of Critical Environmental Concern and,
  200n3; Colorado Plateau and, 36; culture
  of, 302; distribution of land in the West
  and, 245; dynamic partnerships and,
  301–2; environmental lawsuits against,
  7; erosion control and, 64; grazing in
  districts of, xviii–xix, 19–20, 22; health
  of rangelands and, 22; holdings of in
  southeastern Arizona, 228; holdings of in
  southwestern Montana, 228; Lava Lake
  Land & Livestock and, 189, 191, 192, 193;
  management for multiple uses and, 6;
  Northwest Forest Plan and, 160; Rocky

Mountain ecoregion and, 38; scientific
  surveys by, 189; Sierra Pacific Industries
  and, 178; stewardship contracts and agreements and, 121–22, 159, 161–63, 173, 178,
  306; volume of timber production in lands
  of, 160; Weaverville Community Forest
  and, 177, 178–80; Weaverville land swap
  and, x; wilderness areas and, 187; Wilderness Study Areas and, 200n3; Wyden
  Authority and, 170
Bureau of Reclamation, 301–2
Burlington Northern, 126
Burros, Marion, 195
burrowing owl (*Athene cunicularia*), 232
Bush, George W., 109, 115–16
Butz, Earl, 295

cactus ferruginous pygmy owl (*Glaucidium
  brasilianum cactorum*), 247, 254–55
*Calamospiza melancorys* (lark bunting). *See* lark
  bunting (*Calamospiza melanocorys*)
California Climate Reserve, 139
California North Coast Forest Conservation
  Initiative, 155
California State Coastal Conservancy, 156
California State Wildlife Conservation Board,
  156
*Canis lupus* (gray wolf). *See* gray wolf (*Canis
  lupus*)
Canoa Ranch, 253–54, 261
cap-and-trade, 151, 249
carbon credits and carbon sequestration:
  aggregation schemes and, 289; carbon
  emissions cap and trade and, 249; Chicago
  Climate Exchange and, 249, 280–81; Eco
  Products Fund and, 284; as ecosystem
  service, 249, 277, 279–81; environmental
  cobenefits of, 284; Garcia River Forest
  and, 156, 157; greenhouse gas emissions
  trading programs and, 281; income for
  forest owners and, 155; price of meat and,
  286; property rights and, 280; in Redwood
  Region of California, 151–52; third-party
  verification and, 286; value of offsets and,
  283
Carpenter Ranch, 261
Carson National Forest, 89, 90–93
CAR standards. *See* Climate Action Reserve
Carter, Jimmy, 296
CBCCs. *See* community-based conservation
  groups (CBCCs)
CCX. *See* Chicago Climate Exchange (CCX)

Cecil D. Andrus Leadership Award, 191

Center for Biological Diversity, 7, 58–59, 110, 115

Central Idaho Rangelands Network, 186

Centrocercus urophasianus (sage grouse). See sage grouse (Centrocercus urophasianus)

Cervus canadensis (elk). See elk (Cervus canadensis)

Chen caerulescens (snow goose). See snow goose (Chen caerulescens)

Chicago Climate Exchange (CCX), 249, 280–81, 289

Chihuahua bioregion, 61

Chilton, Jim, 58

Chilton family, 59

Chiricahua leopard frog (Lithobates chiricahuenses), 232

Civilian Conservation Corps, 64, 78

Clean Water Act (1970), 89, 277, 281

Climate Action Reserve, 151, 152, 157

climate change: carbon sequestration and, 151–52; desert habitat and, 263; greenhouse-gas emissions regulations and, 157; shifts in ecosystems and, 35; trout populations and, 91

Clinton, Bill, 109, 115–16

Clothier, Van, 93

Clyne Ranch, 261

CNB (Country Natural Beef ). See Country Natural Beef (CNB)

Coalition for Conservation through Ranching, 45

Coalition for Sonoran Desert Protection, 258

coast redwood (Sequoia sempervirens), 138

Coca-Cola, 286

Cochise College, 232

Cochise-Graham Cattle Growers Association, 234

coho salmon (Oncorhynchus kisutch), 148, 156, 161, 170

Colinus virginianus ridwayi (masked bobwhite). See masked bobwhite (Colinus virginianus ridgwayi)

collaborative conservation: alternative dispute resolution and, 55; biodiversity and, xiv; coalitions and, 101; in Colorado Plateau, 38; contradictions and dichotomies to overcome and, 1; as cooperative conservation, 77, 78; core principles for working landscapes and, 108–9; democratic governance and, 72; difficulty of implementing large-scale projects and, 90, 91, 92–93;

diversity of participants and, 91–92; early efforts of, 77–79; ecosystem health and, xiv; forest management conflicts and, 141; funding for, 92; grazing practices and, 213; hard work of, x–xi; importance of innovation and, 92; interdependency and, xvi; at landscape scale, 197–98; Madison Valley Ranchlands Group and, 182; names of, 116; obstacles to, 104; organizations for, 45, 49–50; origins of, 55; participants in, xvi; partners versus stakeholders in, 302–3; place-based strategies and, 102–3; policy environment supportive of, 110–11; presidential initiatives and, 116; reasons for emergence of, 8; rural-urban divide and, xvii; rural voices coming together in, 103–5; stewardship contracting and, 104–5, 116n3, 121–22, 163; Swan Valley project and, 134; toolkit for, 247; transjurisdictional conservation and, xv, xvi–xvii, 2, 8–9; transjurisdictional management and, 305; twenty-first-century move toward, 171–73; values and value of, 172; Weaverville Community Forest and, x; working landscapes and, 38. See also community-based conservation groups (CBCCs)

Collaborative Forest Landscape Restoration program, 11–12

Colorado Plateau, 36–38

commons, tragedy of. See tragedy of the commons

community-based conservation groups (CBCCs): adaptive management and, 72; alliances of, 54, 309; coalitions and, 101, 105, 115, 123, 310; common ground of, 50; consensus building and, 58; critical challenges for, 69–70; dissension within, 70; economic diversification and, 64–66; economic goals of, 50; erosion control and, 62–64; examples of successful groups and, 53; finding and protecting common ground and, 247; fire management and, 60–62, 70; flood control and, 63; forest restoration and, 110; as growing trend, 53–54; initiation of, 49; land trusts and, 243–44; membership of, 69; participants' feelings about, 112–13; partnerships based on trust and, 69; producer-led and government-led, 54; producers' role in, 69–70; restoration of arroyos and, 63; self-definition of, 69; as self-organizing,

community-based conservation groups
(*continued*)
68; Siuslaw Stewardship Group and, 167;
tasks of, 56; typologies of, 54; umbrella
organizations and, 49–50; urban sprawl
and, 66–68; versus traditional interest
groups, 50, 54, 55–56, 57–58, 104; volunteers versus paid staff in, 70–71; watershed groups and, 56–60; working forests
and, 177; working with the elements and,
63, 89, 93. *See also specific groups*
Community-Based Forest and Public Lands
Restoration Act (proposed in 2002), 110–11
Conant, Jeff, 278
conservation: business of, 151, 200; causes of
land degradation and, 231–32; conservation banking and, 282–83; of conservation
knowledge, 227; conservation ranching
and, 231–35; conservation settlements
and, 310n10; experimentation and innovation in, xv–xvi; interdependence
of public and private lands and, 43; land
acquisition for, 124, 127–28, 132–34, 185;
landscape-scale, 185, 198–99; long-term
commitments to, 303–4; multispecies
habitat conservation and, 251; of old-
growth forest habitat, 43; private-sector,
209, 211; protection tools and, 33;
public-private partnerships in, 198–99,
227; resistance to development pressure
and, 124, 128–29, 132, 133; of Rio Grande
cutthroat trout, 90–91; science and
politics in, 255–57; working landscapes
as core areas of, 44. *See also* collaborative
conservation; conservation easements;
economics of conservation
Conservation Alliance, 285–86
conservation easements: Arcata Community
Forest (ACF) and, 137; donation or purchase of, 217; ecosystem services and, 285;
estate tax and, 272; 47 Ranch and, 183, 229,
235–38; Garcia River Forest and, 156, 157;
habitat protection and, 235; income from,
237; land valuations and, 152; Laramie
Foothills Group and, 68; Lava Lake Land &
Livestock and, 196–97; in Madison Valley,
Montana, 212, 216–17, 224; negotiations
for, 235–36; private land conservation
trends and, 241–42, 242; property appraisal and, 236; sale of development
rights, 197; silver bullet solutions and, 278;
Sonoran Desert Conservation Plan and,

260; Sun Ranch and, 216–17, 219, 220; in
Swan Valley, Montana, 128–29, 133, 134; tax
incentives and, 217, 221n7, 236–37, 241, 245;
tax policy and, 267; watershed stewardship and, 284; wording of, 239n3
Conservation Fund (TCF), 86, 120–21, 151,
155–58, 197
Conservation Reserve Program, 295–99
Coon Valley initiative, 77–79
cooperative conservation. *See* collaborative
conservation
Corbett, Jim, 61
Coronado National Forest, 58, 255
Coulee Region Organic Produce Pool
(CROPP), 79
Country Natural Beef (CNB): chal-
lenges of, 203–6; establishment of,
x; landscape-scale conservation and,
199; ranchers as marketers and, 181;
third-party certification of, 287
Craig, Larry, 110
Craters of the Moon National Monument and
Preserve, 187, 188
CROPP. *See* Coulee Region Organic Produce
Pool (CROPP)
Crown of the Continent Ecosystem, 124, 133
*Cygnus buccinator* (trumpeter swan). *See* trum-
peter swan (*Cygnus buccinator*)

Daggett, Dan, xiii, 44
Danforth, Fred, 282
David and Lucile Packard Foundation, 156
Dawes Act (1887), 5
Dean, Sandy, 146
Death Tax Repeal Permanency Act (proposed), 268
deBuys, Bill, xiii, 86–87
Declaration of Interdependency, xiii
Defenders of Wildlife, 198
*Dendrocygna autumnalis* (black-bellied whistling duck). *See* black-bellied whistling
duck (*Dendrocygna autumnalis*)
desert ironwood (*Olneya desota*), 254
developers and development: as beneficiaries
of conflict, 8; biodiversity and, 33, 39–40,
44; bulldozer in the countryside mentality
and, 254; community-based conservation
groups (CBCCs) and, 67–68; conservation easements and, 217; conservation
settlements and, 310n10; definition of
developed land and, 311n28; development
mitigation fees and, 258–60; ecologi-

cally sustainable, 216–17; environmental concerns about, 213, 215–16; estate tax and, 248, 267–73; fire management and, 69–70; fragmentation of ecosystems and, 69–70, 212; government funding in face of high development pressure and, 123; jobs versus environmental protection and, 253; in Madison Valley, Montana, 212–13; in Pioneer Mountains and Craters of the Moon region, 187; planned versus wildcat, 251, 253; population growth and, 24; prevention of, 50; public hearings and, 253, 254; ranchers' income and, 95; ranchettes and, 212, 213, 217; rapid growth and, 13; rate of conversion to developed land and, 24, 96; resistance to development pressure and, 155; rising land values and, 126; Sonoran Desert Conservation Plan and, 257–58, 271; Sun Ranch and, 215–16, 220; in Swan Valley, Montana, 124, 128–29, 132, 133; urban sprawl and, 248, 251–53, 262; working landscapes and, xv

DeVoto, Bernard, 6

Diablo Burger (Flagstaff, Arizona), 66, 75n38, 287

Diablo Trust: awards for, 67; Diablo Burger and, 66, 75n38, 287; ecosystem services and, 290; formation of, 66; goals of, 67; as pioneers in collaborative conservation, 78; as successful CBCC, 53

Diamond Bell Ranch, 261

Dine Bei'ina, 288

Double Check Ranch, 249

Douglas fir (*Pseudotsuga menziesii*), 138, 151, 156

Durham, William, 214

Earth First!, 58, 143–44

EBI. *See* Environmental Benefit Index (EBI)

Ecoasis, 234

ecological health. *See* ecosystem health

economic collapse of 2008: Chicago Climate Exchange and, 280–81; ecosystem services and, 277; Garcia River Forest and, 156; innovations in ranching and, 181; Lava Lake Land & Livestock and, 196; real estate prices before and after, 261–62; Redwood region and, 151, 152; stagnation in construction industry and, 244; Sun Ranch and, 182–83, 211, 215; Sun Ranch Settlements and, 218–19

Economic Growth and Tax Relief Reconciliation Act (2001), 270, 272

economics of conservation: Beginning Farmer-Rancher Loans and, 233; boom-and-bust levels of support and, 116; carbon credits and sequestration and, 120–21, 152; community forest bonds and, 130; compensation for conservation work and, 90, 92; conservation banking and, 282–83; conservation bonds and, 259–60, 261, 262; core principles for working landscapes and, 108–9; costs of inputs versus outputs and, 268; costs of producing for niche markets and, 288–89; creative financing strategies and, xvii–xviii, 156; decline of government funding and, 241; diversification and, 64–66, 182–83, 214, 227, 307; federal investment and, 116; financial structures for community-supported agriculture and, 248; Forest Legacy Program and, 129; forest projects from public lands and, 133; funding cuts and, 242–43, 245; government bonds and, 132, 134; government contracting and, 116–17nn2–4; government demands on private organizations and, 81, 92–93; Habitat Conservation Plans (HCPs) and, 258–60; incentives for grass-fed production and, 97–99; jobs versus environmental protection and, 253; land-rich, cash-poor ranchers and, 214; land values and development and, 126; legislative developments and, 112; locavesting and, 249; management agreements with former ranch owners and, 264; marijuana production and, 154; as market frontier, 249; mitigation banking and, 277; nonreliance on grants and subsidies and, 81; outside investors in ranching and, 182; patient capital and, 276; payments for ecosystem services and, 275–78; Preservation Capital and, 218–19; presidential initiatives and, 116; profitable conservation and, 182; public-private partnerships in, 247; Qualified Forestry Bonds program and, 132; resilience and, 209; resistance to development pressure and, 123, 124; restoration economy and, 248–49, 275–76; revenue generated from timber production and, 137; slow money and, 249, 276; species banks and, 283; staffing of conservation organizations and, 114; stewardship contracting and, 116n3, 121–22, 159; variability in, 234. *See also* economic collapse of 2008; ecosystem services; taxation

Eco Products Fund, 284
ecosystem health, xiv–xv, 88–89
ecosystem services: beneficiaries of, 284–86, 290–91; bundling of, 283–84; carbon sequestration and, 249, 279–81; categories of, 279; colocation and, 283–84; conservation easements and, 285; Conservation Reserve Program, 295; cost and benefit sharing for producers and, 288–90; examples of, 276; habitat preservation as, 279, 281–83; information for participants and, 298; landscape-scale, 288–90, 307; Millennium Ecosystem Assessment and, 278–79; mitigation banking and, 249; payment for, 248–49, 275–79, 283, 284–91, 295, 297–99; philosophical and practical concerns about, 278, 291; prices of eco-labeled products and, 286–87; protection of endangered species and, 249; silver bullet solutions and, 278; supply and demand and, 275, 276; voluntary provision of, 277; water and wetlands and, 249, 279, 281–82
Ecosystem Workforce Program, 103
ecotourism, 182, 213–15, 217–18, 220, 224
Ecotrust, 283–84
elk (Cervus canadensis), 186, 212, 223–24
Elk Creek Conservation Area, 130
Elk Creek Land Exchange, 127–28
Emerson, Kirk, 59
Empirita Ranch, 261
endangered species and Endangered Species Act (1973): cactus ferruginous pygmy owl and, 247; communities disadvantaged by, 172–73; conservation banking and, 282–83; conservation easements and, 236; Habitat Conservation Plans (HCPs) and, 254–58; habitat projects and, 234, 247–48, 249; Incidental Take Permits and, 254; noncompliance lawsuits and, 7, 69; northern spotted owl and, 160; passage of, 7; Section 10 permits and, 254, 263; Sonoran Desert Conservation Plan and, 254; U.S. Forest Service priorities and, 160
Environmental Benefit Index (EBI), 297–98
Environmental Defense Fund, 269
Environmental Protection Agency (EPA). See EPA (Environmental Protection Agency)
EPA (Environmental Protection Agency), 89, 281
erosion control, 62–64, 77–78, 89, 279, 296

Fairfax, Sally, 278
Family Farm Preservation and Conservation Estate Tax Act, 272
Farm Bill, 106, 132, 296
Federal Forest Health Advisory Council, 109–10
federal lands: changing uses of, 19–24; declining, 19; overworked land managers and, 49; as proportion of lands in West, 19; recreational use of, 20–21; role of in Pacific Northwest, 164–65; as working landscapes, 19–24. See also U.S. Forest Service
fire management: biodiversity and, 62; CBCCs and, 60–62, 70; Collaborative Forest Landscape Restoration program and, 111–12; development and, 69–70, 270; Healthy Forest Restoration Act (2003) and, 109–10; Malpai Borderlands Group and, 60–61; management plans and, 74n18; National Fire Plan and, 109–10; necessity of fire and, 8; overplanting of trees and, 166; overzealous fire suppression and, 60; Rural Voices for Conservation Coalition (RVCC) and, 109–10; Weaverville Community Forest and, 178–79
First United Realty, 262
Fisher family, 144
Flathead National Forest, 126, 131, 133
Food Alliance, 204, 287
Food Security Act (1985), 296
Forbes, Peter, xiii
Ford, Jerome, 141
Forest Guild, 139
Forest Landscape Restoration Act (2009), 111
Forest Legacy Program, 129
forest management: checkerboard ownership versus single owner, 119–20, 122; collaborative conservation and, x–xi, 122; in community forests, 121; computer landscape modeling and, 152–53; cost of forest certification and, 148; elimination of bad practices versus improvement in, 104; forestry infrastructure and, 172; forest thinning and, 121, 157, 166; government role in, 141; maintenance work and, 151; patient, 157; planning processes in, 153; seat at the table for foresters and, 69; silviculture and, 157, 168; socially responsible, 150; sustainable, 141; transjurisdictional, 304–5; triple bottom line of, 119, 122. See also forests and foresters; steward-

ship contracts and agreements; timber
production
Forest Project Protocols, 151
forestry. *See* forest management; forests and
foresters; timber production
forests and foresters: carbon sequestration
and, 279–80; Collaborative Forest Landscape Restoration program and, 111–12;
Community-Based Forest and Public
Lands Restoration Act (proposed in 2002)
and, 110–11; community forests and, 121,
129–30, 137, 140, 177; conflict over use and
care of, 101–2; conservation of old-growth
habitat in, 43; definition of forestland
and, 29n1; definition of timberland and,
29n1; deforestation and, 13; embodied
sense of place and, 71; estate tax and, 271–
73; Farm Bill and, 106; Forest Landscape
Restoration Act (2009) and, 111; forestry
certification systems, 286, 289; fragmentation of, 25–26; multigenerational commitments and, 303–4; nontimber forest
products and, 15–16, 19, 20; in Oregon
Coast Range, 40–41; overexploitation of
land and, 71–72; ownership patterns in,
26–27; parcel size and number in, 25, 26;
predicted future composition of, 42–43;
private, 24–25; private industrial versus
nonindustrial, 41, 47n31; restoration of,
110; as working landscapes, 138. *See also*
fire management; forest management;
timber production
Forest Service. *See* U.S. Forest Service
Forest Stewardship Council (FSC): certification by, 120, 139, 141, 147–48, 157, 277, 287;
group certification and, 289; protection
of old-growth trees and, 150; trademark
of, 147
47 Ranch: conservation easements on, 235–38;
conservation ranching and habitat
conservation at, 231–35; description of,
227–28; diversified enterprises at, 227,
229–35; electricity generation and, 229;
Hay Mountain Watershed Restoration
Project and, 233–34; holistic approach
to land care at, 238; as lab and teaching facility, 232–33; livestock breeds at,
229–31; map of, 228; meat sales and, 233;
olive production at, 231; partnerships of,
233–34; Quivira Coalition and, 232; sale of
development rights and, 183
Framing Our Community, 103

Frank Church–River of No Return Wilderness
Area, 187
Freeport-McMoRan Copper and Gold, 71
Freyfogle, Eric, 249
Friends of the Wild Swan, 127
FSC. *See* Forest Stewardship Council (FSC)
Fussell, Betty, 65

Gaebler, Ted, 153
Garcia River Forest, 120, 151, 155–57
General Homestead Act of 1862, 4
Georgia-Pacific, 145–46, 171
giant sacaton grass (*Sporobolus wrightii*), 63
Glacier National Park, 124
*Glaucidium brasilianum cactorum* (cactus ferru-
ginous pygmy owl). *See* cactus ferrugi-
nous pygmy owl (*Glaucidium brasilianum
cactorum*)
Gleckman, Howard, 269
Glenn, Warner, 61
Glenn, Wendy, 61
Grange, 310
Granite Cattle Company, 211
Grant Family Farms, 248
Grassroots Collaborative Conservation Sur-
vey, 56, 70–71
gray wolf (*Canis lupus*), 123–24, 186, 212–13, 224
grazing. *See* livestock grazing
Great Plains grasslands, 36
Greenpeace, 147
greenwashing, 98
grizzly bear (*Ursus arctos horribills*), 123–24,
212, 224
Grossman, Dan, 269
*Grus canadensis* (sandhill crane). *See* sandhill
crane (*Grus canadensis*)
*Gulo gulo* (wolverine). *See* wolverine (*Gulo gulo*)

Habitat Conservation Plans (HCPs): Endangered Species Act (1973) and, 254–58; Environmental Impact Report and, 149–50;
funding of, 258–60; Mendocino Redwood
Company and, 120, 141, 147, 148–50, 153;
requirements of, 149–50; 2007 Farm Bill
and, 132
habitat preservation: conservation easements
and, 235; ecosystem services and, 249,
279, 281–83; northern spotted owl and, 7;
wetlands mitigation banking and, 281
Hadley family, 62
Hardin, Garrett, 4
*Harper's* (magazine), 6

Hasselstrom, Linda, xiii
Hatfield, Connie, x, 203–4, 205
Hatfield, Doc, x, 203–4, 206
Hayhook Ranch, 260–61
Hay Mountain Watershed Restoration Project, 233–34, 238–39n2
HCPs. *See* Habitat Conservation Plans (HCPs)
Healthy Forest Restoration Act (2003), 109–10
Hirt, Paul, 6
Holly, Diane, 270
Home Depot, 148
Humane Society of the United States, 75n35
Humboldt Redwood Company, 120, 150–53
Humboldt State University, 138
Humphrey, Robert, 60

Idaho Conservation League, 197
Idaho Department of Fish and Game, 189, 192, 193, 198
Idaho National Laboratory, 187
Imhoff, Dan, 44
Indian Reorganization Act (1934), 5
Integrated Systems Information, Inc., 286
International River Foundation, 167
International Thiess Riverprize, 167
invasive species, 213
Inventoried Roadless Areas, 187, 200n3
"Invitation to Join the Radical Center," xiii, 84

Jackson, Andrew, 5
Jacoby Creek Forest, 138
Johnson, Barbara, 83–84
Johnson, Paul, xiii
Jordan, Teresa, xiii
Jornada Experimental Range, 230
Josephy, Alvin, xiii

King, Manuel, 63
King family, 63–64, 260–61
King's 98 Ranch, 260–61
Klyza, Christopher McGrory, 7
Knight, Heather, xiii
Knight, Rick, 67–69

Lake County Resources Initiative, 103
Land and Water Conservation Fund, 126, 242–43
land health, 88–90, 243
Land Trust Alliance, 243, 246n5, 273
land trusts: acquisition of state trust land and, 262; acreage protected by, 243, 244–45; census of land trusts and, 242–43; in Colorado Plateau, 38; definition of, xxi

n1; 47 Ranch and, 183; grazing in, 22; livelihoods and, xv; permanence of, 245; private land conservation trends and, 242–43, 242; rancher-led land trusts and, 199; sale of, xxi n2; success of, 243; in Swan Valley, Montana, 126; tax incentives and, 245
Lang, Roger, 213, 219
Lang family, 182, 214, 215
Laramie Foothills Group, 53, 54, 67–68
Laramie Foothills Mountains to Plains Project, 248
lark bunting (*Calamospiza melanocorys*), 40
Las Cienegas National Conservation Area, 284–85
Lava Lake Institute for Science and Conservation, 186, 191, 197, 198
Lava Lake Lamb, 186, 195–96
Lava Lake Land & Livestock: awards for, 191; business strategies and, 188, 189; conservation and profitability at, 182; conservation easements and, 196–97; conservation goals and, 188; description of, 185; economic strategy of, 194; funding for, 191, 192, 197; grazing and, 188, 189, 191, 192–94; innovative practices of, 186; land acquisition for, 182, 188–89, 196; lessons from, 198–200; map of lands of, 190; meat sales and, 194–96; Nature Conservancy and, 189, 191, 196; organizational structure of, 198; origins of, 188; partners of, 186, 192, 193, 195–96, 197; praise for meat quality and, 195; ranchers as marketers and, 181; reduction in sheep numbers and, 193–94; science and conservation advisory board and, 191, 193; science-based management and, 189, 191–92; transjurisdictional landscape conservation and, 181–82, 189
Lava Lake Ranch, 188
Leopold, Aldo, 40, 77, 78, 88
Limerick, Patricia, xiii
*Lithobates chiricahuenses* (Chiricahua leopard frog). *See* Chiricahua leopard frog (*Lithobates chiricahuenses*)
*Lithocarpus densiflorus* (tan oak). *See* tan oak (*Lithocarpus densiflorus*)
livestock grazing: activism against, 224; advocacy for end of on public lands, 7; in Bureau of Land Management lands, xviii–xix; "Cattle-Free in '93" and, 181; collaborative conservation, 213; conservation ranching and, 234; decline in, 1, 13, 19–20; Lava Lake Land & Livestock and, 188, 189, 191, 192–94; mosaic of private and

public land tenure and, 22–24; in national forests, 5; protection of cattle and, 224; with public grazing leases on private lands, 39; public versus local interest and, 6; scientific management and, 5–6; types of lands used for, 16; in West as proportion of U.S. as a whole, 13, 16; in winter, 214; year-round water source and, 232. *See also* ranching and ranchers; rangelands

Local Burger (Lawrence, Kansas), 287

local food movement, 65–66, 68

logging. *See* timber production

Lolo National Forest, 133

Louisiana-Pacific, 144–46

Louisiana Purchase, 3

Lowe, C. E., 36

lowland leopard frog (*Rana yavapaiensis*), 232

lynx (*Lynx canadensis*), 123–24

*Lynx canadensis* (lynx). *See* lynx (*Lynx canadensis*)

*Lynx rufus* (bobcat). *See* bobcat (*Lynx rufus*)

Madison Valley, Montana: conservation easements in, 212, 216–17, 224; conservation issues and partners in, 211–13; description of, 223; development in, 212–13; elk hunting in, 224; history of, 211; invasive plants in, 224–25; land prices in, 223; ranchettes in, 212, 213, 223; wildlife in, 212, 223–24

Madison Valley Expeditions, 213, 220, 224

Madison Valley Ranchlands Group: activities of, 223; dialogues of, 224; ecotourism and, 220; formation of, 213, 224; Grassroots Collaborative Conservation Survey and, 56; Living with Wildlife workshops of, 224; membership of, 69, 219, 224; purpose of, 223; ranchers' voices and, 181; Range Riders and, 224; rising land values and, 182; staffing of, 71; success of, 213; Sun Ranch and, 213; weed control and, 224–25

Malpai Borderlands Group: authors as participants in, 55; erosion control and, 62–63; formation of, 60–61; grassbanks and, 87; Grassroots Collaborative Conservation Survey and, 56; land health and, 243–44; landscape-scale conservation and, 199; membership of, 69; as pioneers in collaborative conservation, 78; radical center and, 83; as successful CBCC, 53; U.S. Forest Service and, 68; working wilderness and, 61

Mapleton-Deadwood community, 161

marbled murrelet (*Brachyramphus marmoratus*), 41, 170

Margerum, R. D., 54, 69

marijuana production, 154

Marley Ranch, 261, 262

Marshall, John, 5

Marston, Ed, xiii

Martin Cattle Company, 260–61

Mary's Peak Stewardship Group, 168

masked bobwhite (*Colinus virginianus ridgwayi*), 59

May, R. M., 34

Mayfield, Butch, 286

Mayfield Ranch, 286

McDonald, Bill, xiii, 83

McKibben, Bill, xiii

meat industry: antibiotics and hormones in, 203–4; Argentine beef and, 203–4; carbon credits and, 286; consumer-driven marketing and, 203–4; cooperatives in, 195, 203; direct-to-consumer marketing and, 230; fat content and, 230; flavor and tenderness and, 230; fluctuating meat prices in, 205–6; fluctuating prices in, 203; grass-fed and -finished labeling and, 96, 98–99, 98; human animal treatment and, 206; locally branded meat and, 206; local supply and, 251, 264–65; natural food retailers and, 205; niche markets and, 95, 194–96, 205, 206, 288–89; number of cattle in West and, 16, 18; oligopsony in, 65; pricing and, 268, 288–89; production back stories and, 287, 288; raise well and graze well standards and, 203, 204–5; third-party certifiers and, 97; USDA-certified slaughterhouses and, 264; value-based marketing and, 205

Meine, Curt, xiii

Mendocino Forest Products, 145, 148

Mendocino Redwood Company: carbon credits and, 151–52; conservation easements and, 152; early decisions of, 145–47, 150; Environmental Impact Report and, 149–50; environmental stewardship of, 141; Forest Stewardship Council certification for, 147–48; founding of, 144–45; Habitat Conservation Plan for, 147, 148–50, 153; Humboldt Redwood Company and, 150; lands of, 142; Mendocino County history and, 141, 143–44; Natural Community Conservation Plan for, 120, 141, 147, 148, 149–50, 153; Timber Harvesting Plans and, 149–50; timber wars and, 146–47; trust building and, 120

*Mephitis mephitis* (striped skunk). *See* striped
skunk (*Mephitis mephitis*)
Merk, Frederick, 4
Metzger family, 66
migratory birds, 232, 234
Millennium Ecosystem Assessment, 278–79
mineral rights, 211
Montana Fish, Wildlife and Parks, 282
Montana Land Reliance, 127, 282
Montana Legacy Project, 120, 132–33
Montana Logging Association, 127
Montana Partners for Fish and Wildlife,
281–82
Montana Wilderness Association, 127
Moroney family, 183, 228
Morris K. Udall Foundation, 59
mountain goat (*Oreamnos americanus*), 186
mountain lion (*Puma concolor*), 186
Mountain States Lamb Cooperative, 181, 195
Mountain States Rosen, 195–96
Mountain States Transmission Intertie
Project, 187
Muir, John, 5, 8
mule deer (*Odocoileus hemionus*), 186

Nabhan, Gary, xiii
National Cattlemen's Beef Association,
268–69
National Environmental Policy Act (1969):
environmental review processes and, 128;
government demands on private organizations and, 92; noncompliance lawsuits
and, 7, 69; passage of, 7; public comment
requirements and, 57; Valle Grande Grass-
bank and, 87
National Fire Plan, 109–10
National Fish and Wildlife Foundation, 71, 193
National Forest Management Act, 116n2
National Forest System, xviii–xix, 5, 133,
163–64, 172–73
National Marine Fisheries Service, 148
National Oceanic and Atmospheric Administration, 282
National Park Service, 36, 188–89, 245, 301–2
National Partnership for Reinventing Govern-
ment, 67
National Research Council, 88
National Rural Assembly, 115
Native Americans: congressional prohibition
of treaties with, 5; ecosystem services
and, 278; ignoring of land claims of, 4;
legislation affecting landownership of,

5; Madison Valley, Montana and, 211;
recognition of local interests and, 6; Trail
of Tears and, 5; westward removal of, 4–5.
*See also* tribal lands
Natural Community Conservation Plans (NC-
CPs), 147–50, 153
Natural Resources Conservation Service: con-
servation easements and, 236–37; erosion
control and, 63–64; Farm and Ranchland
Protection Program and, 197; Grassland
Reserve Program and, 197; habitat projects
and, 232; Hay Mountain Watershed
Restoration Project and, 233–34; Laramie
Foothills Project and, 68; Pioneers Alliance and, 197; soil conservation and, 77
Natural Resources Defense Council, 147, 192
Natural Resources Inventory, 246n5
Nature Conservancy (TNC): Crown of the Continent Ecosystem and, 124; development
rights, 272–73; ecoregional assessments
by, 37; funding and, 62; Garcia River Forest
and, 156; Hay Mountain Watershed Restoration Project and, 238–39n2; Laramie
Foothills Group and, 54, 67, 68; Laramie
Foothills Mountains to Plains Project and,
248; Lava Lake Land & Livestock and, 189,
191, 196; Montana Legacy Project and,
132–33; Nevada Spring Creek Watershed
and, 282; Pioneers Alliance and, 197;
protected lands in U.S. and, 34; relations
of with ranchers, 59; Sonoran Desert
Conservation Plan and, 259; Sun Ranch
and, 216–17, 220
Nature's Market, 205
Navajo-Churro Sheep Association, 288
Navajo Sheep and Lifeways, 288
NCCPs. *See* Natural Community Conservation
Plans (NCCPs)
Nestlé, 286
New Agrarianism, 249–50
New Belgium Brewing, 286
New Seasons Market, 205
Northern Arizona Collaborative Grassroots
Management Team, 66
Northern Pacific Railroad, 126
northern spotted owl (*Strix occidentalis cau-
rina*): declining timber production and,
41; Endangered Species Act (1973) and,
160; in Garcia River Forest, 156; habitat
restoration for, 161; injunctions against
timber harvesting and, 165; listing of as
endangered, 170; Mendocino Redwood

Company and, 148; preservation of habitat of, 7; resistance to grazing and, 181

Northwest Connections, 103

Northwest Forest Plan, 160, 164, 165–66, 170, 172–73

Obama, Barack, 115–16, 270

*Odocoileus hemionus* (mule deer). *See* mule deer (*Odocoileus hemionus*)

Off the Beaten Path, 214

*Olneya desota* (desert ironwood). *See* desert ironwood (*Olneya desota*)

Omnibus Consolidated Appropriations Act of FY 1999, 116n3

Omnibus Public Land Management Act of 2009, 111

*Oncorhynchus clarkii lewisi* (westslope cutthroat trout). *See* westslope cutthroat trout (*Oncorhynchus clarkii lewisi*)

*Oncorhynchus kisutch* (coho salmon). *See* coho salmon (*Oncorhynchus kisutch*)

*Oncorhynchus mykiss* (steelhead trout). *See* steelhead trout (*Oncorhynchus mykiss*)

*Oreamnos americanus* (mountain goat). *See* mountain goat (*Oreamnos americanus*)

Oregon Coast Range, 40–43

Oregon Country Beef. *See* Country Natural Beef (CNB)

Oregon Natural Resources Council, 171

Oregon Wild, 115, 171

Organic Valley, 77, 79

Osborne, David, 153

Overlook Ranch, 270

Pacific Gas and Electric Company, 151, 157

Pacific Lumber Company, 150

Pacific Northwest Region, 160–61, 164

Packer, Larissa, 278

Papoose Creek Lodge, 213, 214–15

Patagonia, Inc., 285–86

PES (payment for ecosystem services). *See* ecosystem services

Pima County, Arizona. *See* Sonoran Desert Conservation Plan

Pinchot, Gifford, xviii, 5–6, 8

Pioneer Mountains and Craters of the Moon region, 186–87, 189, 192, 193–95, 197–98

Pioneers Alliance, 182, 186, 197

Plum Creek Timber Company: community forest proposal and, 130; conservation easements and, 128–29; conservationists' land purchases from, 131–33; develop-

ers' land purchases from, 119, 126–27, 131; land-use planning and, 131; as largest private landowner in U.S., 126; public land acquisition from, 124, 126, 127–28

pollination of crops, 279

Powell, John Wesley, 56–57, 70, 301

Prescott College, 232

Preservation Capital, 218–19

private lands: biodiversity on, 33, 44, 45; in Colorado Plateau, 37–38; ecological buffer zones in, 33, 39; fire suppression and, 8; forestlands and, 24–25; industrial versus nonindustrial, 41, 47n31; interdependence with public lands and, 43; landowners' conservation experiments and, 44; leasing of for grazing and, 22–23, 39; as neglected geography for conservation, 34–35; noncorporate versus corporate, 25; nonindustrial and noncorporate, 31n37; in Oregon Coast Range, 41–43; parcelization of, 1; quantity of forest and rangeland in U.S. and, 27; researchers' and environmentalists' neglect of, 37–38; respect for landowners and, 68; in Rocky Mountain ecosystem, 38–39; timber industry organization and, 26–27; as working landscapes, 24–28

Progressive Era, scientific management and, 5–6

pronghorn (*Antilocapra americana*), 186, 212

Prosser family, 66

protected lands, 33, 34, 35, 36–37

*Pseudotsuga menziesii* (Douglas fir). *See* Douglas fir (*Pseudotsuga menziesii*)

public lands: access to for grazing, 193; acquisition of in Swan Valley, 132–34; biodiversity limited on, 33, 43–45; collaboration versus confrontation in management of, 302–3; declining grazing on, 1; declining timber production on, 1, 41; interdependence with private lands and, 43; land-use planning and, 130–31; local knowledge in management of, 306–7; in Oregon Coast Range, 41–43; in Pacific Northwest Region, 164–65; ranchers' dependence on, 22–24; in Rocky Mountain ecoregion, 38; timber production on, 137; zoning and, 131

Public Lands Council, 269

public policy: brief history of in West, 3–11; environmental review processes, 128; working landscapes as core conservation areas and, 44

*Puma concolor* (mountain lion). *See* mountain lion (*Puma concolor*)

pygmy owl. *See* cactus ferruginous pygmy owl (*Glaucidium brasilianum cactorum*)

Qualified Forestry Bonds program, 132

Quivira Coalition: accomplishments of, 82; alliances and, 310; collaborative conservation and, 49; Comanche Creek habitat restoration and, 90–93; Declaration of Interdependency, xiii; as do tank, 82; founding of, 83; geographic reach of, 83–84; government demands on private organizations and, 92–93; Grassroots Collaborative Conservation Survey and, 56, 70–71; induced meandering and, 89; innovation at, 82–83; mission of, 82; name of, 83; New Agrarian Program of, 232; New Ranch and, 85–86; no legislation or litigation and, 82, 83; range of activities of, 86, 90, 93; resilience building strategies and, 93–94; riparian restoration and, 89–93; as umbrella organization, 54; Valle Grande Grassbank and, 86–87

radical center: coining of term, 82–85; collaborative conservation organizations and, 45; community good and, ix; as critical to West that works, 302; "Invitation to Join the Radical Center" and, xiii; limited success of, 84–85; Malpai Borderlands Group and, 61. *See also* "Invitation to Join the Radical Center"

*Rana yavapaiensis* (lowland leopard frog). *See* lowland leopard frog (*Rana yavapaiensis*)

ranching and ranchers: adjacent to urban areas, 251–53, 262; carbon credits and, 96; commodity markets and, 95–96; conservation bonds and, 260; conservation ranching and, 231–35; declining acreage in, 267–68; developers buying land from, 95, 96; as difficult way to make a living, 23–24, 64, 198–200, 214; dude ranches and, 65; earnings of per food dollar spent, 96; economic diversification and, 65; economics of grass-fed production and, 97–99; embodied sense of place and, 71; environmentalists' attitude toward, 256; estate tax and, 260, 267–71; federal regulation of, 5; fluctuating land values and, 270; fluctuating meat prices and, 205–6; fragmentation of forestlands and, 248; fragmentation

of ranches and, 187, 270; grassbanks and, 86–87; grass-fed and -finished labeling and, 96, 98–99, 98; green ranching and, x; hunting permits and, 65; innovation and, 95, 96–97, 181, 183, 218–19; intergenerational transfer of lands and, 268–69; as land-rich and cash-poor, 214; livestock breeding and, 229–30, 238n1; loss of access to public rangelands and, 22–24; management agreements with former ranch owners and, 264; meat industry and, 65–66, 75n35; mosaic of private and public land tenure and, 22–24; multigenerational commitments and, 303–4; New Ranch and, 82, 85–88; oligopsony and, 65; overexploitation of land and, 71–72; rancher-led land trusts and, 199; ranchers as "lords of yesterday" and, 7; ranchers' conservationist ethic and, 8; ranchers in CBCCs and, 54; ranchettes and, 269; rural-urban divide and, 262–65; science and research and, 185; seat at the table for, 69; Sonoran Desert Conservation Plan and, 256, 258, 260; stagnant cattle prices and, 223; stewardship by, 40; third-party certifiers and, 97; varieties of cattle and, 229–31, 238n1; wolves and, 198; younger generation and, 233, 268. *See also* livestock grazing; meat industry

Rancho Seco, 260–61, 271

rangelands: animal unit months and, 20, 30n16; bison ranging patterns and, 214; carbon sequestration and, 279–80; conversion of to other uses, 27–28; definition of, 29n10; extent of in West, 16; federal and nonfederal in West, 18; fragmentation of, 27, 28; health of, 22, 28; interspersed landownership patterns and, 185–86, 193; land health and, 88–89; ownership of in West, 16, 17; range wars and, xiii–xiv; Rural Voices for Conservation Coalition and, 106; stewardship contracting in, 162; transjurisdictional management and, 304–5; wildlife corridors and, 182

real estate speculation. *See* developers and development

Recon Environmental, 255

Redwood Forest Foundation, 151

Redwood Region of California, 120, 150–52

Redwood Summer, 143–44

renewable energy, 65

Renewing America's Food Traditions Alliance, 288

restoration economy. *See* economics of conservation

Ricketts, J., 36

riparian restoration, 88–93, 193, 214

Rivers and Harbors Appropriation Act (1899), 281

robin (*Turdus migratorius*), 40

Robles, M. D., 25

Rocky Mountain ecoregion, 36, 38–40

Rocky Mountain Elk Foundation, 128–29

Rodriguez, J. P., 34

Rome, Adam, 254

Roosevelt, Theodore, xviii, 5–6, 8

Roosevelt, Theodore, IV, xiii

Roots of Change, 249

*Roseland Plantation LLC v. United States Fish and Wildlife Service et al.*, 280

Rowley family (of Rancho Seco), 59, 260–61

Rural Revitalization Act (1990), 172

Rural Voices for Conservation Coalition (RVCC): alliances and, 310; approach of to policy change, 107–12; collaborative conservation and, 45, 49–50; core principles for working landscapes and, 108–9; establishment of, 103; estate tax and, 269; fire management and, 109–10; future of, 115–16; growing constituency of, 115; growth and sustenance of, 106; legislative advocacy of, 105–8, 110–12, 113–14; lessons learned by, 112–15; membership of, 106–7; mission of, 103; needs fulfilled by, 104; objects and focus areas of, 105–7; organizational model of, 115; orientation of new members and, 113; policy-issue papers of, 107, 113–14; staffing of, 114; trust building in local communities and, 115; as umbrella organization, 54

RVCC. *See* Rural Voices for Conservation Coalition (RVCC)

Sabores sin Fronteras, 234

Sagebrush Steppe Ecosystem Reserve, 187

sage grouse (*Centrocercus urophasianus*), 186

Saguaro National Park, 255

Salazar, John, 272–73

*Salmo trutta* (brown trout). *See* brown trout (*Salmo trutta*)

*Salvelinus confluentus* (bull trout). *See* bull trout (*Salvelinus confluentus*)

sandhill crane (*Grus canadensis*), 234

Sands Ranch, 261

Sawtooth National Recreation Area, 198

Sayre, Nathan, 62

Schwenessen, Paul, 249

Schwenessen, Sarah, 249

Scott, J. M., 39

Section 347(c)(1) of the Omnibus Consolidated Appropriations Act (1999), 104–5

Seminole people, 5

*Sequoia sempervirens* (coast redwood). *See* coast redwood (*Sequoia sempervirens*)

Shafer, C. L., 34

sheep, number of in West, 16, 18

Sierra Pacific Industries, 178

Sisk, T. D., 38

Siuslaw Basin Partnership, 167

Siuslaw Institute, 161, 167, 171, 173

Siuslaw National Forest (SNF), 121, 161, 164–72

Siuslaw Soil and Water Conservation District, 167

Siuslaw Stewardship Fund, 167

Siuslaw Stewardship Group (SSG), 121, 167, 169–71

Siuslaw Watershed, 165

Siuslaw Watershed Council, 161

Six Bar Ranch, 261

Slow Food International, 288

Slow Money Alliance, 249

Smithsonian Institution, 254

SNF. *See* Siuslaw National Forest (SNF)

Snow, Donald, 55

snow goose (*Chen caerulescens*), 234

Society for Range Management, 234

Soil Conservation Service, 64

soil stability, 89–90

Sonoran Desert Conservation Plan: conservation bonds and, 259–60, 261, 262; conservation easements and, 260, 261; creation and purpose of, 247; Cultural Resources Technical Advisory Team and, 256; description of, 251; educational meetings and, 257–58; erosion control and, 64; funding of, 248, 258–60; goals of, 251–52; initiation of, 251–52, 254; land acquisitions and, 248, 260–62, 271; launch of, 59; map of, 263; multispecies habitat conservation and, 254–58, 262–63; Pima County ranches and, 260–62; prioritizing implementation of, 258–60; Ranch Conservation Technical Advisory Team and, 256; rural-urban divide and, 262–65; Science Technical Advisory Team and, 255–58, 263; scope of, 247–48; setting of, 251–53, 262–65; steering committee and, 257–58

Sonoran Desert ecoregion, 36, 61

Sooes Forest, 283–84

Sópori Ranch, 260, 261, 262, 266n26
Sousa, David, 7
Southern Arizona Home Builders Association, 257
Southern Rocky Mountain ecoregion. *See* Rocky Mountain ecoregion
Southwest Grassfed Livestock Alliance, 234
Soviet Union, 296
Spies, T. A., 42
*Sporobolus wrightii* (giant sacaton grass). *See* giant sacaton grass (*Sporobolus wrightii*)
spotted knapweed, 224–25
spotted owl. *See* northern spotted owl (*Strix occidentalis caurina*)
starling (*Sturnus vulgaris*), 40
steelhead trout (*Oncorhynchus mykiss*), 156
stewardship contracts and agreements: accomplishments of, 168–69; agreements versus contracts and, 178; benefits and challenges of, 169–71, 179–80; collaborative conservation and, 104–5, 116n3; description of, 159, 161–63; forest management and, 159; interpretation of stewardship authority and, 172–73; as land management model, 306–7; number of contracts awarded and, 163–64; pilot projects and, 163, 166; promotion of, 173; receipts from sale of timber and, 166, 167–69, 171, 172; in Siuslaw National Forest, 164–69; values behind, 172; varying viability of, 171; Weaverville Community Forest and, 121–22, 162
striped skunk (*Mephitis mephitis*), 40
*Strix occidentalis caurina* (northern spotted owl). *See* northern spotted owl (*Strix occidentalis caurina*)
*Sturnus vulgaris* (starling). *See* starling (*Sturnus vulgaris*)
subsistence producers, 6
Sun Ranch: carbon credits and, 280; conservation easements and, 216–17, 219, 220; description of, 209, 211; as eco-enterprise model, 219; economic collapse of 2008 and, 218–19; economic diversification at, 182; establishment of, 211; Lodge at, 213, 214–15, 217; Madison Valley Ranchlands Group and, 213; map of, 210; name of, 211; outside investors in ranching and, 182; ownership of, 213; private-sector conservation and, 211; residential development and, 215–16; success of, 220; sustainability strategies and, 214–19; winter and, 214
Sun Ranch Group, 216–17, 218–19

Sun Ranch Institute, 216
Sun Ranch Settlements, 215–18
Sunstein, Cass, 54
Sun Valley resort, 186
Superior Livestock, 286
sustainability: bison ranging patterns and, 214; certification for, 120–21; definition of, xiv; economic, 214, 234; small-scale residential development and, 216–17; Sun Ranch model and, 214–19; triple bottom line and, 119, 283–84
Sustainable Forestry Initiative, 121, 157
Sustainable Northwest: Cecil D. Andrus Award and, 191; Rural Voices for Conservation Coalition and, 103, 105, 106, 107, 114
Sustainable Northwest Wood, Inc., 287
Swan Ecosystem Center, 129, 130
Swan Valley, Montana: checkerboard land ownership pattern in, 119–20, 124, 125, 126, 127, 128, 132, 135; community forest in, 129–30; conservation easements in, 128–29, 133; corporate land divestment in, 131–32; description of, 123–24, 126; factories determining success of land project in, 134; land exchange in, 127–28; land-use planning in, 130–31; Montana Legacy Project and, 132–33; public acquisition of lands in, 126, 131–33; resistance to development pressure in, 124, 128–29, 132; rising land values in, 126, 129, 131, 132, 135; zoning in, 131

take permits, 62, 74n23
Talbert, C. B., 39
tan oak (*Lithocarpus densiflorus*), 151
taxation: conservation easements and, 217, 221n7, 236–37, 241, 245, 267; deferred taxes and, 273; escalating land values and, 267; estate tax and, 248, 260, 267–73; fragmentation of land, 28; land rich, cash poor landowners and, 267; land trusts and, 245; subdivision of land and, 267; tax policy and, 267, 268, 270, 272–73
Tax Policy Center, 269–70, 272
Tax Relief, Unemployment Insurance Reauthorization and Job Creation Act (2010), 270
Taylor Grazing Act (1934), 5, 6
TCF. *See* Conservation Fund (TCF)
TCRCD. *See* Trinity County Resource Conservation District (TCRCD)
Thompson, Mike, 272–73
Thompson-Fisher Conservation Easement, 128–29, 272

Thune, John, 268

timber production: advocacy for end of on public lands, 7; Arcata Community Forest (ACF) and, 139; benefits of to community, 154; versus carbon sequestration, 152; changing context of, 119; corporate versus local benefit of, 165–66; decline in, 1, 13, 19, 21, 25, 41, 160–61; definition of timberland and, 29n1; employment in, 15, 16, 19, 20, 21, 160–61; environmental activism against, 143–44; federal regulation of, 5; foresters' conservationist ethic and, 8; foresters in CBCCs and, 54; forest thinning and, 157; harvests by ownership type, 20; industry organization and, 26–27; from larger- to smaller-diameter timber, 171; loggers as "lords of yesterday" and, 7; market demand for wood products and, 21; mill infrastructure and, 21–22, 21, 23; in national forests, 5; ownership of timberlands and, 14–15, 15; patience during forest restoration and, 155; on private lands, 24–25; public versus local interest and, 6; revenue for conservation from, 137; sales volume and, 27; scientific management and, 5–6; service contracts and, 116n2; in Siuslaw National Forest, 164; sustainable forestry and, 141; third-party certifiers and, 286; timber wars and, 146–47; traditional contracts in, 104; unsustainable, 120; unsustainable harvesting and, 143; value-added niche markets and, 288; in West as proportion of U.S. whole, 13, 15. See also forest management; forests and foresters; stewardship contracts and agreements

TNC. See Nature Conservancy (TNC)
tragedy of the commons, 4
Trail of Tears, 5
Transition Colorado, 248
Treaty of Guadalupe-Hidalgo, 4, 5
tribal lands: biodiversity on, 33, 44, 45; in Colorado Plateau, 36, 37–38; ecological buffer zones in, 33; Elk Creek Conservation Area and, 130; establishment of reservations and, 5; jurisdictional complexity in, 19; as neglected geography for conservation, 34–35; researchers' and environmentalists' neglect of, 37–38; Sonoran Desert Conservation Plan and, 255; status of as unthreatened, xix. See also Native Americans

Trinity County Resource Conservation District (TCRCD), 178–80
trumpeter swan (Cygnus buccinator), 234
Trust for Public Land, 126, 127, 128–29, 132–33
Turdus migratorius (robin). See robin (Turdus migratorius)
Turner, Ray, 61

Udall, Stewart, xiii
Udall Center for Studies in Public Policy, 59
Union Lumber, 145–46
University of Arizona, 230, 232, 238, 238–39n2
University of Wisconsin, 78
urban and exurban areas: alliances with, 308–10; biodiversity and, 39–40; CBBCs and, 66–68; collaborative conservation and, xvii; developers and, 248, 251, 251–53, 262; local food movement and, 309–10; rural-urban divide and, 262–65
Urban Institute, 269
Ursus americanus (black bear). See black bear (Ursus americanus)
Ursus arctos horribills (grizzly bear). See grizzly bear (Ursus arctos horribills)
Usal Redwood Forest, 151
U.S. Army Corps of Engineers, 281, 282
USDA. See U.S. Department of Agriculture
U.S. Department of Agriculture: Beginning Farmer-Rancher Loans and, 233; Conservation Fund and, 298; Forest Service of, 127; Jornada Experimental Range and, 230; slaughterhouses and, 65, 264; stewardship contracting and, 167. See also Agriculture Research Service; Natural Resources Conservation Service
U.S. Department of the Interior, 167
U.S. Endangered Species Act, 74n23
U.S. Fish and Wildlife Service: conservation banking and, 282; conservation easements and, 236–37; dynamic partnerships and, 301–2; endangered species and, 254; environmental review processes and, 128; fire management and, 74n18; Habitat Conservation Plans and, 147, 257, 263; Incidental Take Permits and, 254; Landowner Incentive Program of, 236–37; Lava Lake Land & Livestock and, 192; prescribed burns and, 62; Roseland Plantation LLC v. United States Fish and Wildlife Service et al. and, 280; Swan Valley conservation project and, 127; wetlands mitigation banking and, 281–82
U.S. Forest Capital, 130

U.S. Forest Service: authority of, 172–73; Buenos Aires National Wildlife Refuge and, 59; changing mission and focus of, 159–60; collaborative conservation and, 171–73; Collaborative Forest Landscape Restoration program and, 111; Colorado Plateau and, 36; conservation easements and, 217, 219; culture of, 302; demands of on private organizations, 92; Diablo Trust and, 66; distribution of land in the West and, 245; dynamic partnerships and, 301–2; environmental lawsuits against, 7; Forest Legacy Program and, 129; habitat projects and, 232; higher mountain ranges and, 57; Inventoried Roadless Areas and, 200n3; land exchanges and, 128; Lava Lake Land & Livestock and, 189, 191, 192; leased grazing allotments and, 209; local interests and, 6; Malpai Borderlands Group and, 68; management for multiple uses and, 6; Mendocino Redwood Company and, 148; Montana Legacy Project and, 133; Northwest Forest Plan and, 160; overzealous fire suppression and, 61; prescribed burns and, 61–62; protection of old-growth trees and, 150; rangeland health and, 22; Rocky Mountain ecoregion and, 38; Rural Voices for Conservation Coalition and, 105–6; scientific surveys by, 189; Siuslaw National Forest and, 161; stewardship contracting and, 121–22, 159, 163, 166–70, 306; timber sales and service contracts and, 116n2; top-down approach and, 68; Valle Grande Grassbank and, 86, 87; volume of timber production in lands of, 160; Weaverville Community Forest and, 177, 179–80; Western range wars and, 58; wilderness areas and, 187; Wood River Wolf Project and, 198; Wyden Amendment and, 106; Wyden Authority and, 170

U.S. Institute for Environmental Conflict Resolution, 59

U.S. Soil Erosion Service, 77, 78

Valle Grande Grassbank, 86–88
VerifiedGreen, 286

Walbert, David, 249
Wallowa Resources, 103
water quality and quantity: beverage companies and, 286; conservation easements and, 284; as ecosystem service, 249, 276, 279, 281–82; in Garcia River Forest, 91; riparian restoration and, 91; stewardship contracting and, 163. See also watersheds

water rights, 211
Watershed Research and Training Center, 103
Watershed Restoration and Enhancement Agreement Authority. See Wyden Authority

watersheds: conservation easements and, 284; downstream beneficiaries of ecosystem services and, 284–85; forest restoration and, 121; Hay Mountain Watershed Restoration Project and, 233–34, 238–39n2; irrigation districts and, 56–57; land health and, 89; Nevada Spring Creek Watershed and, 281–82; Sonoran Desert Conservation Plan and, 260; transjurisdictional, 131; watershed groups in CBCCs and, 58, 68; Wyden Amendment and, 106. See also water quality and quantity; water rights

Weaverville 1,000, 178
Weaverville Community Forest, ix–x, 121–22, 162, 177–79

Wentworth, Rand, 273
West, the: adaptive management needed in, 305–6; building permits issued in, 244; complex land tenure in, 8–9; culture of public lands management and, 302–3; declining acreage of farm and ranch lands in, 267–68; distribution of land types in, 244–45, 244; dude ranching in, 215; dynamic partnerships needed in, 301–2; ecoregions in, 35; extent of, 16; family forestland in, 272; federal and nonfederal rangelands in, 18; federal government as dominant resource manager in, 57; grass-fed labeling hotspots in, 98; land uses in, 24; long-term commitments to, 303–4; mosaic of private and public land tenure and, 49; number of cattle and sheep in, 16, 18; ownership of rangelands in, 16, 17; ownership of timberlands in, 14–15, 15; polarization of interest groups in, 49; private land conservation trends in, 241–45, 242–43, 246n5; proportion of lands in federal ownership and, 19; "rangeland conflict" in, 6; rectilinear land parcels in, 57; urbanization of, 6–7, 8; Western Week in Washington, DC, and, 107

Western Environmental Law Center, 115
Western Governors' Association, 109–10

Westland Resources, 74n29
westslope cutthroat trout (*Oncorhynchus clarkii lewisi*), 281
wetlands. *See* habitat preservation
Whipple Observatory, 254
White, Courtney, xiii, 83–84
Whole Foods Market, 205
Wilcove, D. S., 34
Wild at Heart, 234
Wilderness Society, 110, 115, 147
Wilderness Study Areas, 187, 200n3
Wildlands Project, 37
Wildlife Services, 198
Wilkinson, Charles, xiii, 7
Willamette Partnership, 283
William and Flora Hewlett Foundation, 197
Winder, Jim, 83–84
wolf. *See* gray wolf (*Canis lupus*)
wolverine (*Gulo gulo*), 186
Wood River Land Trust, 197
Wood River Wolf Project, 186, 198
working landscapes: alternatives to conservation of, xviii–xix; biodiversity in, xiv–xv, 37–38, 39; broad toolboxes for conservation of, 308; collaborative conservation in, 38, 247; community forests and, 243; as core conservation areas, 44; core principles supporting, 108–9; definition of, xiv; difficulty maintaining livelihoods in, 28–29; ecosystem health and, xiv–xv; ecosystem services of, 278–83; embodied sense of place and, 71; exemption from estate tax and, 269, 272; federal lands as, 19–24; land acquisition for conservation of, 124; map of case studies in, xx; private lands as, 24–28; protection of, 33; versus real-estate market, 126; real estate speculation versus, xv; rural-urban divide and, 264–65

World Resources Institute, 286
World Wildlife Fund, 147
Worster, Donald, 49
Wyden Amendment, 106, 169
Wyden Authority: content of, 174n10; enactment of, 162; stewardship contracting and, 162, 167, 168, 169, 170

Yellowstone National Park, 212, 224

Zeedyk, Bill, 63, 70, 89, 93